# Boundary integral and singularity methods for linearized viscous flow

*Cambridge texts in applied mathematics*

Maximum and minimum principles: a unified approach with applications
M. J. SEWELL

Introduction to numerical linear algebra and optimization
P. G. CIARLET

Solitons: an introduction
P. G. DRAZIN AND R. S. JOHNSON

The kinematics of mixing: stretching, chaos and transport
J. M. OTTINO

Integral equations: a practical treatment, from spectral theory to applications
D. PORTER AND D. S. G. STIRLING

The thermomechanics of plasticity and fracture
G. A. MAUGIN

Boundary integral and singularity methods for linearized viscous flow
C. POZRIKIDIS

# *Boundary Integral and Singularity Methods for Linearized Viscous Flow*

C. POZRIKIDIS
*University of California, San Diego*

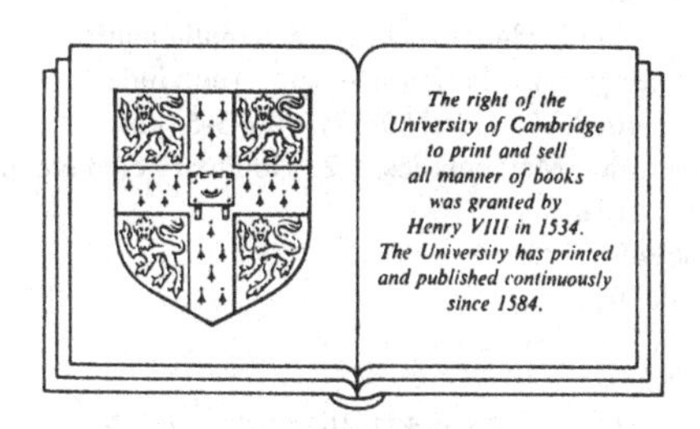


*CAMBRIDGE UNIVERSITY PRESS*
*Cambridge*
*New York Port Chester*
*Melbourne Sydney*

CAMBRIDGE UNIVERSITY PRESS
Cambridge, New York, Melbourne, Madrid, Cape Town, Singapore, São Paulo

Cambridge University Press
The Edinburgh Building, Cambridge CB2 2RU, UK

Published in the United States of America by Cambridge University Press, New York

www.cambridge.org
Information on this title: www.cambridge.org/9780521405027



First published 1992

*A catalogue record for this publication is available from the British Library*

*Library of Congress Cataloguing in Publication data*

Pozrikidis, C.
Boundary integral and singularity methods for linearized viscous flow / C. Pozrikidis.
p. cm. – (Cambridge texts in applied mathematics; 8)
Includes bibliographical references (p. ) and index.
ISBN 0-521-40502-5. – ISBN 0-521-40693-5 (pbk.)
1. Viscous flow – Mathematics. 2. Boundary element methods.
I. Title. II. Series.
TA357.5.V56P69 1991
620.1′064 – dc20 91-28067
CIP

ISBN-13 978-0-521-40502-7 hardback
ISBN-10 0-521-40502-5 hardback

ISBN-13 978-0-521-40693-2 paperback
ISBN-10 0-521-40693-5 paperback

Transferred to digital printing 2005

'For what shall it profit a man,
if he shall gain the whole world,
and lose his own soul?'

Mark 8:36

# *Contents*

# *Preface*

The goal of this book is to bring together classical and recent developments in the field of boundary integral and singularity methods for steady and unsteady Stokes flow. The targeted audience includes graduate students of engineering and applied mathematics, as well as academic and industrial researchers in the field of fluid mechanics. The material was selected so that the book may serve both as a reference monograph and as a textbook in an advanced course of fluid mechanics or computational fluid dynamics. The prerequisites are introductory fluid mechanics, real analysis, and numerical methods. Each section of every chapter is followed by a number of theoretical or computer problems whose objectives are to complement the theory, indicate extensions, and provide further insights. The references were chosen so as to provide the reader with a convenient entry to the immense literature of boundary integral, boundary element, and singularity methods.

The author would like to express his appreciation to Professor Sangtae Kim of the University of Wisconsin for giving him access to his recent work. Insightful discussions with Dr Mark Kennedy provided the motivation for developing and restating a number of results. The typescript benefitted in many different ways from the superb work of Dr Susan Parkinson, technical editor of the Cambridge University Press. Thanks are due to Dr Seppo Karilla for offering constructive comments on the early typescript.

The author would like to take this opportunity to thank Ms Audrey Hill for her indispensable encouragement and support. Finally, the author wishes to dedicate this book to each one of the members of his family – Aris, Chrisi, Savas, and Kyriakos for their love and for providing a shelter all these years.

# 1

# Preliminaries

## 1.1 Linearization of the equations of fluid flow

Referring to a frame of reference which is either fixed or translated in space, let us consider the flow of an incompressible Newtonian fluid. The motion of the fluid is governed by the continuity equation for the flow velocity $\mathbf{u}$:

$$\nabla \cdot \mathbf{u} = 0 \tag{1.1.1}$$

which expresses conservation of mass, and the Navier–Stokes equation

$$\rho\left(\frac{\partial \mathbf{u}}{\partial t} + \mathbf{u} \cdot \nabla \mathbf{u}\right) = -\nabla P + \mu \nabla^2 \mathbf{u} + \rho \mathbf{b} \tag{1.1.2}$$

which expresses Newton's second law for a small parcel of fluid. In (1.1.2) $\rho$ and $\mu$ are the density and the viscosity of the fluid, and $\mathbf{b}$ is a body force which, for simplicity, we shall assume to be constant. Inspecting the flow we find a characteristic length $L$ related to the size of the boundaries, a characteristic velocity $U$ determined by the particular mechanism driving the flow, and a characteristic time $T$ that is either imposed by external forcing or simply defined as $L/U$. We scale the velocity by $U$, all lengths by $L$, time by $T$, and each term on the right-hand side of (1.1.2) by $\mu U/L^2$. Then, we introduce the dimensionless variables $\mathbf{u}' = \mathbf{u}/U$, $\mathbf{x}' = \mathbf{x}/L$, $t' = t/T$, $P' = PL/\mu U$, and write (1.1.2) in the dimensionless form

$$\beta \frac{\partial \mathbf{u}'}{\partial t'} + \mathrm{Re}\, \mathbf{u}' \cdot \nabla' \mathbf{u}' = -\nabla' P' + \nabla'^2 \mathbf{u}' + \frac{Re}{Fr} \frac{\mathbf{b}}{|\mathbf{b}|} \tag{1.1.3}$$

We identify the first term on the left-hand side of (1.1.3) as the inertial acceleration term, and the second term as the inertial convective term.

Three dimensionless numbers appear in (1.1.3): the frequency parameter $\beta = L^2/\nu T$, the Reynolds number $Re = UL/\nu$, and the Froude number $Fr = U^2/|\mathbf{b}|L$, where $\nu = \mu/\rho$ is the kinematic viscosity of the fluid. The frequency parameter $\beta$ expresses the magnitude of inertial acceleration forces relative to viscous forces, or equivalently, the ratio between the characteristic time of diffusion of vorticity $L^2/\nu$ and the characteristic time of flow $T$. The Reynolds number expresses the magnitude of inertial convective forces relative to viscous forces, or equivalently, the ratio between the characteristic time of diffusion of vorticity $L^2/\nu$ and the convective time $L/U$. The Froude number expresses the magnitude of inertial convective forces relative to body forces. Finally, the group $Re/Fr = L^2|\mathbf{b}|/\nu U$ expresses the magnitude of body forces relative to viscous forces. It should be noted that in the absence of external forcing, $T$ may be defined as $L/U$ in which case $\beta$ reduces to $Re$, and the dimensionless Navier–Stokes equation (1.1.3) involves only two independent parameters, namely $Re$ and $Fr$.

Now, when $Re, \beta \ll 1$, all terms on the left-hand side of (1.1.3) are small compared with those on the right-hand side and thus may be neglected. Reverting to dimensional variables we find that the flow is governed by the *Stokes equation*

$$-\nabla P + \mu\nabla^2\mathbf{u} + \rho\mathbf{b} = 0 \tag{1.1.4}$$

which states that pressure, viscous, and body forces balance at any instant in time even though the flow may be unsteady. The instantaneous structure of the flow depends solely on the present boundary configuration and boundary conditions, and is independent of the history of motion. To be more precise, the history of motion enters the problem only by determining the current location of the boundaries.

When $Re \ll 1$ but $\beta \sim 1$, the inertial convective term on the left-hand side of (1.1.3) is small compared with the rest of the terms and thus may be neglected. Again reverting to dimensional variables we find that the flow is governed by the *unsteady Stokes equation* or *linearized Navier–Stokes equation*

$$\rho\frac{\partial\mathbf{u}}{\partial t} = -\nabla P + \mu\nabla^2\mathbf{u} + \rho\mathbf{b} \tag{1.1.5}$$

Because of the presence of the acceleration term, the instantaneous structure of the flow depends not only on the instantaneous boundary configuration and boundary conditions but also on the history of motion. Physically, the unsteady Stokes equation is valid for flows that are characterized by sudden acceleration or deceleration, such as those occurring during hydrodynamic braking, during the impact of a particle on a solid surface, or during the initial stages of the flow due to a particle settling from rest.

**Problem**

1.1.1 Consider a rigid body that is convected by a steady flow, and write the Navier–Stokes equation in an accelerating but not rotating frame of reference that is attached to the body. Identify $\beta$, *Re*, and *Fr*, and establish conditions for the steady or unsteady Stokes equation to be valid in the vicinity of the body.

## 1.2 The equations of Stokes flow

We have seen that the flow of a Newtonian fluid at small values of the frequency parameter $\beta$ and Reynolds number *Re* is governed by the continuity equation $\nabla\cdot\mathbf{u}=0$ and the Stokes equation

$$-\nabla P+\mu\nabla^2\mathbf{u}+\rho\mathbf{b}=\nabla\cdot\boldsymbol{\sigma}+\rho\mathbf{b}=0 \tag{1.2.1}$$

where $\boldsymbol{\sigma}$ is the stress tensor defined as follows:

$$\sigma_{ij}=-P\delta_{ij}+\mu\left(\frac{\partial u_i}{\partial x_j}+\frac{\partial u_j}{\partial x_i}\right)=-P\delta_{ij}+2\mu e_{ij} \tag{1.2.2}$$

and

$$e_{ij}=\frac{1}{2}\left(\frac{\partial u_i}{\partial x_j}+\frac{\partial u_j}{\partial x_i}\right) \tag{1.2.3}$$

is the rate of deformation tensor. Taking the divergence of (1.2.1) and using the continuity equation we find that the pressure is a harmonic function, i.e.

$$\nabla^2 P=0 \tag{1.2.4}$$

Taking the Laplacian of the Stokes equation and using (1.2.4) we find that the velocity satisfies the vectorial biharmonic equation, i.e.

$$\nabla^4\mathbf{u}=0 \tag{1.2.5}$$

Finally, taking the curl of the Stokes equation and using the vector identity $\nabla\times\nabla\mathrm{F}=0$, valid for any twice differentiable function F, we find that the vorticity $\boldsymbol{\omega}=\nabla\times\mathbf{u}$ is a harmonic function, i.e.

$$\nabla^2\boldsymbol{\omega}=0 \tag{1.2.6}$$

Invoking the general properties of harmonic functions we deduce that there is no intense concentration of vorticity in Stokes flow and furthermore, that the vorticity attains extreme values at the boundaries of the flow. The onset of regions of recirculating fluid does not imply the presence of compact vortices, as it does in the case of inviscid flow.

Integrating (1.2.1) over a volume $V$ that resides within the domain of flow and is bounded by the closed surface $D$, and applying the divergence theorem to convert the volume integral involving the stress into a surface integral, we find

$$\mathbf{F}=\int_D \boldsymbol{\sigma}\cdot\mathbf{n}\,\mathrm{d}S\equiv\int_D \mathbf{f}\,\mathrm{d}S=-\int_V \rho\mathbf{b}\,\mathrm{d}V \tag{1.2.7}$$

where $\mathbf{F}$ is the *hydrodynamic force* exerted on $D$, $\mathbf{f} = \boldsymbol{\sigma}\cdot\mathbf{n}$ is the *surface force* or *traction* exerted on $D$, and $\mathbf{n}$ is the unit normal vector pointing outside $V$. Working in a similar manner we obtain

$$\mathbf{L} = \int_D \mathbf{x} \times (\boldsymbol{\sigma}\cdot\mathbf{n})\,dS \equiv \int_D \mathbf{x} \times \mathbf{f}\,dS = -\int_V \rho\mathbf{x} \times \mathbf{b}\,dV \tag{1.2.8}$$

where $\mathbf{L}$ is the *hydrodynamic torque* exerted on $D$. Equations (1.2.7) and (1.2.8) imply that in the absence of a body force, the force and torque exerted on any volume of pure fluid are equal to zero, and furthermore, the force and torque exerted on any two reducible surfaces have the same values. It will be noted that the force exerted on a surface that encloses a boundary or a singular point may be finite, but must be equal to that exerted on the boundary or on a small surface that encloses the singular point; similarly for the torque.

For convenience, in the ensuing discussion we shall incorporate the effect of the body force $\mathbf{b}$ into a modified pressure defined as

$$P^{MOD} = P - \rho\mathbf{b}\cdot\mathbf{x} \tag{1.2.9}$$

and thus, we shall consider the body-force-free Stokes equation written in terms of $P^{MOD}$. The distinction between the regular and modified pressure will be relevant only when we consider boundary conditions for the surface force.

To facilitate the analysis of two-dimensional flow it is often helpful to introduce the stream function $\Psi$, defined by the equation

$$\mathbf{u} = \nabla \times (\mathbf{k}\Psi) \tag{1.2.10}$$

where $\mathbf{k}$ is the unit vector perpendicular to the $(x, y)$ plane of the flow. The underlying motivation for introducing the stream function is that the continuity equation is satisfied for any choice of $\Psi$ and thus may be overlooked. Explicitly, the $x$ and $y$ Cartesian components of the velocity are given by

$$u = \frac{\partial \Psi}{\partial y} \qquad v = -\frac{\partial \Psi}{\partial x} \tag{1.2.11}$$

The radial and angular components of the velocity are given by

$$u_r = \frac{1}{r}\frac{\partial \Psi}{\partial \theta} \qquad u_\theta = -\frac{\partial \Psi}{\partial r} \tag{1.2.12}$$

In terms of the stream function, the vorticity is equal to

$$\boldsymbol{\omega} = \nabla \times \mathbf{u} = -\mathbf{k}\nabla^2\Psi \tag{1.2.13}$$

Recalling that the vorticity is a harmonic function, we find that the stream function satisfies the biharmonic equation

$$\nabla^4\Psi = 0 \tag{1.2.14}$$

Now, using the identity $\nabla^2\mathbf{u} = -\nabla \times \boldsymbol{\omega}$ and the Stokes equation $\nabla^2\mathbf{u} = (1/\mu)\nabla P$, we find $\mu\nabla \times \boldsymbol{\omega} = -\nabla P$, which suggests that the magnitude of the vorticity and the pressure satisfy the Cauchy–Riemann equations

$$\frac{\partial \omega}{\partial x} = \frac{1}{\mu}\frac{\partial P}{\partial y} \qquad \frac{\partial \omega}{\partial y} = -\frac{1}{\mu}\frac{\partial P}{\partial x} \tag{1.2.15}$$

As a result, the complex function

$$f(z) = \omega + \frac{\mathrm{i}}{\mu}P \tag{1.2.16}$$

is an analytic function of $z = x + \mathrm{i}y$, where i is the square root of minus one. This observation allows us to study two-dimensional Stokes flow within the general framework of the theory of complex functions (Mikhlin 1957, Chapter 5; Langlois 1964, Chapter 7).

In an alternative formulation, we express a two-dimensional Stokes flow in terms of the Airy stress function $\Phi$, defined by the equations

$$\sigma_{xx} = \frac{\partial^2 \Phi}{\partial y^2} \qquad \sigma_{xy} = \sigma_{yx} = -\frac{\partial^2 \Phi}{\partial x \partial y} \qquad \sigma_{yy} = \frac{\partial^2 \Phi}{\partial x^2} \tag{1.2.17}$$

It will be noted that the Stokes equation is satisfied for any choice of $\Phi$. Using the continuity equation and recalling that the pressure is a harmonic function, we deduce that $\Phi$ satisfies the biharmonic equation. Writing out the three independent components of the stress tensor in terms of the pressure and the stream function, and using (1.2.17) to eliminate the pressure, we obtain

$$\frac{\partial^2 \Phi}{\partial x^2} - \frac{\partial^2 \Phi}{\partial y^2} = -4\mu\frac{\partial^2 \Psi}{\partial x \partial y} \qquad \frac{\partial^2 \Psi}{\partial x^2} - \frac{\partial^2 \Psi}{\partial y^2} = \frac{1}{\mu}\frac{\partial^2 \Phi}{\partial x \partial y} \tag{1.2.18}$$

Furthermore, using (1.2.18), we find that the complex function $\chi = \Phi - \mathrm{i}\,2\mu\Psi$ satisfies the equation

$$\left(\frac{\partial^2 x}{\partial z^{*2}}\right)_z = 0 \tag{1.2.19}$$

where $z = x + \mathrm{i}y$, and an asterisk indicates the complex conjugate (Coleman 1980). Integrating (1.2.19) we find

$$\chi(z) = z^*\chi_1(z) + \chi_2(z) \tag{1.2.20}$$

where $\chi_1$ and $\chi_2$ are two arbitrary analytic functions of $z$. Different selections for $\chi_1$ and $\chi_2$ produce various types of two-dimensional flow (problem 1.2.3).

Next, we switch to cylindrical polar coordinates $(x, \sigma, \phi)$ and consider an axisymmetric flow in which neither the velocity nor the pressure depend on the azimuthal angle $\phi$. To facilitate the analysis, it is convenient to introduce the Stokes stream function $\psi$ defined by the equation

$$\mathbf{u} = \nabla \times \left(0, 0, \frac{\psi}{\sigma}\right) \tag{1.2.21}$$

It will be noted that the continuity equation is satisfied for any choice of $\psi$. The $x$ and $\sigma$ components of the velocity are given explicitly by

$$u_x = \frac{1}{\sigma}\frac{\partial \psi}{\partial \sigma} \qquad u_\sigma = -\frac{1}{\sigma}\frac{\partial \psi}{\partial x} \tag{1.2.22}$$

Switching temporarily to spherical polar coordinates $(r, \theta, \phi)$, we find that the radial and meridional components of the velocity are given by

$$u_r = \frac{1}{r^2 \sin\theta}\frac{\partial \psi}{\partial \theta} \qquad u_\theta = -\frac{1}{r \sin\theta}\frac{\partial \psi}{\partial r} \tag{1.2.23}$$

where $\sigma = r\sin\theta$ and $x = r\cos\theta$. The vorticity is equal to

$$\boldsymbol{\omega} = -\frac{1}{\sigma}E^2\psi \mathbf{e}_\phi \tag{1.2.24}$$

where $\mathbf{e}_\phi$ is the unit vector in the azimuthal direction, and $E^2$ is a differential operator defined as

$$E^2 \equiv \frac{\partial^2}{\partial x^2} + \frac{\partial^2}{\partial \sigma^2} - \frac{1}{\sigma}\frac{\partial}{\partial \sigma} = \frac{\partial^2}{\partial r^2} + \frac{\sin\theta}{r^2}\frac{\partial}{\partial \theta}\left(\frac{1}{\sin\theta}\frac{\partial}{\partial \theta}\right) = \frac{\partial^2}{\partial r^2} + \frac{1}{r^2}\frac{\partial^2}{\partial \theta^2} - \frac{\cot\theta}{r^2}\frac{\partial}{\partial \theta} \tag{1.2.25}$$

Noting that the vorticity is a harmonic function, we deduce that $\psi$ satisfies the fouth-order differential equation

$$\mathrm{E}^4\psi = 0 \tag{1.2.26}$$

Turning our attention next to axisymmetric swirling flow (such as that produced by the axial rotation of a prolate spheroid), we find it convenient to introduce the *swirl* $\Omega(x, \sigma)$, defined by the equation

$$u_\phi = \frac{\Omega}{\sigma} \tag{1.2.27}$$

The vorticity associated with the swirling flow is given by

$$\boldsymbol{\omega} = \frac{1}{\sigma}\frac{\partial \Omega}{\partial \sigma}\mathbf{i} \tag{1.2.28}$$

where $\mathbf{i}$ is the unit vector in the $x$ direction, and the pressure is constant. Substituting (1.2.27) into the Stokes equation, we find that the swirl satisfies the second-order differential equation

$$E^2\Omega = 0 \tag{1.2.29}$$

In an alternative formulation, we express a swirling flow in terms of a single scalar function $\chi$ defined by the equations

$$\sigma_{x\phi} = \mu \frac{1}{\sigma^2} \frac{\partial \chi}{\partial \sigma} \qquad \sigma_{\sigma\phi} = -\mu \frac{1}{\sigma^2} \frac{\partial \chi}{\partial x} \tag{1.2.30}$$

(Love 1944, p. 325). It will be noted that the Stokes equation is satisfied for any choice of $\chi$. The reader may verify that $\chi$ is constant along a line of vanishing surface force. Recalling that

$$\sigma_{x\phi} = \mu \frac{\partial u_\phi}{\partial x} \qquad \sigma_{\sigma\phi} = \mu \sigma \frac{\partial}{\partial \sigma}\left(\frac{u_\phi}{\sigma}\right) \tag{1.2.31}$$

and using (1.2.30) we find that $\chi$ satisfies the differential equation

$$\frac{\partial^2 \chi}{\partial x^2} + \frac{\partial^2 \chi}{\partial \sigma^2} - \frac{3}{\sigma} \frac{\partial \chi}{\partial \sigma} = 0 \tag{1.2.32}$$

Furthermore, using (1.2.32) we find that the function $\Phi = \chi \cos 2\phi / \sigma^2$ satisfies the three-dimensional Laplace equation

$$\nabla^2 \Phi = 0. \tag{1.2.33}$$

**Problems**

1.2.1 Show that the integral of the stress tensor over a volume $V$ that encloses pure fluid is

$$\int_V \sigma_{ij}\,dV = \int_D \sigma_{ik} n_k x_j\,dS - \int_V \frac{\partial \sigma_{ik}}{\partial x_k} x_j\,dV$$

where $D$ is the boundary of $V$ and $\mathbf{n}$ is the normal vector pointing outside $V$.

1.2.2 Establish the validity of the following reciprocal identity

$$\int_D \left( u'_i \frac{\partial u_k}{\partial x_i} - u_i \frac{\partial u'_k}{\partial x_i} \right) n_k\,dS = 0$$

where $\mathbf{u}$ and $\mathbf{u}'$ are two incompressible vector fields and $D$ is the boundary of an arbitrary volume of fluid $V$.

1.2.3 Show that setting

$$\chi = \tfrac{1}{2}\mu[z^2(z^* - \tfrac{1}{3}z) - 4z]$$

in (1.2.20) produces two-dimensional Poiseuille flow through a slot of unit half-width (Coleman 1980).

1.2.4 Show that the derivative of the function $\chi$ normal to a line $C$ that rotates as a rigid body, i.e. $u_\phi = \Omega\sigma$ over $C$ where $\Omega$ is the angular velocity of rotation, is equal to zero.

1.2.5 Prove (1.2.33).

## 1.3 Reversibility of Stokes flow

Let us assume that $\mathbf{u}$ and $P$ form a pair of velocity and pressure fields that satisfy the equations of Stokes flow. Clearly, $-\mathbf{u}$ and $-P$ satisfy the equations of Stokes flow as well, thereby implying that reversed flow is a mathematically acceptable and physically viable solution. It should be noted that the direction of the force and torque acting on any surface are also reversed when the signs of $\mathbf{u}$ and $P$ are switched. The property of reversibility is not shared by flow at finite Reynolds numbers for in that case the non-linear term $\mathbf{u} \cdot \nabla \mathbf{u}$ maintains its sign when the sign of the velocity is reversed.

The reversibility of Stokes flow may be invoked to derive a number of interesting and useful results. Consider, for instance, a solid sphere moving under the action of a shear flow in the vicinity of a plane wall. In principle, the hydrodynamic force acting on the sphere may have a component perpendicular to the wall and a component parallel to the wall. Let us assume for a moment that the component of the force perpendicular to the wall pushes the sphere away from the wall. Reversing the direction of the shear flow must reverse the direction of this force and thus must push the sphere towards the wall. Such an anisotropy, however, is physically unacceptable in view of the fore-and-aft symmetry of the domain of flow. We must conclude that the normal component of the force on the sphere is equal to zero, implying that the sphere must keep moving parallel to the wall.

Using the concept of reversibility we may infer that the streamline pattern around an axisymmetric and fore-and-aft symmetric object that moves along its axis must also be axisymmetric and fore-and-aft symmetric. The streamline pattern over a two-dimensional rectangular cavity must be symmetric with respect to the mid-plane of the cavity. A neutrally buoyant spherical particle that is convected by a parabolic flow in a cylindrical pipe may not move towards the center of the pipe or the wall, but must maintain its initial radial position. As an example of a more subtle situation, consider a buoyant drop, in the shape of a ring, rising under the action of gravity in an infinite ambient fluid. Reversibility requires that if the diameter of the ring is increasing, the cross-sectional shape of the ring may not have a fore-and-aft symmetry (Kojima, Hinch & Acrivos 1984).

### Problem

1.3.1 Show that the force on a solid sphere that is rotating in the vicinity of a plane wall does not have a component perpendicular to the wall.

## 1.4 The reciprocal identity

Let us assume that $\mathbf{u}$ and $\mathbf{u}'$ are two solutions of the equations of Stokes flow with associated stress tensors $\boldsymbol{\sigma}$ and $\boldsymbol{\sigma}'$ respectively, and compute

$$u'_i \frac{\partial \sigma_{ij}}{\partial x_j} = \frac{\partial}{\partial x_j}(u'_i \sigma_{ij}) - \sigma_{ij}\frac{\partial u'_i}{\partial x_j} = \frac{\partial}{\partial x_j}(u'_i \sigma_{ij}) - \left[ -P\delta_{ij} + \mu\left(\frac{\partial u_i}{\partial x_j} + \frac{\partial u_j}{\partial x_i}\right)\right]\frac{\partial u'_i}{\partial x_j}$$

$$= \frac{\partial}{\partial x_j}(u'_i \sigma_{ij}) - \mu\left(\frac{\partial u_i}{\partial x_j} + \frac{\partial u_j}{\partial x_i}\right)\frac{\partial u'_i}{\partial x_j} \tag{1.4.1}$$

Note that we have used the continuity equation to eliminate the pressure. Interchanging the roles of $\mathbf{u}$ and $\mathbf{u}'$ we obtain

$$u_i \frac{\partial \sigma'_{ij}}{\partial x_j} = \frac{\partial}{\partial x_j}(u_i \sigma'_{ij}) - \mu\left(\frac{\partial u'_i}{\partial x_j} + \frac{\partial u'_j}{\partial x_i}\right)\frac{\partial u_i}{\partial x_j} \tag{1.4.2}$$

Subtracting (1.4.2) from (1.4.1) we find

$$\frac{\partial}{\partial x_j}(u'_i \sigma_{ij} - u_i \sigma'_{ij}) = u'_i \frac{\partial \sigma_{ij}}{\partial x_j} - u_i \frac{\partial \sigma'_{ij}}{\partial x_j} \tag{1.4.3}$$

If the flows $u$ and $u'$ are regular, i.e. they contain no singular points, the right-hand side of (1.4.3) vanishes, yielding the reciprocal identity

$$\nabla \cdot (\mathbf{u}' \cdot \boldsymbol{\sigma} - \mathbf{u} \cdot \boldsymbol{\sigma}') = 0 \tag{1.4.4}$$

due to Lorentz (1907). In problems 1.4.1 and 1.4.2 we shall discuss a generalization of the reciprocal identity for fluids with different viscosities, and an alternative expression in terms of the velocity and the pressure (Happel & Brenner 1973, pp. 80, 85). To place the reciprocal identity into a more general perspective, it will be useful to note that (1.4.4) is the counterpart of Green's second identity in the theory of potential flow (Kellogg 1954, p. 215), and Betti's formula in the theory of linear elastostatics (Love 1944, p. 173).

A useful form of the reciprocal identity emerges by integrating (1.4.4) over a volume of fluid $V$ that is bounded by the closed surface $D$ and then using the divergence theorem to convert the volume integral into a surface integral over $D$. In this manner we obtain

$$\int_D \mathbf{u}' \cdot \mathbf{f} \, dS = \int_D \mathbf{f}' \cdot \mathbf{u} \, dS \tag{1.4.5}$$

where $\mathbf{f} = \boldsymbol{\sigma} \cdot \mathbf{n}$ and $\mathbf{f}' = \boldsymbol{\sigma}' \cdot \mathbf{n}$ are the surface forces exerted on $D$, and $\mathbf{n}$ is the unit normal vector pointing outside $V$.

The major strength of the reciprocal identity is that it allows us to obtain information about a flow without having to solve the equations of motion explicitly, but merely by using information about another flow.

To illustrate the resulting simplifications, we proceed now to discuss several applications in the field of particulate flows.

Let us consider a solid particle that is held stationary in an infinite incident ambient flow $\mathbf{u}^\infty$. The presence of the particle causes a disturbance flow $\mathbf{u}^D$ which is added to the ambient flow to give the total flow $\mathbf{u} = \mathbf{u}^\infty + \mathbf{u}^D$. Turning to the reciprocal theorem, we identify $\mathbf{u}'$ with the velocity produced when the particle translates with velocity $\mathbf{V}$. Exploiting the linearity of the Stokes equation we write the corresponding surface force exerted on the particle in the form

$$\mathbf{f}^T = -\mu \mathcal{G}^T \cdot \mathbf{V} \tag{1.4.6}$$

where $\mathcal{G}^T$ is the *translational surface force resistance matrix*, and the superscript T indicates translation. We select a control volume V that is enclosed by the surface of the particle $S_P$ and by a surface $S_\infty$ of large radius, and apply (1.4.5) for the pair $\mathbf{u}^T$ and $\mathbf{u}^D$ obtaining

$$\int_{S_\infty, S_P} \mathbf{u}^T \cdot \mathbf{f}^D \, dS = \int_{S_\infty, S_P} \mathbf{u}^D \cdot \mathbf{f}^T \, dS \tag{1.4.7}$$

Letting the radius of $S_\infty$ tend to infinity we find that the surface integrals over $S_\infty$ vanish, for the velocity at infinity decays at least as fast as the inverse of the distance $r$ from the particle, and the surface force decays at least as fast as $r^{-2}$ (see discussion of unbounded flow in section 2.3). Equation (1.4.7) then reduces to

$$\mathbf{V} \cdot \mathbf{F}^D = \int_{S_P} \mathbf{u}^D \cdot \mathbf{f}^T \, dS \tag{1.4.8}$$

where $\mathbf{F}^D$ is the disturbance force exerted on the particle, defined as

$$\mathbf{F}^D = \int_{S_P} \mathbf{f}^D \, dS \tag{1.4.9}$$

Applying the boundary condition $\mathbf{u} = 0$ or $\mathbf{u}^D = -\mathbf{u}^\infty$ on $S_P$, substituting (1.4.6) into (1.4.8), and noting that in the absence of a body force the disturbance force $\mathbf{F}^D$ is equal to the total force $\mathbf{F}$, we finally obtain

$$\mathbf{F} = \mu \int_{S_P} \mathbf{u}^\infty \cdot \mathcal{G}^T dS \tag{1.4.10}$$

(Brenner 1964b). Equation (1.4.10) provides us with an expression for the force simply in terms of the values of the incident velocity $\mathbf{u}^\infty$ over the surface of the particle and the surface force resistance matrix for translation. It will be noted that if the resistance matrix $\mathcal{G}^T$ happens to be constant, as is the case for a spherical particle, the force on the particle is simply proportional to the average value of the incident velocity over the surface of the particle.

Next, we consider the torque exerted on a particle that is held stationary in an infinite ambient flow $\mathbf{u}^{\infty}$. Following the above procedures, we introduce the *rotary surface force resistance matrix* $\mathcal{G}^{\mathrm{R}}$ defined by the equation

$$\mathbf{f}^{\mathrm{R}} = -\mu \mathcal{G}^{\mathrm{R}} \cdot \mathbf{\Omega} \tag{1.4.11}$$

where $\mathbf{f}^{\mathrm{R}}$ is the surface force acting on a particle when it rotates with angular velocity $\mathbf{\Omega}$, and find

$$\mathbf{L} = \mu \int_{S_{\mathrm{P}}} \mathbf{u}^{\infty} \cdot \mathcal{G}^{\mathrm{R}} \mathrm{d}S \tag{1.4.12}$$

where

$$\mathbf{L} = \int_{S_{\mathrm{P}}} \mathbf{x} \times \mathbf{f} \, \mathrm{d}S \tag{1.4.13}$$

is the torque acting on the particle. Equation (1.4.12) provides us with an expression for the torque simply in terms of the value of the incident velocity $\mathbf{u}^{\infty}$ over the surface of the particle, and the surface force resistance matrix for rotation.

Equations (1.4.10) and (1.4.12) constitute one version of the Faxen relations for the force and the torque, discussed in more detail in section 2.5.

As a specific application of (1.4.10) and (1.4.12), we note that the translational and rotary surface force resistance matrices for a rigid spherical particle are given by

$$\mathcal{G}^{\mathrm{T}}_{ij} = \frac{3}{2a} \delta_{ij} \qquad \mathcal{G}^{R}_{ij} = \frac{3}{a} \varepsilon_{ijk} \hat{x}_k \tag{1.4.14}$$

where $a$ is the radius of the sphere, $\hat{\mathbf{x}} = \mathbf{x} - \mathbf{x}_0$, and $\mathbf{x}_0$ is the center of the sphere (see (7.3.8), (7.3.21) and (7.3.27)). Substituting (1.4.14) into (1.4.10) and (1.4.12) we obtain

$$\mathbf{F} = \frac{3}{2} \frac{\mu}{a} \int_{S_{\mathrm{P}}} \mathbf{u}^{\infty} \mathrm{d}S \qquad \mathbf{L} = 3 \frac{\mu}{a} \int_{S_{\mathrm{P}}} \hat{\mathbf{x}} \times \mathbf{u}^{\infty} \, \mathrm{d}S \tag{1.4.15}$$

Now, as a further application of the reciprocal theorem, we consider a suspended rigid particle that translates and rotates under the action of an incident flow $\mathbf{u}^{\infty}$. The presence of the particle causes a disturbance flow $\mathbf{u}^{\mathrm{D}}$ which is added to the incident flow to give the total flow $\mathbf{u} = \mathbf{u}^{\mathrm{D}} + \mathbf{u}^{\infty}$. Applying the reciprocal identity for the disturbance flow and the flow produced when the particle translates in an otherwise quiescent fluid, we obtain (1.4.8). Requiring that on the surface of the particle $\mathbf{u} = \mathbf{U} + \boldsymbol{\omega} \times \mathbf{x}$, or equivalently, $\mathbf{u}^{\mathrm{D}} = \mathbf{U} + \boldsymbol{\omega} \times \mathbf{x} - \mathbf{u}^{\infty}$ where $\mathbf{U}$ and $\boldsymbol{\omega}$ are the velocities of translation and rotation respectively, and recalling that in the absence of a body force the disturbance force is equal to the total force, we obtain

$$\mathbf{U}\cdot\mathbf{F}^{\mathrm{T}}+\boldsymbol{\omega}\cdot\mathbf{L}^{\mathrm{T}}=\int_{S_P}\mathbf{u}^{\infty}\cdot\mathbf{f}^{\mathrm{T}}\,dS+\mathbf{V}\cdot\mathbf{F} \tag{1.4.16}$$

where $\mathbf{F}^{\mathrm{T}}$ and $\mathbf{L}^{\mathrm{T}}$ are the force and torque acting on the particle when it translates with velocity $\mathbf{V}$. Working in a similar manner we obtain

$$\mathbf{U}\cdot\mathbf{F}^{\mathrm{R}}+\boldsymbol{\omega}\cdot\mathbf{L}^{\mathrm{R}}=\int_{S_P}\mathbf{u}^{\infty}\cdot\mathbf{f}^{\mathrm{R}}\,dS+\boldsymbol{\Omega}\cdot\mathbf{L} \tag{1.4.17}$$

where $\mathbf{F}^{\mathrm{R}}$ and $\mathbf{L}^{\mathrm{R}}$ are the force and torque acting on the particle when it rotates with angular velocity $\boldsymbol{\Omega}$. Now, exploiting the linear nature of the equations of Stokes flow, we write

$$\mathbf{F}^{\mathrm{T}}=-\mu\mathbf{X}\cdot\mathbf{V}\qquad\mathbf{L}^{\mathrm{T}}=-\mu\mathbf{P}'\cdot\mathbf{V} \tag{1.4.18}$$

$$\mathbf{F}^{\mathrm{R}}=-\mu\mathbf{P}\cdot\boldsymbol{\Omega}\qquad\mathbf{L}^{\mathrm{R}}=-\mu\mathbf{Y}\cdot\boldsymbol{\Omega} \tag{1.4.19}$$

where $\mathbf{X},\mathbf{P}'$ are resistance matrices for translation and $\mathbf{P},\mathbf{Y}$ are resistance matrices for rotation respectively. Using (1.4.6), (1.4.11), (1.4.18), and (1.4.19), we write (1.4.16) and (1.4.17) in the equivalent forms

$$\mathbf{U}\cdot\mathbf{X}+\boldsymbol{\omega}\cdot\mathbf{P}'=\int_{S_P}\mathbf{u}^{\infty}\cdot\mathscr{G}^{\mathrm{T}}\,dS-\frac{1}{\mu}\mathbf{F} \tag{1.4.20}$$

and

$$\mathbf{U}\cdot\mathbf{P}+\boldsymbol{\omega}\cdot\mathbf{Y}=\int_{S_P}\mathbf{u}^{\infty}\cdot\mathscr{G}^{\mathrm{R}}\,dS-\frac{1}{\mu}\mathbf{L} \tag{1.4.21}$$

Given the force $\mathbf{F}$ and torque $\mathbf{L}$ exerted on the particle, as well as the resistance matrices for translation and rotation, we obtain a system of linear algebraic equations for the translational and rotational velocities $\mathbf{U}$ and $\boldsymbol{\omega}$.

Next, we consider the specific case of a force-free and torque-free particle for which $\mathbf{F}=\mathbf{L}=0$ and take a linear combination of (1.4.20) and (1.4.21), obtaining

$$\mathbf{U}\cdot(\mathbf{X}\cdot\boldsymbol{\alpha}+\mathbf{P}\cdot\boldsymbol{\beta})+\boldsymbol{\omega}\cdot(\mathbf{P}'\cdot\boldsymbol{\alpha}+\mathbf{Y}\cdot\boldsymbol{\beta})=\int_{S_P}\mathbf{u}^{\infty}\cdot(\mathscr{G}^{\mathrm{T}}\cdot\boldsymbol{\alpha}+\mathscr{G}^{\mathrm{R}}\cdot\boldsymbol{\beta})\,dS \tag{1.4.22}$$

where $\boldsymbol{\alpha}$ and $\boldsymbol{\beta}$ are two arbitrary constant vectors. It is certainly possible to tune the values of $\boldsymbol{\alpha}$ and $\boldsymbol{\beta}$ so that one of the two terms in the parentheses on the left-hand side of (1.4.22) is equal to zero. When the second term is made to vanish we obtain

$$\mathbf{U}\cdot\mathscr{F}=\int_{S_P}\mathbf{u}^{\infty}\cdot\mathbf{f}^{\mathrm{F}}\,dS \tag{1.4.23}$$

where

$$\mathscr{F}=-\mu(\mathbf{X}\cdot\boldsymbol{\alpha}+\mathbf{P}\cdot\boldsymbol{\beta}),\qquad\mathbf{f}^{\mathrm{F}}=-\mu(\mathscr{G}^{\mathrm{T}}\cdot\boldsymbol{\alpha}+\mathscr{G}^{\mathrm{R}}\cdot\boldsymbol{\beta}) \tag{1.4.24}$$

Referring back to (1.4.6), (1.4.11), and (1.4.18), we find that $\boldsymbol{\alpha}$ and $\boldsymbol{\beta}$ are the translational and angular velocities of a torque-free particle that moves under the influence of the force $\mathscr{F}$. When the terms in the first parentheses on the left-hand side of (1.4.22) are set equal to zero we obtain

$$\boldsymbol{\omega}\cdot\mathscr{M} = \int_{S_P} \mathbf{u}^{\infty}\cdot\mathbf{f}^{\mathrm{M}}\,\mathrm{d}S \tag{1.4.25}$$

where

$$\mathscr{M} = -\mu(\mathbf{P}'\cdot\boldsymbol{\alpha} + \mathbf{Y}\cdot\boldsymbol{\beta}), \qquad \mathbf{f}^{\mathrm{M}} = -\mu(\mathscr{G}^{\mathrm{T}}\cdot\boldsymbol{\alpha} + \mathscr{G}^{\mathrm{R}}\cdot\boldsymbol{\beta}) \tag{1.4.26}$$

and find that $\boldsymbol{\alpha}$ and $\boldsymbol{\beta}$ are the translational and angular velocities of a force-free particle that moves under the influence of the torque $\mathscr{M}$.

The important message delivered by (1.4.23) and (1.4.25) is the following: the translational and rotational velocities of a force-free and torque-free particle that is immersed in an arbitrary ambient flow may be computed from a knowledge of the values of the incident velocity over the surface of the particle and of the surface force that is exerted on such a particle when it moves under the influence of an external force or torque.

## Problems

1.4.1 Assume that $\mathbf{u}$ and $\mathbf{u}'$ are two solutions to the equations of Stokes flow with associated stress tensors $\boldsymbol{\sigma}$ and $\boldsymbol{\sigma}'$ and corresponding viscosities $\mu$ and $\mu'$. Then, prove the generalized reciprocal relation (Happel & Brenner 1973, p. 86)

$$\frac{\partial}{\partial x_j}(\mu' u'_i \sigma_{ij} - \mu u_i \sigma'_{ij}) = \mu' u'_i \frac{\partial \sigma_{ij}}{\partial x_j} - \mu u_i \frac{\partial \sigma'_{ij}}{\partial x_j}$$

1.4.2 Show that two regular Stokes flows $\mathbf{u}$ and $\mathbf{u}'$ with corresponding pressure fields $P$ and $P'$ satisfy the reciprocal relation

$$\frac{\partial}{\partial x_k}\left[u'_i\left(-\delta_{ik}P + \mu\frac{\partial u_i}{\partial x_k}\right) - u_i\left(-\delta_{ik}P' + \mu\frac{\partial u'_i}{\partial x_k}\right)\right] = 0$$

1.4.3 Noting that the velocity field of a Stokes flow satisfies the biharmonic equation, prove the identities (2.5.39) and (2.5.40). (Hint: expand the velocity field in a Taylor series about the center of the sphere.)

1.4.4 Using (1.4.15) show that the force exerted on a spherical particle that is placed at the axis of the paraboloidal flow $U[(y/a)^2 + (z/a)^2, 0, 0]$ is equal to $\mathbf{F} = 4\pi\mu U a\,[1, 0, 0]$, where $a$ is the radius of the particle.

1.4.5 Using the reciprocal theorem show that the resistance matrices $\mathbf{X}$ and $\mathbf{Y}$ defined in (1.4.18) and (1.4.19) are symmetric, whereas $\mathbf{P}'$ is the transpose of $\mathbf{P}$ (Brenner 1963, 1964a).

1.4.6 Using the reciprocal theorem show that if we know the surface force on a rigid particle that translates or rotates within a tube, we can compute the force and torque on the particle when it is held stationary in an incident flow through the tube.

## 1.5 Uniqueness of solution, significance of homogeneous boundary conditions, and further properties of Stokes flow

Let us assume that $\mathbf{u}$ is a complex Stokes velocity field, in the sense that it is composed of a real and an imaginary part which, together with the associated pressure fields, satisfy the equations of Stokes flow. Incidentally, we note that this extension to complex flow fields is motivated by the analysis of the Fredholm integral equations of the second kind arising from boundary integral representations, as will be discussed in Chapter 4. Now, let $V$ be an arbitrary, possibly multiply-connected, volume of fluid that is bounded by the smooth surface $D$. Using the divergence theorem, the continuity equation, and the Stokes equation, we compute

$$\begin{aligned}\int_D u_i^* \sigma_{ik} n_k \, dS &= \int_D u_i^*(-P\delta_{ik} + 2\mu e_{ik}) n_k \, dS \\ &= -\int_V \left[ -u_i^* \frac{\partial P}{\partial x_i} + 2\mu \frac{\partial}{\partial x_k}(u_i^* e_{ik}) \right] dV \\ &= -\int_V \left[ -\mu u_i^* \frac{\partial u_i}{\partial x_k \partial x_k} + 2\mu \frac{\partial}{\partial x_k}(u_i^* e_{ik}) \right] dV \\ &= -2\mu \int_V \frac{\partial u_i^*}{\partial x_k} e_{ik} \, dV = -2\mu \int_V e_{ik}^* e_{ik} \, dV \end{aligned} \tag{1.5.1}$$

where the unit normal vector $\mathbf{n}$ points *into* the control volume $V$, $\mathbf{e}$ is the rate of deformation tensor defined in (1.2.3), and an asterisk indicates the complex conjugate. Briefly, we have

$$\int_D u_i^* f_i \, dS = -2\mu \int_V e_{ik}^* e_{ik} \, dV \tag{1.5.2}$$

We note that since the integral on the right-hand side of (1.5.2) is real and non-negative, the integral on the left-hand side must be real and non-positive.

Restricting ourselves to real velocity fields, we identify the right-hand side of (1.5.2) with the rate of dissipation of mechanical energy within the control volume $V$, and the left-hand side with the rate of working of the surface force on $D$. Under this light, we recognize that (1.5.2) provides us with a statement of conservation of energy. Because the fluid possesses no momentum and hence, no inertial mass, the rate of accumulation of kinetic energy into the control volume is equal to zero.

Now, let us assume that over the boundary $D$ either the velocity $\mathbf{u}$ vanishes, or the surface force $\mathbf{f}$ vanishes, or the velocity is normal to the surface force $\mathbf{u}\cdot\mathbf{f}=0$. Under these circumstances, the left-hand side of (1.5.2) will be equal to zero, requiring that the rate of deformation tensor $\mathbf{e}$ vanishes throughout $V$ and hence, that $\mathbf{u}$ expresses rigid body motion, i.e. translation and rotation (problem 1.5.1). An important consequence of this result is the uniqueness of solution of the equations of Stokes flow, first noted by Helmholtz in 1868 (see Batchelor 1967, p. 227). Assume for a moment that there are two solutions corresponding to a given set of boundary conditions on $D$, either for the velocity or for the surface force. The flow expressed by the difference between these solutions has homogeneous boundary conditions on $D$ and thus it must express rigid body motion. If the boundary conditions specify the velocity over a portion of $D$, rigid body motion is not permissible, the difference flow must vanish, and the solution is unique. If, on the other hand, the boundary conditions specify the surface force exclusively, any solution may be augmented with the addition of an arbitrary rigid body motion.

To present a further application of (1.5.2), we consider the flow produced by a rigid particle that translates and rotates in an infinite ambient fluid. We select a control volume $V$ that is confined by the surface of the particle $S_P$ and by a surface $S_\infty$ of large radius, and apply (1.5.2) with $D$ equal to $S_P$ plus $S_\infty$. Letting the radius of $S_\infty$ tend to infinity, we find that the corresponding surface integral on the left-hand side makes a vanishing contribution (see section 2.3). Requiring the boundary condition $\mathbf{u}=\mathbf{U}+\boldsymbol{\omega}\times\mathbf{x}$ on $S_P$, where $\mathbf{U}$ and $\boldsymbol{\omega}$ are the linear and angular velocities of motion, we obtain

$$\mathbf{U}\cdot\mathbf{F}+\boldsymbol{\omega}\cdot\mathbf{L}=-2\mu\int_V e_{ik}e_{ik}\,\mathrm{d}V \tag{1.5.3}$$

where $\mathbf{F}$ and $\mathbf{L}$ are the force and torque exerted on the particle. Equation (1.5.3) expresses a balance between the rate of supply of mechanical energy necessary to sustain the motion of the particle, and the rate of viscous dissipation within the flow. Evidently, the left-hand side of (1.5.3) must be non-positive for any combination of $\mathbf{U}$ and $\boldsymbol{\omega}$, requiring that $\mathbf{U}\cdot\mathbf{F}\leqslant 0$ and $\boldsymbol{\omega}\cdot\mathbf{L}\leqslant 0$, independently. Physically, these inequalities imply that the drag force exerted on a translating particle and the torque exerted on a rotating particle must resist the motion of the particle.

One further theorem of Stokes flow, due to Helmholtz, states that the rate of viscous dissipation in a Stokes flow is lower than that in any other incompressible flow that has the same boundary values of the velocity (see Batchelor 1967, p. 227). Hill & Power (1956) and several subsequent workers derived upper and lower bounds of the rate of energy dissipation in a Stokes flow, and used them to estimate the force and torque exerted on a particle that is suspended in a Stokes flow (see Happel & Brenner 1973, p. 91, and Kim & Karrila 1991, Chapter 2).

**Problems**

1.5.1 Show that if the rate of deformation tensor $\mathbf{e}$ vanishes throughout a flow, the flow must represent rigid body motion.

1.5.2 Using the minimum energy dissipation theorem show that the magnitude of the drag force and torque exerted on a rigid particle that moves steadily in a fluid under conditions of Stokes flow are lower than those for flow at finite Reynolds numbers.

## 1.6 Unsteady Stokes flow

In section 1.1 we saw that when the frequency parameter $\beta$ is of order unity, but the Reynolds number is much less than unity, the flow is governed by the continuity equation $\nabla \cdot \mathbf{u} = 0$ and the unsteady Stokes equation

$$\rho \frac{\partial \mathbf{u}}{\partial t} = -\nabla P + \mu \nabla^2 \mathbf{u} + \rho \mathbf{b} \tag{1.6.1}$$

Taking the divergence of (1.6.1) and using the continuity equation we find that the pressure is a harmonic function. Furthermore, taking the curl of (1.6.1) we find that the vorticity satisfies the vectorial unsteady heat conduction equation

$$\frac{\partial \boldsymbol{\omega}}{\partial t} = \nu \nabla^2 \boldsymbol{\omega} \tag{1.6.2}$$

To simplify the analysis, it is useful to take an integral transform of (1.6.1) with respect to time. When the flow has started impulsively from rest, i.e. $\mathbf{u}(t = \bar{0}) = 0, P(t = \bar{0}) = 0$, it is convenient to take the Laplace transform obtaining

$$\rho s \hat{\mathbf{u}} = -\nabla \hat{P} + \mu \nabla^2 \hat{\mathbf{u}} + \rho \hat{\mathbf{b}} \tag{1.6.3}$$

where

$$[\hat{\mathbf{u}}, \hat{P}, \hat{\mathbf{b}}] = \int_{0^-}^{\infty} [\mathbf{u}, P, \mathbf{b}] \exp(-st)\,dt \tag{1.6.4}$$

For a general unsteady flow, it is convenient to take the Fourier transform obtaining

$$- \mathrm{i}\omega\rho \mathcal{u} = - \nabla\mathcal{P} + \mu\nabla^2 \mathcal{u} + \rho\mathcal{b} \tag{1.6.5}$$

where

$$[\mathcal{u}, \mathcal{P}, \mathcal{b}] = \int_{-\infty}^{\infty} [\mathbf{u}, P, \mathbf{b}] \exp(\mathrm{i}\omega t)\, \mathrm{d}t \tag{1.6.6}$$

The inverse transform gives

$$[\mathbf{u}, P, \mathbf{b}] = \frac{1}{2\pi} \int_{-\infty}^{\infty} [\mathcal{u}, \mathcal{P}, \mathcal{b}] \exp(-\mathrm{i}\omega t)\, \mathrm{d}\omega \tag{1.6.7}$$

For an oscillatory flow, it is convenient to express the velocity and pressure as well as the body force in the form

$$[\mathbf{u}, P, \mathbf{b}] = [\mathcal{u}, \mathcal{P}, \mathcal{b}] \exp(-\mathrm{i}\omega t) \tag{1.6.8}$$

where $\omega$ is the angular velocity of the oscillations. Substituting (1.6.8) into (1.6.1) we obtain (1.6.5).

Now, non-dimensionalizing (1.6.3) and (1.6.5) as indicated in section 1.1, we obtain the dimensionless unsteady Stokes equation

$$(\nabla'^2 - \lambda^2)\mathbf{u}' = \nabla' P' - \mathbf{b}' \tag{1.6.9}$$

where in the case of Laplace transform $\lambda^2 = sL^2/\nu$, whereas in the case of Fourier transform or oscillatory flow $\lambda^2 = -\mathrm{i}\omega L^2/\nu$. It will be noted that (1.6.9) is identical with Brinkman's equation describing flow in porous media, in which case $\lambda^2$ plays the role of the medium's resistance to flow (Howell 1974).

For simplicity, in the ensuing discussion we shall drop the primes, but shall tacitly assume that we are dealing with non-dimensional variables. In addition, we shall accommodate the effect of the body force $\mathbf{b}$ into the modified pressure defined in (1.2.9).

The equations of unsteady Stokes flow share several of the properties of the equations of steady or quasi-steady Stokes flow, in particular, the property of reversibility and those properties expressed in the reciprocal identities (1.4.3), (1.4.4), and (1.4.5) (problem 1.6.1). The counterpart of the energy equation (1.5.2) is

$$\int_D u_i^* f_i \,\mathrm{d}S = -\lambda^2 \int_V u_i^* u_i \,\mathrm{d}V - 2 \int_V e_{ik}^* e_{ik} \,\mathrm{d}V \tag{1.6.10}$$

expressing a balance between the rate of working of hydrodynamic forces on the surface $D$, the rate of change of the kinetic energy of the fluid within $V$, and the rate of viscous dissipation within $V$ (problem 1.6.2). Note that an asterisk in (1.6.10) indicates the complex conjugate. To demonstrate that the solution to the equations of unsteady Stokes flow is unique, we assume that there are two solutions with identical boundary values of the velocity or surface force over $D$ and apply (1.6.10) for the flow expressed by the difference between the two solutions. Because this difference flow has homogeneous boundary conditions over $D$, the integral on the left-hand side of (1.6.10) vanishes yielding

$$\lambda^2 \int_{\mathscr{V}} u_i^* u_i \, \mathrm{d}V = -2 \int_{\mathscr{V}} e_{ik}^* e_{ik} \, \mathrm{d}V \qquad (1.6.11)$$

Noting that both integrals in (1.6.11) are real and non-negative and recalling that $\lambda^2$ is either purely imaginary or real and positive we require that the difference velocity vanishes throughout the domain of flow.

One strategy for computing a general unsteady Stokes flow is to take the Laplace or Fourier transform of the boundary conditions, to solve the transformed equations of unsteady Stokes flow for several values of $\lambda^2$ subject to the transformed boundary conditions, and then to compute the inverse transform using a numerical method. Unless the geometry of the boundaries is very simple, however, this procedure requires a considerable analytical and computational effort. Accordingly, previous work on unsteady Stokes flow has focused mainly on oscillatory flow and simple cases of impulsively started flow (Landau & Lifschitz 1987, section 24; Lawrence & Weinbaum 1988; Pozrikidis 1989a, b).

It should be noted that an alternative strategy for computing a general unsteady Stokes flow is to express the solution as a convolution integral in time using the space–time fundamental solution of the transformed equations of unsteady Stokes flow (Banerjee & Butterfield 1981, Chapter 10). A discussion of this method falls outside the scope of this book.

## Problems

1.6.1 Show that the reciprocal identities (1.4.3), (1.4.4), and (1.4.5) are also valid for unsteady Stokes flow.

1.6.2 Prove the energy balance (1.6.10).

1.6.3 Show that an unsteady Stokes flow with homogeneous boundary conditions for the velocity or surface force must necessarily vanish.

# 2

# *The boundary integral equations*

## 2.1 Green's functions of Stokes flow

The Green's functions of Stokes flow represent solutions of the continuity equation $\nabla \cdot \mathbf{u} = 0$ and the singularly forced Stokes equation

$$-\nabla P + \mu \nabla^2 \mathbf{u} + \mathbf{g}\delta(\mathbf{x} - \mathbf{x}_0) = 0 \tag{2.1.1}$$

where $\mathbf{g}$ is an arbitrary constant, $\mathbf{x}_0$ is an arbitrary point, and $\delta$ is the three-dimensional delta function. Introducing the Green's function $\mathbf{G}$, we write the solution of (2.1.1) in the form

$$u_i(\mathbf{x}) = \frac{1}{8\pi\mu} G_{ij}(\mathbf{x}, \mathbf{x}_0) g_j \tag{2.1.2}$$

where $\mathbf{x}_0$ is the *pole* or the *source point*, and $\mathbf{x}$ is the *observation* or *field point*. Physically, (2.1.2) expresses the velocity field due to a concentrated point force of strength $\mathbf{g}$ placed at the point $\mathbf{x}_0$, and may be identified with the flow produced by the slow settling of a small particle. In the literature of boundary integral methods, the Green's function may appear under the names *fundamental solution* or *propagator*.

It is convenient to classify the Green's functions into three categories depending on the topology of the domain of flow. First, we have the free-space Green's function for infinite unbounded flow; second, the Green's functions for infinite or semi-infinite flow that is bounded by a solid surface; and third, the Green's functions for internal flow that is completely confined by solid surfaces. The Green's functions in the second and third categories are required to vanish over the internal or external boundaries of the flow. As the observation point $\mathbf{x}$ approaches the pole $\mathbf{x}_0$ all Green's functions exhibit singular behaviour and, to leading order, behave like the free-space Green's function. The Green's functions for infinite unbounded or bounded flow are required to decay at infinity at a rate equal to or lower than that of the free-space Green's function.

Taking the divergence of (2.1.2) and using the continuity equation we find

$$\frac{\partial G_{ij}}{\partial x_i}(\mathbf{x}, \mathbf{x}_0) = 0 \tag{2.1.3}$$

Integrating (2.1.3) over a volume of fluid that is bounded by the surface $D$ and using the divergence theorem, we find

$$\int_D G_{ij}(\mathbf{x}, \mathbf{x}_0) n_i(\mathbf{x})\, dS(\mathbf{x}) = 0 \tag{2.1.4}$$

independently of whether the pole $\mathbf{x}_0$ is located inside, right on, or outside $D$.

The vorticity, pressure, and stress fields associated with the flow (2.1.2) may be presented in the corresponding forms:

$$\omega_i(\mathbf{x}) = \frac{1}{8\pi\mu} \Omega_{ij}(\mathbf{x}, \mathbf{x}_0) g_j \tag{2.1.5}$$

$$P(\mathbf{x}) = \frac{1}{8\pi} p_j(\mathbf{x}, \mathbf{x}_0) g_j \tag{2.1.6}$$

$$\sigma_{ik}(\mathbf{x}) = \frac{1}{8\pi} T_{ijk}(\mathbf{x}, \mathbf{x}_0) g_j \tag{2.1.7}$$

where $\Omega$, $\mathbf{p}$, and $\mathbf{T}$ are the vorticity tensor, pressure vector, and stress tensor associated with the Green's function. The stress tensor $\mathbf{T}$, in particular, is defined as

$$T_{ijk}(\mathbf{x}, \mathbf{x}_0) = -\delta_{ik} p_j(\mathbf{x}, \mathbf{x}_0) + \frac{\partial G_{ij}}{\partial x_k}(\mathbf{x}, \mathbf{x}_0) + \frac{\partial G_{kj}}{\partial x_i}(\mathbf{x}, \mathbf{x}_0) \tag{2.1.8}$$

It will be noted that $T_{ijk} = T_{kji}$ as required by the symmetry of the stress tensor $\boldsymbol{\sigma}$. When the domain of flow is infinite, we require that all $\boldsymbol{\Omega}$, $\mathbf{p}$, and $\mathbf{T}$ vanish as the observation point is moved to infinity.

Substituting (2.1.2), (2.1.6), and (2.1.8) into (2.1.1) we obtain the equations

$$-\frac{\partial p_j}{\partial x_k}(\mathbf{x}, \mathbf{x}_0) + \nabla^2 G_{kj}(\mathbf{x}, \mathbf{x}_0) = -8\pi\delta_{kj}\delta(\mathbf{x} - \mathbf{x}_0) \tag{2.1.9}$$

and

$$\frac{\partial T_{ijk}}{\partial x_i}(\mathbf{x}, \mathbf{x}_0) = \frac{\partial T_{jki}}{\partial x_i}(\mathbf{x}, \mathbf{x}_0) = -8\pi\delta_{kj}\delta(\mathbf{x} - \mathbf{x}_0) \tag{2.1.10}$$

Furthermore, using (2.1.10) we find

$$\frac{\partial}{\partial x_k}[\varepsilon_{ilm} x_l T_{mjk}(\mathbf{x}, \mathbf{x}_0)] = -8\pi\varepsilon_{ilj} x_l \delta(\mathbf{x} - \mathbf{x}_0) \tag{2.1.11}$$

Integrating (2.1.10) and (2.1.11) over the volume of fluid enclosed by the smooth surface $D$ and using the divergence theorem to convert the volume integral into a surface integral, we obtain the identities

$$\int_D T_{ijk}(\mathbf{x}, \mathbf{x}_0) n_i(\mathbf{x})\, dS(\mathbf{x}) = \int_D T_{kji}(\mathbf{x}, \mathbf{x}_0) n_i(\mathbf{x})\, dS(\mathbf{x}) = -\begin{bmatrix} 8\pi \\ 4\pi \\ 0 \end{bmatrix} \delta_{jk} \quad (2.1.12)$$

$$\varepsilon_{ilm} \int_D x_l T_{mjk}(\mathbf{x}, \mathbf{x}_0) n_k(\mathbf{x})\, dS(\mathbf{x}) = -\begin{bmatrix} 8\pi \\ 4\pi \\ 0 \end{bmatrix} \varepsilon_{ilj} x_{0,l} \quad (2.1.13)$$

where the unit normal vector $\mathbf{n}$ is directed outside the control volume, and $x_{0,l}$ on the right-hand side of (2.1.13) indicates the $l$ component of $\mathbf{x}_0$. The values $-8\pi$, $-4\pi$, and 0 on the right-hand sides of (2.1.12) and (2.1.13) apply when the point $\mathbf{x}_0$ is located respectively inside, right on, or outside $D$. When $\mathbf{x}_0$ is right on $D$, the integrals in (2.1.12) and (2.1.13) are improper but convergent (see discussion at the end of section 2.3).

In section 3.2 we shall see that the pressure vector $\mathbf{p}$ and the stress tensor $\mathbf{T}$ associated with a Green's function for infinite unbounded or bounded flow represent two fundamental solutions of Stokes flow. Specifically, we shall show that $\mathbf{p}(\mathbf{x}, \mathbf{x}_0)$ represents the velocity field at the point $\mathbf{x}_0$, due to a point source of strength $-8\pi$ with pole at $\mathbf{x}$. Furthermore, we shall show that

$$u_j(\mathbf{x}_0) = T_{ijk}(\mathbf{x}, \mathbf{x}_0) q_{ik} \quad (2.1.14)$$

where $\mathbf{q}$ is a constant matrix, represents the velocity field due to a singularity called the stresslet with pole at $\mathbf{x}$. The pressure field corresponding to (2.1.14) may be conveniently expressed in terms of a pressure matrix $\mathbf{\Pi}$ as

$$P(\mathbf{x}_0) = \mu \Pi_{ik}(\mathbf{x}_0, \mathbf{x}) q_{ik} \quad (2.1.15)$$

The precise definition and further properties of $\mathbf{\Pi}$ will be discussed in section 3.2.

Adding a number of Green's functions with different poles $\mathbf{x}_n$ we can devise a Green's function with multiple poles, namely

$$\mathbf{G} = \sum_{n=1}^{N} \mathbf{G}(\mathbf{x}, \mathbf{x}_n) \quad (2.1.16)$$

In the limiting case where an infinite number of poles are placed exceedingly close to each other, the sum in (2.1.16) reduces to an integral yielding a line, surface, or volume distribution of point forces. Differentiating the Green's function with respect to the pole $\mathbf{x}_0$ we can derive differential singular solutions representing multipoles of the point force (see section 7.2). For instance, differentiating the Green's function once, we obtain the point force doublet that represents the flow produced by two point forces with opposite strengths and indistinguishable poles.

## 2.2 The free-space Green's function

To compute the free-space Green's function we replace the delta function on the right-hand side of (2.1.1) with the equivalent expression

$$\delta(\hat{\mathbf{x}}) = -\frac{1}{4\pi}\nabla^2\left(\frac{1}{r}\right) \tag{2.2.1}$$

where $r = |\hat{\mathbf{x}}|$, $\hat{\mathbf{x}} = \mathbf{x} - \mathbf{x}_0$. Recalling that the pressure is a harmonic function, and balancing the dimensions of the pressure term with those of the delta function in equation (2.1.1), we set

$$P = -\frac{1}{4\pi}\mathbf{g}\cdot\nabla\left(\frac{1}{r}\right) \tag{2.2.2}$$

Substituting (2.2.1) and (2.2.2) into (2.1.1) we obtain

$$\mu\nabla^2\mathbf{u} = -\frac{1}{4\pi}\mathbf{g}\cdot(\nabla\nabla - \mathbf{I}\nabla^2)\left(\frac{1}{r}\right) \tag{2.2.3}$$

Next, we express the velocity in terms of a scalar function $H$ as

$$\mathbf{u} = \frac{1}{\mu}\mathbf{g}\cdot(\nabla\nabla - \mathbf{I}\nabla^2)H \tag{2.2.4}$$

It will be noted that the continuity equation is satisfied for any choice of $H$. Substituting (2.2.4) into (2.2.3) and discarding the arbitrary constant $\mathbf{g}$ we obtain

$$(\nabla\nabla - \mathbf{I}\nabla^2)\left(\nabla^2 H + \frac{1}{4\pi r}\right) = 0 \tag{2.2.5}$$

Clearly, (2.2.5) is satisfied by any solution of Poisson's equation, $\nabla^2 H = -1/(4\pi r)$. Using (2.2.1) we find that $H$ is, in fact, the fundamental solution of the biharmonic equation $\nabla^4 H = \delta(\hat{\mathbf{x}})$. Thus

$$H = -\frac{r}{8\pi} \tag{2.2.6}$$

Substituting (2.2.6) into (2.2.4) we find

$$u_i(\mathbf{x}) = \frac{1}{8\pi\mu}\mathcal{S}_{ij}(\hat{\mathbf{x}})g_j \tag{2.2.7}$$

where

$$\mathcal{S}_{ij}(\hat{\mathbf{x}}) = \frac{\delta_{ij}}{r} + \frac{\hat{x}_i\hat{x}_j}{r^3} \tag{2.2.8}$$

is the free-space Green's function, also called the *Stokeslet*, or the

*Oseen–Burgers tensor.* The vorticity, pressure, and stress fields associated with the flow (2.2.7) may be written in the standard forms (2.1.5), (2.1.6), and (2.1.7) where

$$\Omega_{ij}(\hat{\mathbf{x}}) = 2\varepsilon_{ijl}\frac{\hat{x}_l}{r^3} \tag{2.2.9}$$

and

$$p_i(\hat{\mathbf{x}}) = 2\frac{\hat{x}_i}{r^3} \tag{2.2.10}$$

Substituting (2.2.7) and (2.2.10) into (2.1.8) we obtain the stress tensor

$$T_{ijk}(\hat{\mathbf{x}}) = -6\frac{\hat{x}_i\hat{x}_j\hat{x}_k}{r^5} \tag{2.2.11}$$

As mentioned in section 2.1, $\mathbf{p}$ and $\mathbf{T}$ represent two fundamental solutions of Stokes flow. Specifically, $\mathbf{p}$ represents the velocity at the point $\mathbf{x}$ due to a point source of strength $8\pi$ with pole at $\mathbf{x}_0$, or, equivalently, the velocity at $\mathbf{x}_0$ due to a point source of strength $-8\pi$ with pole at $\mathbf{x}$, whereas

$$u_j(\mathbf{x}_0) = T_{ijk}(\mathbf{x} - \mathbf{x}_0)q_{ik} = -T_{ijk}(\mathbf{x}_0 - \mathbf{x})q_{ik} \tag{2.2.12}$$

where $\mathbf{q}$ is a constant matrix, represents the velocity field due to a stresslet with pole at $\mathbf{x}$. Using the results of section 7.2 we find that the pressure field corresponding to the flow (2.2.12) is given by (2.1.15) where

$$\Pi_{ik}(\mathbf{x}_0, \mathbf{x}) = 4\left(-\frac{\delta_{ik}}{r^3} + 3\frac{\hat{x}_i\hat{x}_k}{r^5}\right) \tag{2.2.13}$$

The associated stress field will be discussed in problem 2.2.2.

Now, as an exercise, we shall compute the surface force exerted on a fluid sphere of radius $r$ centered at the pole of a point force. Using (2.1.7) and (2.2.11) we find

$$f_i(\mathbf{x}) = \sigma_{ik}(\mathbf{x})n_k(\mathbf{x}) = \frac{1}{8\pi}T_{ijk}(\mathbf{x}, \mathbf{x}_0)n_k(\mathbf{x})g_j = -\frac{3}{4\pi}\frac{\hat{x}_i\hat{x}_j}{r^4}g_j \tag{2.2.14}$$

The force acting on the sphere is

$$F_i = \int_{\text{sphere}} f_i(\mathbf{x})\,\mathrm{d}S(\mathbf{x}) = -\frac{3}{4\pi}g_j\frac{1}{r^4}\int_{\text{sphere}} \hat{x}_i\hat{x}_j\,\mathrm{d}S(\mathbf{x}) \tag{2.2.15}$$

Using the divergence theorem we compute

$$\int_{\text{sphere}} \hat{x}_i\hat{x}_j\,\mathrm{d}S(\mathbf{x}) = r\int_{\text{sphere}} \hat{x}_i n_j\,\mathrm{d}S(\mathbf{x}) = r\int_{\text{sphere}} \frac{\partial \hat{x}_i}{\partial \hat{x}_j}\,\mathrm{d}V(\mathbf{x}) = \delta_{ij}\tfrac{4}{3}\pi r^4 \tag{2.2.16}$$

Combining (2.2.15) and (2.2.16) we find $\mathbf{F} = -\mathbf{g}$ independently of the radius of the sphere, in agreement with our previous discussion in section 1.2. The torque with respect to the pole of a point force on any surface that encloses the pole of the point force is equal to zero (see problem 2.2.3).

### Problems

2.2.1 An alternative method for deriving the free-space Green's function is by using Fourier transforms. Take the three-dimensional complex Fourier transform of (2.1.1) and the continuity equation to find

$$\hat{G}_{ij} = \frac{4}{(2\pi)^{1/2}} \frac{1}{|\mathbf{k}|^2}\left(\delta_{ij} - \frac{k_i k_j}{|\mathbf{k}|^2}\right) \qquad \hat{p}_j = -\frac{4i}{(2\pi)^{1/2}} \frac{k_j}{|\mathbf{k}|^2} \tag{1}$$

where the three-dimensional complex Fourier transform of a function $f(\mathbf{x})$ is defined as

$$\hat{f}(\mathbf{k}) = \frac{1}{(2\pi)^{3/2}} \int_{\substack{\text{whole}\\ \text{space}}} f(\mathbf{x}) \exp(-i\mathbf{k}\cdot\mathbf{x})\, d\mathbf{x} \tag{2}$$

Next, invert (1) using

$$f(\mathbf{x}) = \frac{1}{(2\pi)^{3/2}} \int_{\substack{\text{whole}\\ \text{space}}} \hat{f}(\mathbf{k}) \exp(i\mathbf{x}\cdot\mathbf{k})\, d\mathbf{k} \tag{3}$$

to obtain the Stokeslet (Ladyzhenskaya 1969, p. 50).

2.2.2 Show that the stress field associated with the flow (2.2.12) is given by $\sigma_{ik} = 2\mu T^{\mathrm{STR}}_{ijlk}(\hat{\mathbf{x}}) q_{jl}$, where the stress tensor $\mathbf{T}^{\mathrm{STR}}$ is given in (7.2.26).

2.2.3 Using (2.2.14) show that the torque with respect to the pole of a point force on any surface that encloses the pole of the point force is equal to zero. What is the torque with respect to another point in space?

## 2.3 The boundary integral equation

It is well known that the solution of linear, elliptic, and homogeneous boundary value problems may be represented in terms of boundary integrals involving the boundary values of the unknown function and its derivatives (Stakgold 1968). One example of a boundary integral representation is Green's third identity for harmonic functions (Kellogg 1954, p. 219). Another example is Somigliana's identity for the displacement field in linear elastostatics (Love 1944, p. 245). In the case of Stokes flow, we obtain a boundary integral representation involving the boundary values of the velocity and surface force.

A convenient starting point for deriving the boundary integral equation is the Lorentz reciprocal identity (1.4.4) stating that for any two non-singular (regular) flows $\mathbf{u}$ and $\mathbf{u}'$ with corresponding stress tensors $\boldsymbol{\sigma}$ and $\boldsymbol{\sigma}'$,

$$\frac{\partial}{\partial x_k}(u_i' \sigma_{ik} - u_i \sigma_{ik}') = 0 \tag{2.3.1}$$

Identifying $\mathbf{u}'$ with the flow due to a point force with strength $\mathbf{g}$ located at the point $\mathbf{x}_0$, we obtain

$$u_i'(\mathbf{x}) = \frac{1}{8\pi\mu} G_{ij}(\mathbf{x}, \mathbf{x}_0) g_j \qquad \sigma_{ik}'(\mathbf{x}) = \frac{1}{8\pi} T_{ijk}(\mathbf{x}, \mathbf{x}_0) g_j \tag{2.3.2}$$

Substituting (2.3.2) into (2.3.1), and discarding the arbitrary constant $\mathbf{g}$, we obtain

$$\frac{\partial}{\partial x_k}[G_{ij}(\mathbf{x}, \mathbf{x}_0)\sigma_{ik}(\mathbf{x}) - \mu u_i(\mathbf{x}) T_{ijk}(\mathbf{x}, \mathbf{x}_0)] = 0 \tag{2.3.3}$$

Now, we select a control volume $V$ that is bounded by the closed (simply- or multiply-connected) surface $D$, as illustrated in Figure 2.3.1. Note that $D$ may be composed of fluid surfaces, fluid interfaces, or solid surfaces. In addition, we select a point $\mathbf{x}_0$ outside $V$. Noting that the function within the square bracket in (2.3.3) is regular throughout $V$, integrating (2.3.3) over $V$, and using the divergence theorem to convert the volume integral over $V$ into a surface integral over $D$, we obtain

$$\int_D [G_{ij}(\mathbf{x}, \mathbf{x}_0)\sigma_{ik}(\mathbf{x}) - \mu u_i(\mathbf{x}) T_{ijk}(\mathbf{x}, \mathbf{x}_0)] n_k(\mathbf{x}) \, dS(\mathbf{x}) = 0 \tag{2.3.4}$$

In (2.3.4) as well as in all subsequent equations, the normal vector $\mathbf{n}$ is directed into the control volume $V$.

Next we select a point $\mathbf{x}_0$ in the interior of $V$, and define a small spherical volume $V_\varepsilon$ of radius $\varepsilon$ centered at $\mathbf{x}_0$. The function within the square bracket

*Figure 2.3.1.* A control volume $V$ within the domain of a flow.

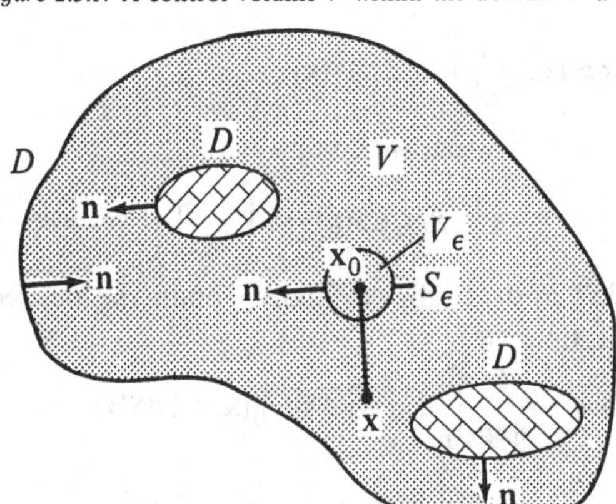

in (2.3.3) is regular throughout the reduced volume $V - V_\varepsilon$. Integrating (2.3.3) over $V - V_\varepsilon$ and using the divergence theorem to convert the volume integral into a surface integral, we obtain

$$\int_{D,S_\varepsilon} [G_{ij}(\mathbf{x}, \mathbf{x}_0)\sigma_{ik}(\mathbf{x}) - \mu u_i(\mathbf{x}) T_{ijk}(\mathbf{x}, \mathbf{x}_0)] n_k(\mathbf{x})\, dS(\mathbf{x}) = 0 \tag{2.3.5}$$

where $S_\varepsilon$ is the spherical surface enclosing $V_\varepsilon$ as indicated in Figure 2.3.1. Letting the radius $\varepsilon$ tend to zero we find that over $S_\varepsilon$, to leading order in $\varepsilon$, the tensors $\mathbf{G}$ and $\mathbf{T}$ reduce to the Stokeslet and its associated stress tensor, respectively, i.e.

$$G_{ij} \approx \frac{\delta_{ij}}{\varepsilon} + \frac{\hat{x}_i \hat{x}_j}{\varepsilon^3} \qquad T_{ijk} \approx -6\frac{\hat{x}_i \hat{x}_j \hat{x}_k}{\varepsilon^5} \tag{2.3.6}$$

where $\hat{\mathbf{x}} = \mathbf{x} - \mathbf{x}_0$. Over $S_\varepsilon$, $\mathbf{n} = \hat{\mathbf{x}}/\varepsilon$ and $dS = \varepsilon^2\, d\Omega$, where $\Omega$ is the differential solid angle. Substituting these expressions along with (2.3.6) into (2.3.5) we obtain

$$\int_D [G_{ij}(\mathbf{x}, \mathbf{x}_0)\sigma_{ik}(\mathbf{x}) - \mu u_i(\mathbf{x}) T_{ijk}(\mathbf{x}, \mathbf{x}_0)] n_k(\mathbf{x})\, dS(\mathbf{x})$$

$$= -\int_{S_\varepsilon} \left[\left(\delta_{ij} + \frac{\hat{x}_i \hat{x}_j}{\varepsilon^2}\right)\sigma_{ik}(\mathbf{x}) + 6\mu u_i(\mathbf{x})\frac{\hat{x}_i \hat{x}_j \hat{x}_k}{\varepsilon^4}\right]\hat{x}_k\, d\Omega \tag{2.3.7}$$

As $\varepsilon \to 0$, the values of $\mathbf{u}$ and $\boldsymbol{\sigma}$ over $S_\varepsilon$ tend to their corresponding values at the center of $V_\varepsilon$, i.e. to $\mathbf{u}(\mathbf{x}_0)$ and $\boldsymbol{\sigma}(\mathbf{x}_0)$, respectively. Since $\hat{\mathbf{x}}$ decreases linearly with $\varepsilon$, as $\varepsilon \to 0$ the contribution of the stress term within the integral on the right-hand side of (2.3.7) decreases linearly in $\varepsilon$, whereas the contribution of the velocity term tends to a constant value. Thus, in the limit $\varepsilon \to 0$, (2.3.7) reduces to

$$\int_D [G_{ij}(\mathbf{x}, \mathbf{x}_0)\sigma_{ik}(\mathbf{x}) - \mu u_i(\mathbf{x}) T_{ijk}(\mathbf{x}, \mathbf{x}_0)] n_k(\mathbf{x})\, dS(\mathbf{x})$$

$$= -6\mu u_i(\mathbf{x}_0)\frac{1}{\varepsilon^4}\int_{S_\varepsilon} \hat{x}_i \hat{x}_j\, dS(\mathbf{x}) \tag{2.3.8}$$

Using the divergence theorem we compute

$$\int_{S_\varepsilon} \hat{x}_i \hat{x}_j\, dS(\mathbf{x}) = \varepsilon \int_{S_\varepsilon} \hat{x}_i n_j\, dS(\mathbf{x}) = \varepsilon \int_{V_\varepsilon} \frac{\partial \hat{x}_i}{\partial \hat{x}_j}\, dV(\mathbf{x}) = \delta_{ij}\tfrac{4}{3}\pi\varepsilon^4 \tag{2.3.9}$$

Substituting (2.3.9) into (2.3.8) we finally obtain the desired boundary integral representation

$$u_j(\mathbf{x}_0) = -\frac{1}{8\pi\mu}\int_D \sigma_{ik}(\mathbf{x}) n_k(\mathbf{x}) G_{ij}(\mathbf{x}, \mathbf{x}_0)\, dS(\mathbf{x})$$

$$+ \frac{1}{8\pi}\int_D u_i(\mathbf{x}) T_{ijk}(\mathbf{x}, \mathbf{x}_0) n_k(\mathbf{x})\, dS(\mathbf{x}) \tag{2.3.10}$$

It will be convenient to introduce the surface force $\mathbf{f} = \boldsymbol{\sigma} \cdot \mathbf{n}$ and rewrite (2.3.10) in the equivalent form

$$u_j(\mathbf{x}_0) = -\frac{1}{8\pi\mu} \int_D f_i(\mathbf{x}) G_{ij}(\mathbf{x}, \mathbf{x}_0) \, dS(\mathbf{x}) + \frac{1}{8\pi} \int_D u_i(\mathbf{x}) T_{ijk}(\mathbf{x}, \mathbf{x}_0) n_k(\mathbf{x}) \, dS(\mathbf{x}) \qquad (2.3.11)$$

Equation (2.3.11) provides us with a representation of a flow in terms of two boundary distributions involving the Green's function $\mathbf{G}$ and the associated stress tensor $\mathbf{T}$. The densities of these distributions are proportional to the boundary values of the surface force and velocity. The first distribution on the right-hand of (2.3.11) is termed the *single-layer potential*, whereas the second distribution is termed the *double-layer potential*. A detailed discussion of the significance and properties of these potentials will be deferred until Chapter 4.

Now, viewing the double-layer potential as a mere mathematical function, we compute its limiting values as the point $\mathbf{x}_0$ approaches the boundary $D$ either from the internal or from the external side, and obtain two different values. Specifically, if $D$ is a Lyapunov surface, i.e. it has a continuously varying normal vector (see Jaswon & Symm 1977), and the velocity over $D$ varies in a continuous manner, we find

$$\lim_{\mathbf{x}_0 \to D} \int_D u_i(\mathbf{x}) T_{ijk}(\mathbf{x}, \mathbf{x}_0) n_k(\mathbf{x}) \, dS(\mathbf{x}) = \pm 4\pi u_j(\mathbf{x}_0) + \int_D^{\mathscr{PV}} u_i(\mathbf{x}) T_{ijk}(\mathbf{x}, \mathbf{x}_0) n_k(\mathbf{x}) \, dS(\mathbf{x}) \qquad (2.3.12)$$

where the plus sign applies when the point $\mathbf{x}_0$ approaches $D$ from the side of the flow (indicated by the direction of the normal vector), and the minus sign otherwise (see section 4.3). The superscript $\mathscr{PV}$ indicates the principal value of the double-layer potential, defined as the value of the improper double-layer integral when the point $\mathbf{x}_0$ is right on $D$. Substituting (2.3.12) with the plus sign into (2.3.11) or with the minus sign into (2.3.4), we find that for a point $\mathbf{x}_0$ that is located right on the boundary $D$,

$$u_j(\mathbf{x}_0) = -\frac{1}{4\pi\mu} \int_D f_i(\mathbf{x}) G_{ij}(\mathbf{x}, \mathbf{x}_0) \, dS(\mathbf{x}) + \frac{1}{4\pi} \int_D^{\mathscr{PV}} u_i(\mathbf{x}) T_{ijk}(\mathbf{x}, \mathbf{x}_0) n_k(\mathbf{x}) \, dS(\mathbf{x}) \qquad (2.3.13)$$

In summary, equations (2.3.4), (2.3.11), and (2.3.13) are valid when the point $\mathbf{x}_0$ is located outside, inside, or right on the boundary of a selected volume of flow.

In section 3.1 we shall show that the Green's functions satisfy the symmetry property

$$G_{ij}(\mathbf{x}, \mathbf{x}_0) = G_{ji}(\mathbf{x}_0, \mathbf{x}) \tag{2.3.14}$$

which allows us to switch the order of the indices as long as we also switch the order of the arguments, i.e. the location of the observation point and the pole. Substituting (2.3.14) into (2.3.11) we obtain

$$u_j(\mathbf{x}_0) = -\frac{1}{8\pi\mu}\int_D G_{ji}(\mathbf{x}_0, \mathbf{x}) f_i(\mathbf{x})\,\mathrm{d}S(\mathbf{x}) + \frac{1}{8\pi}\int_D u_i(\mathbf{x}) T_{ijk}(\mathbf{x}, \mathbf{x}_0) n_k(\mathbf{x})\,\mathrm{d}S(\mathbf{x}) \tag{2.3.15}$$

Clearly, the single-layer potential on the right-hand side of (2.3.15) represents a boundary distribution of point forces with strength $-\mathbf{f}$. To understand the significance of the double-layer potential, we decompose the stress tensor $\mathbf{T}$ into its constituents using (2.1.8). Exploiting (2.3.14) we obtain

$$\int_D u_i(\mathbf{x}) T_{ijk}(\mathbf{x}, \mathbf{x}_0) n_k(\mathbf{x})\,\mathrm{d}S(\mathbf{x}) = -\int_D p_j(\mathbf{x}, \mathbf{x}_0) u_i(\mathbf{x}) n_i(\mathbf{x})\,\mathrm{d}S(\mathbf{x})$$

$$+ \int_D \frac{\partial G_{ji}(\mathbf{x}_0, \mathbf{x})}{\partial x_k}(u_i n_k + u_k n_i)(\mathbf{x})\,\mathrm{d}S(\mathbf{x}) \tag{2.3.16}$$

In section 3.2 we shall see that when $\mathbf{p}$ corresponds to a Green's function of infinite unbounded or bounded flow, the first integral on the right-hand side of (2.3.16) represents a distribution of point sources. The density of this distribution vanishes over a solid surface or stationary fluid interface where $\mathbf{u} = 0$ or $\mathbf{u} \cdot \mathbf{n} = 0$ respectively. The second integral on the right-hand side of (2.3.16) represents a distribution of symmetric point force dipoles.

Now, inspecting (2.3.15) suggests an expression for the pressure in terms of two boundary distributions corresponding to the single-layer and double-layer potential, namely

$$P(\mathbf{x}_0) = -\frac{1}{8\pi}\int_{\mathscr{C}} p_i(\mathbf{x}_0, \mathbf{x}) f_i(\mathbf{x})\,\mathrm{d}l(\mathbf{x}) + \frac{\mu}{8\pi}\int_{\mathscr{C}} u_i(\mathbf{x}) \Pi_{ik}(\mathbf{x}_0, \mathbf{x}) n_k(\mathbf{x})\,\mathrm{d}l(\mathbf{x}) \tag{2.3.17}$$

where $\mathbf{p}$ and $\mathbf{\Pi}$ express the pressure corresponding to the Green's function and its associated stress tensor defined in (2.1.6) and (2.1.15), respectively.

It will be instructive to apply the boundary integral equation for certain simple flows that are known to be exact solutions to the equations of Stokes flow. For instance, if we are considering rigid body motion then $\mathbf{u} = \mathbf{U} + \boldsymbol{\omega} \times \mathbf{x}$; setting $\mathbf{f} = -P\mathbf{n}$, where $P$ is the constant pressure, and using (2.3.4), (2.3.11), (2.3.13), and (2.1.4) we find

$$\int_D T_{ijk}(\mathbf{x}, \mathbf{x}_0) n_k(\mathbf{x})\,\mathrm{d}S(\mathbf{x}) = \begin{bmatrix} 8\pi \\ 4\pi \\ 0 \end{bmatrix} \delta_{ij} \tag{2.3.18}$$

and

$$\varepsilon_{ilm} \int_D x_m T_{ijk}(\mathbf{x}, \mathbf{x}_0) n_k(\mathbf{x}) \, \mathrm{d}S(\mathbf{x}) = \begin{bmatrix} 8\pi \\ 4\pi \\ 0 \end{bmatrix} \varepsilon_{jlm} x_{0,m} \tag{2.3.19}$$

for a point $\mathbf{x}_0$ located inside, right on, or outside $D$, respectively (in the second case the integrals should be interpreted in the principal value sense).

The reader will note that (2.3.18) and (2.3.19) are identical to (2.1.12) and (2.1.13) with the exception of a minus sign due to the opposite orientation of the normal vector (the normal vector in (2.1.12) and (2.1.13) is directed outside the control volume). Two sets of identities similar to (2.3.18) and (2.3.19) may be derived by applying the boundary integral equations for linear and parabolic flow (problem 2.3.4).

To derive the above boundary integral equations, we used the reciprocal identity (1.4.4). Had we used the alternative reciprocal identity discussed in problem (1.4.2), we would have obtained a different but equivalent set of equations. Specifically, for a point $\mathbf{x}_0$ that is located within a selected volume of flow we would have obtained

$$u_j(\mathbf{x}_0) = -\frac{1}{8\pi\mu} \int_D G_{ji}(\mathbf{x}_0, \mathbf{x}) \left( -P n_i + \frac{\partial u_i}{\partial x_k} n_k \right)(\mathbf{x}) \, \mathrm{d}S(\mathbf{x})$$
$$+ \frac{1}{8\pi} \int_D u_i(\mathbf{x}) \left[ -p_j(\mathbf{x}, \mathbf{x}_0) n_i(\mathbf{x}) + \frac{\partial G_{ji}(\mathbf{x}_0, \mathbf{x})}{\partial x_k} n_k(\mathbf{x}) \right] \mathrm{d}S(\mathbf{x}) \tag{2.3.20}$$

which is the counterpart of (2.3.15) (Happel & Brenner 1973, p. 81). Due to the more direct physical significance of the density of the single-layer potential, equation (2.3.15) is preferable to (2.3.20) in theoretical analyses as well as numerical implementations.

### *Infinite flow*

A number of problems involve flow in completely unbounded or partially bounded domains. Two examples are flow due to the motion of a small particle in an infinitely dilute suspension, and semi-infinite shear flow over a wall containing a depression or projection. In these cases, in order to apply the boundary integral equation, we select a control volume that is confined by a solid or fluid boundary $S_B$ and a large spherical surface $S_\infty$ extending to infinity. If the fluid at infinity is quiescent, the velocity must decay at least as fast as $1/r$, whereas the pressure and stress must decay at least as fast as $1/r^2$, where $r$ is a typical distance from $S_B$. These scalings become evident by expanding the far pressure field in terms of spherical harmonics, requiring that the pressure at infinity tends to a constant value, and inspecting the corresponding velocity (Lamb 1932, section 335; see also Chapter 7 of this volume). We note that the Green's function decays at least as fast as $1/r$, and that its associated stress tensor decays at least as fast as $1/r^2$, and this suggests that as the radius of $S_\infty$ tends to infinity both the single-layer and double-layer potentials over $S_\infty$ make vanishing contributions. As a result, the domain $D$ of the boundary integral equation is conveniently reduced to $S_B$.

### *Simplification by the use of proper Green's functions*

The domain of the single-layer and double-layer potentials in the boundary integral equation consists of all fluid or solid surfaces that enclose an arbitrarily selected volume of flow. If the velocity happens to vanish over a portion of the boundary, the corresponding double-layer integral makes a vanishing contribution to the double-layer potential. Similarly, if the surface force happens to vanish over another portion of the boundary, the corresponding single-layer integral makes a vanishing contribution to the single-layer potential. A further reduction in the domain of integration may be effected by using a Green's function that observes the topology, symmetry, and character of the flow, as discussed below.

First, let us consider a flow that is bounded internally or externally by a stationary solid boundary $S_B$, as shown in Figure 2.3.2(*a*). Requiring that the velocity vanishes over $S_B$ ensures that the double-layer potential over $S_B$ is equal to zero. If we use a Green's function that vanishes over

*Figure 2.3.2.* The domain of the boundary integral equation may be conveniently reduced by using a proper Green's function that observes the topology, symmetry, or other special feature of a flow. (*a*) Flow confined by a rigid boundary $S_B$: the Green's function vanishes over $S_B$. (*b*) Periodic flow: the Green's function is periodic in the streamwise direction and may vanish over one or all of the boundaries of the flow.

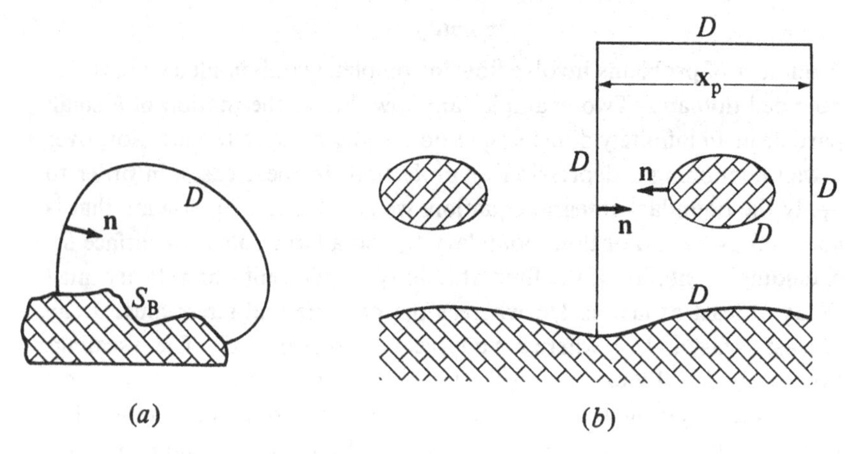

$S_B$, i.e. $\mathbf{G}(\mathbf{x}, \mathbf{x}_0) = 0$ when $\mathbf{x}$ is on $S_B$, we shall find that the single-layer potential over $S_B$ is also equal to zero. Thus, we will reduce the domain of the boundary integral equation from $D$ to $D - S_B$. Green's functions that vanish over several solid boundaries will be discussed extensively in Chapter 3.

When the flow is periodic in one or more directions, it is convenient to use a periodic Green's function that conforms with the periodicity of the flow. Physically, this Green's function represents the flow induced by a single or multiple array of point forces. To be more specific, let us assume that the flow satisfies the periodicity condition

$$\mathbf{u}(\mathbf{x}) = \mathbf{u}(\mathbf{x} + \mathbf{x}_p) \tag{2.3.21}$$

where the vector $\mathbf{x}_p$ indicates the direction and wavelength of perodicity. We introduce a periodic Green's function such that

$$\mathbf{G}(\mathbf{x}, \mathbf{x}_0) = \mathbf{G}(\mathbf{x} + \mathbf{x}_p, \mathbf{x}_0) \tag{2.3.22}$$

$$\mathbf{T}(\mathbf{x}, \mathbf{x}_0) = \mathbf{T}(\mathbf{x} + \mathbf{x}_p, \mathbf{x}_0) \tag{2.3.23}$$

and identify the control volume with one period of the flow as illustrated in Figure 2.3.2(*b*). The double-layer integrals over the two vertical sides of the indicated boundary make equal and opposite contributions and, hence, a vanishing net contribution. If there is no pressure drop in the streamwise direction, the surface forces over the two vertical sides are equal and opposite, and the corresponding single-layer integrals make a vanishing net contribution. In summary, both the single-layer and double-layer integrals over the vertical sides may be overlooked. If we use a periodic Green's function that vanishes over one or all of the boundaries of the flow (indicated in Figure 2.3.2(*b*)), we will find that the corresponding single-layer integrals are equal to zero, yielding further simplification.

A periodic Green's function representing a three-dimensional array of point forces was derived by Hasimoto (1959). Several examples of periodic Green's functions for two-dimensional flow will be discussed in section 3.5.

### *Eliminating the single-layer or double-layer potential*

In a number of theoretical and computational applications, it is convenient to simplify the boundary integral equation by eliminating either the single-layer or the double-layer potential. In this manner, we obtain *indirect* or *generalized boundary integral representations* involving hydrodynamic potentials with modified density functions.

Let us first consider the elimination of the double-layer potential. Assuming that the domain of the flow $\mathbf{u}$ is the interior of the closed boundary $D$, we introduce a complementary flow $\mathbf{u}'$ in the exterior of $D$. We stipulate that $\mathbf{u}'$ vanishes at infinity and has the same boundary values as $\mathbf{u}$ on $D$. Correspondingly, if the domain of the flow $\mathbf{u}$ is the exterior of the boundary $D$, we introduce a complementary flow $\mathbf{u}'$ in the interior of $D$ so that $\mathbf{u}'$ has the same boundary values as $\mathbf{u}$ on $D$. In either case, we select a point $\mathbf{x}_0$ inside the domain of $\mathbf{u}$, and use (2.3.4) to obtain

$$\int_D f_i'(\mathbf{x})G_{ij}(\mathbf{x},\mathbf{x}_0)\,dS(\mathbf{x}) - \mu\int_D u_i'(\mathbf{x})T_{ijk}(\mathbf{x},\mathbf{x}_0)n_k(\mathbf{x})\,dS(\mathbf{x}) = 0 \qquad (2.3.24)$$

Combining (2.3.24) with (2.3.11) we find

$$u_j(\mathbf{x}_0) = -\frac{1}{8\pi\mu}\int_D q_i(\mathbf{x})G_{ij}(\mathbf{x},\mathbf{x}_0)\,dS(\mathbf{x}) \qquad (2.3.25)$$

where $\mathbf{q} = \mathbf{f} - \mathbf{f}'$.

Effectively, we have managed to express the flow in terms of a single-layer potential with modified density $\mathbf{q}$. It should be emphasized that in the above derivation we have assumed that apart from the pole, the Green's function $\mathbf{G}$ does not have any singular points within the internal or external domains of flow.

Now, the validity of (2.3.25) hinges upon the existence of the complementary flow $\mathbf{u}'$. In section 4.2 we shall see that as long as the boundary values of $\mathbf{u}$ satisfy the integral constraint

$$\int_D \mathbf{u}\cdot\mathbf{n}\,dS = 0 \qquad (2.3.26)$$

it will always be possible to find the required external or internal flow $\mathbf{u}'$. This implies that the generalized single-layer representation (2.3.25) will be successful in describing any internal flow and those external flows not violating (2.3.26).

To present a specific application, we consider the flow produced by the translation or rotation of a rigid body, and apply the boundary integral equation (2.3.11) in a volume of fluid that is confined by the surface of the body $S_B$ and a large surface extending to infinity. Noting that the velocity on the surface of the body is $\mathbf{u} = \mathbf{V} + \mathbf{\Omega}\times\mathbf{x}$, where $\mathbf{V}$ and $\mathbf{\Omega}$ are the linear and angular velocities of motion, and using (2.3.18) and (2.3.19), we find that the double-layer integral over the surface of the body is equal to zero. Neglecting the integrals at infinity we obtain a representation in terms of a single-layer potential

$$u_j(\mathbf{x}_0) = -\frac{1}{8\pi\mu}\int_{S_B} f_i(\mathbf{x})G_{ij}(\mathbf{x},\mathbf{x}_0)\,dS(\mathbf{x}) \qquad (2.3.27)$$

It will be noted that (2.3.27) is valid not only in the interior of the flow but also on the surface of the body (problem 2.3.8).

As a second application, we consider a flow past a rigid stationary or moving boundary. It will be useful to decompose the total flow $\mathbf{u}$ into an undisturbed component $\mathbf{u}^\infty$ that would prevail in the absence of the boundary, and a disturbance component $\mathbf{u}^D$ owing to the boundary. We apply (2.3.11) for the disturbance flow $\mathbf{u}^D$, and require that the total velocity of the fluid on the boundary is $\mathbf{u} = \mathbf{V} + \boldsymbol{\Omega} \times \mathbf{x}$ or equivalently $\mathbf{u}^D = -\mathbf{u}^\infty + \mathbf{V} + \boldsymbol{\Omega} \times \mathbf{x}$, where $\mathbf{V}$ and $\boldsymbol{\Omega}$ are the velocities of translation and rotation of the boundary. Assuming that the boundary is a closed surface, neglecting the integrals at infinity, and using (2.3.18) and (2.3.19) we obtain

$$u_j^D(\mathbf{x}_0) = -\frac{1}{8\pi\mu}\int_D f_i^D(\mathbf{x})G_{ij}(\mathbf{x},\mathbf{x}_0)\,dS(\mathbf{x}) - \frac{1}{8\pi}\int_D u_i^\infty(\mathbf{x})T_{ijk}(\mathbf{x},\mathbf{x}_0)n_k(\mathbf{x})\,dS(\mathbf{x}) \tag{2.3.28}$$

Next, recalling that the point $\mathbf{x}_0$ is located outside $D$, we apply the reciprocal identity (2.3.4) for the incident flow $\mathbf{u}^\infty$, obtaining

$$\mu\int_D u_i^\infty(\mathbf{x})T_{ijk}(\mathbf{x},\mathbf{x}_0)n_k(\mathbf{x})\,dS(\mathbf{x}) = \int_D f_i^\infty(\mathbf{x})G_{ij}(\mathbf{x},\mathbf{x}_0)\,dS(\mathbf{x}) \tag{2.3.29}$$

Using (2.3.29) to eliminate the double-layer integral on the right-hand side of (2.3.28) and adding the incident velocity field $\mathbf{u}^\infty$ to both sides of the resulting equation we obtain the simplified representation

$$u_j(\mathbf{x}_0) = u_j^\infty(\mathbf{x}_0) - \frac{1}{8\pi\mu}\int_D f_i(\mathbf{x})G_{ij}(\mathbf{x},\mathbf{x}_0)\,dS(\mathbf{x}) \tag{2.3.30}$$

It will be observed that (2.3.30) is valid not only within the domain of flow, but also on the boundary $D$ (problem 2.3.8).

The reader will note that in order to derive (2.3.30) we had to assume that the boundary $D$ is a closed surface. This, however, is not an intrinsic restriction. Consider, for instance, a shearing flow over an infinite plane wall that contains a projection, as illustrated in Figure 2.3.3. Physically, the projection may be identified with the surface of a drop or cell that is adhering to the wall. We consider the boundary integral equation (2.3.11) for the disturbance flow due to the projection, and identify $D$ with the surface of the projection $\mathscr{D}$, the uncovered surface of the plane wall $A$, and a large surface extending to infinity. Noting that $\mathbf{u}^D = -\mathbf{u}^\infty$ on $\mathscr{D}$ and $\mathbf{u}^D = 0$ on $A$, and neglecting the integrals at infinity, we obtain

$$u_j^D(\mathbf{x}_0) = -\frac{1}{8\pi\mu}\int_{A,\mathscr{D}} f_i^D(\mathbf{x})G_{ij}(\mathbf{x},\mathbf{x}_0)\,dS(\mathbf{x}) - \frac{1}{8\pi}\int_{\mathscr{D}} u_i^\infty(\mathbf{x})T_{ijk}(\mathbf{x},\mathbf{x}_0)n_k(\mathbf{x})\,dS(\mathbf{x}) \tag{2.3.31}$$

Next, we apply the reciprocal identity (2.3.4) for the flow $\mathbf{u}^\infty$ incident upon a control volume enclosed by the projection and the wall. Noting that $\mathbf{u}^\infty = 0$ on $A$, we find

$$\mu \int_{\mathscr{D}} u_i^\infty(\mathbf{x}) T_{ijk}(\mathbf{x}, \mathbf{x}_0) n_k(\mathbf{x}) \, dS(\mathbf{x}) = \int_{S,\mathscr{D}} f_i^\infty(\mathbf{x}) G_{ij}(\mathbf{x}, \mathbf{x}_0) \, dS(\mathbf{x}) \quad (2.3.32)$$

where $S$ is the area of the wall covered by the projection. Using (2.3.32) to eliminate the double-layer integral on the right-hand side of (2.3.31), and adding the incident velocity field $\mathbf{u}^\infty$ to both sides of the resulting equation we obtain a simplified representation in terms of a single-layer potential, namely

$$u_j(\mathbf{x}_0) = u_j^\infty(\mathbf{x}_0) - \frac{1}{8\pi\mu} \int_A f_i^D(\mathbf{x}) G_{ij}(\mathbf{x}, \mathbf{x}_0) \, dS(\mathbf{x}) - \frac{1}{8\pi\mu} \int_{\mathscr{D}} f_i(\mathbf{x}) G_{ij}(\mathbf{x}, \mathbf{x}_0) \, dS(\mathbf{x})$$

$$+ \frac{1}{8\pi\mu} \int_S f_i^\infty(\mathbf{x}) G_{ij}(\mathbf{x}, \mathbf{x}_0) \, dS(\mathbf{x}) \quad (2.3.33)$$

If we use a Green's function that vanishes on the plane wall, i.e. $\mathbf{G}(\mathbf{x}, \mathbf{x}_0) = 0$ when $\mathbf{x}$ is on $S$ or $A$, we will find that the first and third integrals on the right-hand side of (2.3.33) vanish yielding the simplified representation (2.3.30).

Proceeding, we consider the elimination of the single-layer potential from the general boundary integral equation (2.3.11). For this purpose, we introduce a complementary flow $\mathbf{u}'$ that has the same values of the surface force as the flow $\mathbf{u}$ over the boundary $D$, i.e. $\mathbf{f}' = \mathbf{f}$ over $D$. Using (2.3.24), we derive the simplified representation

$$u_j(\mathbf{x}_0) = \frac{1}{8\pi} \int_D q_i(\mathbf{x}) T_{ijk}(\mathbf{x}, \mathbf{x}_0) n_k(\mathbf{x}) \, dS(\mathbf{x}) \quad (2.3.34)$$

*Figure 2.3.3.* Flow over a plane wall containing a projection; $S$ is the area of the wall occupied by the projection.

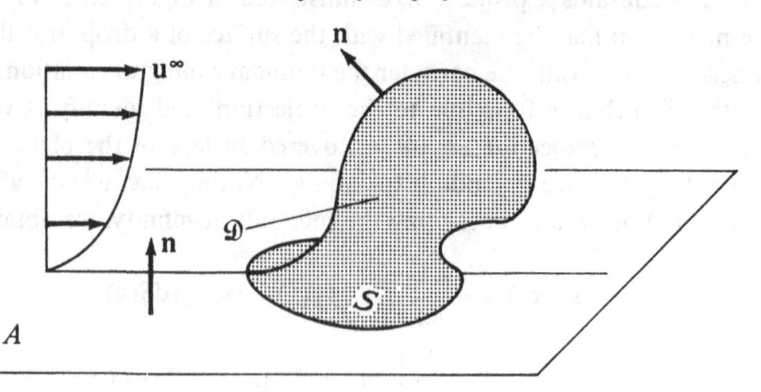

expressing the flow solely in terms of a double-layer potential with density $\mathbf{q} = \mathbf{u} - \mathbf{u}'$. Implicit in this derivation is the assumption that apart from the pole, the Green's function $\mathbf{G}$ does not have any singular points within the internal or external domains of flow.

Addressing the validity of (2.3.34), we inquire whether it is feasible to find the required complementary flow $\mathbf{u}'$. In section 4.2 we shall see that as long as $\mathbf{u}$ satisfies (2.3.26), it will always be possible to find an external flow $\mathbf{u}'$ that yields a prescribed boundary surface force $\mathbf{f}'$. The force and torque exerted on $D$ due to $\mathbf{u}'$ are seen to vanish, for $\mathbf{f}'$ represents the boundary surface force of a regular internal flow. Thus, the double-layer potential will successfully represent any internal flow. In section 4.2 we shall also see that unless the force and torque on $D$ due to $\mathbf{f}$ happen to vanish, it will be impossible to find an internal flow $\mathbf{u}'$ with a prescribed boundary surface force $\mathbf{f}'$. This constraint places severe limitations on the ability of the double-layer potential to represent an arbitrary external flow. These issues will be discussed in more detail in Chapter 4.

In problem 2.3.6 we shall discuss a double-layer representation of a flow in terms of the stress tensor of a Green's function that vanishes over the boundaries of the flow.

### *The boundary integral equation as a Fredholm integral equation*

The boundary integral equation (2.3.13) imposes a mathematical constraint between the distributions of boundary velocity and surface force. Physically, this constraint implies that the boundary velocity and surface force may not be specified independently in an arbitrary manner, but must be prescribed in such a way that (2.3.13) is fulfilled. Indeed, in practice, when stating a particular problem we require boundary conditions either for the velocity or for the surface force but not for both. Typically, we require that the velocity over a solid boundary is equal to the velocity of the boundary, the tangential component of the surface force over an isothermal free surface that contains no impurities or surfactants is equal to zero, and the normal component of the surface force over a free surface is balanced by surface tension. Boundary conditions along a fluid interface will be discussed in detail in Chapter 5.

If we prescribe the boundary velocity $\mathbf{u}$ over $D$, we will reduce equation (2.3.13) to a *Fredholm integral equation of the first kind* for the boundary surface force $\mathbf{f}$, namely

$$\int_D f_i(\mathbf{x}) G_{ij}(\mathbf{x}, \mathbf{x}_0)\, \mathrm{d}S(\mathbf{x}) = -4\pi\mu u_j(\mathbf{x}_0) + \mu I_j^D(\mathbf{x}_0) \qquad (2.3.35)$$

where $\mathbf{I}^D$ represents the known double-layer potential

$$I_j^D(\mathbf{x}_0) \equiv \int_D^{\mathscr{PV}} u_i(\mathbf{x})\, T_{ijk}(\mathbf{x}, \mathbf{x}_0) n_k(\mathbf{x})\, dS(\mathbf{x}) \tag{2.3.36}$$

Alternatively, if we prescribe the boundary surface force $\mathbf{f}$ over $D$, we will reduce equation (2.3.13) to a *Fredholm integral equation of the second kind* for the boundary velocity $\mathbf{u}$, namely

$$u_j(\mathbf{x}_0) = \frac{1}{4\pi} \int_D^{\mathscr{PV}} u_i(\mathbf{x})\, T_{ijk}(\mathbf{x}, \mathbf{x}_0) n_k(\mathbf{x})\, dS(\mathbf{x}) - \frac{1}{4\pi\mu} I_j^S(\mathbf{x}_0) \tag{2.3.37}$$

where $\mathbf{I}^S$ represents the known single-layer potential

$$I_j^S(\mathbf{x}_0) \equiv \int_D f_i(\mathbf{x}) G_{ij}(\mathbf{x}, \mathbf{x}_0)\, dS(\mathbf{x}) \tag{2.3.38}$$

If we prescribe the velocity over a portion of $D$ and the surface force over the remaining portion of $D$ (as we do when the flow is bounded by solid boundaries and free surfaces), we will obtain a Fredholm integral equation of mixed type for the unknown boundary distributions. Once these equations are solved, the velocity, pressure, and stress fields may be computed using the boundary integral representations (2.3.11) and (2.3.17).

Now, inspecting the integral equations (2.3.35) and (2.3.37) we observe that the corresponding kernels $\mathbf{G}(\mathbf{x}, \mathbf{x}_0)$ and $T_{ijk}(\mathbf{x}, \mathbf{x}_0) n_k(\mathbf{x})$ become singular as the observation point $\mathbf{x}$ approaches the pole $\mathbf{x}_0$. In fact, close inspection reveals that the singularities of the kernels are not square integrable. This behaviour not only raises computational difficulties but also prevents us from studying (2.3.35) and (2.3.37) in the context of the theories of Fredholm and Hilbert–Schmidt, which are applicable for integral equations with square integrable kernels (Mikhlin 1957, Pogorzelski 1966). Fortunately, when the boundary $D$ is a Lyapunov surface the kernels are weakly singular, thereby implying that both the single-layer and the double-layer integral are compact linear operators for which the Fredholm theory remains applicable (Kress 1989; Pogorzelski 1966, Chapter 3). A detailed discussion of the properties of (2.3.35) and (2.3.37) will be deferred until sections 4.2, 4.4.

It must be noted that the principal value of the double-layer potential (defined as the value of the improper double-layer integral arising when $\mathbf{x}_0$ is on $D$) is different from the usual Cauchy principal value of a singular one-dimensional integral. Indeed, the latter requires that the improper integration is effected by excluding two small intervals on either side of the singularity from the domain of integration and taking the limit as the size of these intervals tends to zero in a simultaneous manner. On the contrary, since the kernel of the double-layer integral is weakly singular (when the domain is a Lyapunov surface), the principal value of the double-layer integral exists in the usual sense of an improper integral.

**Problems**

2.3.1 Prove the identity

$$\frac{\partial}{\partial x_k}[(\mu u_i(\mathbf{x})T_{ilk}(\mathbf{x},\mathbf{x}_0)-G_{il}(\mathbf{x},\mathbf{x}_0)\sigma_{ik}(\mathbf{x})]=-8\pi\mu u_l\delta(\mathbf{x}-\mathbf{x}_0)$$

where $\mathbf{u}$ is a regular flow field with associated stress field $\boldsymbol{\sigma}$, and $\delta$ is the three-dimensional delta function. Integrating this identity over a selected control volume, and using the divergence theorem, derive the boundary integral equation (2.3.11).

2.3.2 Following the limiting procedure outlined in the beginning of this section, derive the boundary integral equation (2.3.13) for a point at a smooth boundary $D$. (Hint: consider a point $\mathbf{x}_0$ on $D$, define a control volume that is enclosed by $D$ but excludes the volume of a half-sphere with radius $\varepsilon$ centered at $\mathbf{x}_0$, and let $\varepsilon$ tend to zero).

2.3.3 Following the procedure of problem 2.3.2 derive a boundary integral equation for a point at a boundary corner. Specifically, show that for a point $\mathbf{x}_0$ right on a corner, equation (2.3.13) remains valid provided that $u_j(\mathbf{x}_0)$ on the left-hand side is replaced by $u_i(\mathbf{x}_0)c_{ij}(\mathbf{x}_0)$, where $c_{ij}(\mathbf{x}_0)$ is a matrix whose value depends on the detailed geometry of the corner. Evaluate $c_{ij}(\mathbf{x}_0)$ on the corner of a two-dimensional wedge.

2.3.4 Apply the boundary integral equations (2.3.4), (2.3.11), and (2.3.13) for unidirectional linear and parabolic flow to derive two sets of identities similar to (2.3.18) and (2.3.19).

2.3.5 Consider the flow produced by the motion of a piece of paper of infinitesimal thickness. Show that the double-layer integral in the boundary integral equation may be eliminated in a natural manner, and that as a result the flow may be represented solely in terms of a single-layer integral whose density possesses a simple physical interpretation.

2.3.6 Consider an internal or external flow bounded by the closed surface $D$. Show that by using a Green's function that vanishes over $D$ we can express the flow simply in terms of a double-layer potential. In this manner we can express a velocity field in terms of a prescribed boundary velocity.

2.3.7 Consider an infinite incident flow $\mathbf{u}^\infty$ of a fluid with viscosity $\mu_1$ past a liquid drop with viscosity $\mu_2$. Show that the velocity field in the exterior and interior of the drop may be represented as

$$u_j(\mathbf{x}_0)=u_j^\infty(\mathbf{x}_0)-\frac{1}{8\pi\mu_1}\int_{D^+}f_i(\mathbf{x})G_{ij}(\mathbf{x},\mathbf{x}_0)\,\mathrm{d}S(\mathbf{x})$$
$$+\frac{1}{8\pi}\int_D u_i(\mathbf{x})T_{ijk}(\mathbf{x},\mathbf{x}_0)n_k(\mathbf{x})\,\mathrm{d}S(\mathbf{x}) \qquad (1)$$

and

$$u_j(\mathbf{x}_0) = \frac{1}{8\pi\mu_2}\int_{D^-} f_i(\mathbf{x})G_{ij}(\mathbf{x},\mathbf{x}_0)\,dS(\mathbf{x}) - \frac{1}{8\pi}\int_D u_i(\mathbf{x})T_{ijk}(\mathbf{x},\mathbf{x}_0)n_k(\mathbf{x})\,dS(\mathbf{x}) \quad (2)$$

where $D^+$ and $D^-$ are the external and internal sides of the drop respectively, and $\mathbf{n}$ is the unit normal vector pointing into the ambient fluid. Show that as the viscosity of the drop increases and the drop becomes a stationary or moving rigid particle (1) reduces to (2.3.30). Finally, show that for a point $\mathbf{x}_0$ on the surface of the drop

$$u_j(\mathbf{x}_0) = 2u_j^{\infty}(\mathbf{x}_0) - \frac{1}{4\pi\mu_1}\int_{D^+} f_i(\mathbf{x})G_{ij}(\mathbf{x},\mathbf{x}_0)\,dS(\mathbf{x})$$
$$+\frac{1}{4\pi}\int_D^{\mathscr{PV}} u_i(\mathbf{x})T_{ijk}(\mathbf{x},\mathbf{x}_0)n_k(\mathbf{x})\,dS(\mathbf{x})$$

2.3.8 Show that (2.3.27) and (2.3.30) are valid in the bulk of the flow as well as on the surface of the body.

2.3.9 Consider an ambient incident flow past a stationary spherical particle. Using (2.3.30) and requiring that the velocity vanishes on the surface of the particle we obtain

$$u_j^{\infty}(\mathbf{x}_0) = \frac{1}{8\pi\mu}\int_{S_P} f_i(\mathbf{x})\mathscr{S}_{ij}(\mathbf{x}-\mathbf{x}_0)\,dS(\mathbf{x}) \quad (1)$$

where $\mathscr{S}$ is the Stokeslet and $S_P$ is the surface of the particle. Integrating this equation over the surface of the particle, and using the identity

$$\int_{S_P} \mathscr{S}_{ij}(\mathbf{x}-\mathbf{x}_0)\,dS(\mathbf{x}) = \tfrac{16}{3}\pi a\delta_{ij} \quad (2)$$

where $a$ is the radius of the particle and $\mathbf{x}_0$ is on the surface of the particle (Batchelor 1972), show that the force exerted on the particle is given by the first of the Faxen relations (1.4.15).

2.3.10 Following a procedure similar to that outlined in problem 2.3.9 derive the Faxen relation for the torque on a spherical particle given by the second equation of (1.4.15). Note that it will be necessary to use the identity

$$\varepsilon_{ilm}\int_{S_P} (x_m - x_{c,m})\mathscr{S}_{ij}(\mathbf{x}-\mathbf{x}_0)\,dS(\mathbf{x}) = \tfrac{8}{3}\pi a\varepsilon_{jlm}(x_{0,m} - x_{c,m}) \quad (1)$$

where $a$ is the radius of the particle, $\mathbf{x}_0$ is on the surface of the particle, and $\mathbf{x}_c$ is the centre of the particle.

## 2.4 Flow in an axisymmetric domain

In this section we develop simplified versions of the boundary integral equation for axisymmetric flow, as well as for three-dimensional flow in an axisymmetric domain. Our goal is to reduce the boundary integral equation to a one-dimensional equation, or a system of one-dimensional equations, over the trace of the boundaries in an azimuthal plane.

Considering first axisymmetric flow with no swirling motion, we observe that in cylindrical coordinates, none of the boundary variables is a function of the azimuthal angle $\phi$ (Figure 2.4.1). To reduce the number of unknowns, we note that $f_\phi = u_\phi = n_\phi = 0$, express the Cartesian components of the surface force, velocity, and normal vector in terms of the corresponding polar cylindrical components by writing, for instance, $[f_x, f_y, f_z] = [f_x, f_\sigma \cos\phi, f_\sigma \sin\phi]$, and substitute into (2.3.11). Furthermore, we write $dS = \sigma\, d\phi\, dl$, where $dl$ is the differential arc length of the trace of the boundary $C$ in the $x$–$y$ azimuthal plane, and apply (2.3.11) at a point $\mathbf{x}_0$ in the $x$–$y$ plane (Figure 2.4.1). Performing the boundary integrations in the azimuthal direction, we obtain

$$u_\alpha(\mathbf{x}_0) = -\frac{1}{8\pi\mu}\int_C M_{\alpha\beta}(\mathbf{x}_0, \mathbf{x}) f_\beta(\mathbf{x})\, dl(\mathbf{x}) + \frac{1}{8\pi}\int_C q_{\alpha\beta\gamma}(\mathbf{x}_0, \mathbf{x}) u_\beta(\mathbf{x}) n_\gamma(\mathbf{x})\, dl(\mathbf{x}) \tag{2.4.1}$$

where the Greek subscripts $\alpha, \beta, \gamma$ are either $x$ or $\sigma$, indicating the axial and radial components respectively. The matrices $\mathbf{M}$ and $\mathbf{q}$ on the right-hand side of (2.4.1) are given by

$$\mathbf{M}(\mathbf{x}_0, \mathbf{x}) = \begin{bmatrix} M_{xx} & M_{x\sigma} \\ M_{\sigma x} & M_{\sigma\sigma} \end{bmatrix}(\mathbf{x}_0, \mathbf{x}) = \sigma \int_0^{2\pi} \begin{bmatrix} G_{xx} & G_{yx}\cos\phi + G_{zx}\sin\phi \\ G_{xy} & G_{yy}\cos\phi + G_{zy}\sin\phi \end{bmatrix} d\phi \tag{2.4.2}$$

$$\begin{bmatrix} q_{xxx} & q_{xx\sigma} \\ q_{x\sigma x} & q_{x\sigma\sigma} \end{bmatrix}(\mathbf{x}_0, \mathbf{x}) =$$
$$\sigma \int_0^{2\pi} \begin{bmatrix} T_{xxx} & T_{yxx}\cos\phi + T_{zxx}\sin\phi \\ T_{xxy}\cos\phi + T_{xxz}\sin\phi & T_{yxy}\cos^2\phi + T_{zxz}\sin^2\phi + 2T_{yxz}\cos\phi\sin\phi \end{bmatrix} d\phi$$

*Figure 2.4.1.* Schematic illustration of an axisymmetric boundary with reference to a polar cylindrical coordinate system $(x, \sigma, \phi)$.

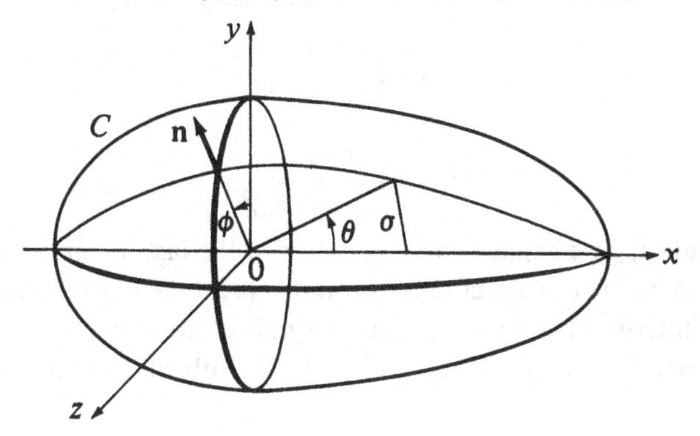

$$\begin{bmatrix} q_{\sigma xx} & q_{\sigma x\sigma} \\ q_{\sigma\sigma x} & q_{\sigma\sigma\sigma} \end{bmatrix}(\mathbf{x}_0, \mathbf{x}) =$$

$$\sigma \int_0^{2\pi} \begin{bmatrix} T_{xyx} & T_{yyx}\cos\phi + T_{zyx}\sin\phi \\ T_{xyy}\cos\phi + T_{xyz}\sin\phi & T_{yyy}\cos^2\phi + T_{zyz}\sin^2\phi + 2T_{yyz}\cos\phi\sin\phi \end{bmatrix} \mathrm{d}\phi \tag{2.4.3}$$

and the arguments of **G** and **T** are $(\mathbf{x}, \mathbf{x}_0)$.

It will be helpful to note that physically,

$$u_\alpha(\mathbf{x}_0) = \frac{1}{8\pi\mu} M_{\alpha\beta}(\mathbf{x}_0, \mathbf{x}) f_\beta \tag{2.4.4}$$

represents the velocity at the point $(x_0, \sigma_0)$ due to a ring of point forces with strength per unit length **f** passing through $(x, \sigma)$.

Now, in order to make the expressions for **M** and **q** more specific we identify **G** with the free-space Green's function, namely the Stokeslet. Substituting $y = \sigma\cos\phi$, and $z = \sigma\sin\phi$ into the right-hand sides of (2.4.2) and (2.4.3) we find

$$\mathbf{M} = \sigma \begin{bmatrix} I_{10} + \hat{x}^2 I_{30} & \hat{x}(\sigma I_{30} - \sigma_0 I_{31}) \\ \hat{x}(\sigma I_{31} - \sigma_0 I_{30}) & I_{11} + (\sigma^2 + \sigma_0{}^2) I_{31} - \sigma\sigma_0(I_{30} + I_{32}) \end{bmatrix} \tag{2.4.5}$$

$$\begin{bmatrix} q_{xxx} \\ q_{xx\sigma} = q_{x\sigma x} \\ q_{x\sigma\sigma} \end{bmatrix} = -6\sigma\hat{x} \begin{bmatrix} \hat{x}^2 I_{50} \\ \hat{x}(\sigma I_{50} - \sigma_0 I_{51}) \\ \sigma_0^2 I_{52} + \sigma^2 I_{50} - 2\sigma\sigma_0 I_{51} \end{bmatrix}$$

and

$$\begin{bmatrix} q_{\sigma xx} \\ q_{\sigma x\sigma} = q_{\sigma\sigma x} \\ q_{\sigma\sigma\sigma} \end{bmatrix} = -6\sigma \begin{bmatrix} \hat{x}^2(\sigma I_{51} - \sigma_0 I_{50}) \\ \hat{x}[(\sigma^2 + \sigma_0{}^2) I_{51} - \sigma\sigma_0(I_{50} + I_{52})] \\ \sigma^3 I_{51} - \sigma^2\sigma_0(I_{50} + 2I_{52}) + \sigma\sigma_0^2(I_{53} + 2I_{51}) - \sigma_0^3 I_{52} \end{bmatrix} \tag{2.4.6}$$

where $\hat{x} = x - x_0$. For convenience, we have introduced the integrals

$$I_{mn}(\hat{x}, \sigma, \sigma_0) \equiv \int_0^{2\pi} \frac{\cos^n\omega}{[\hat{x}^2 + \sigma^2 + \sigma_0^2 - 2\sigma\sigma_0\cos\omega]^{m/2}} \mathrm{d}\omega$$
$$= \frac{4k^m}{(4\sigma\sigma_0)^{m/2}} \int_0^{\pi/2} \frac{(2\cos^2\omega - 1)^n}{(1 - k^2\cos^2\omega)^{m/2}} \mathrm{d}\omega \tag{2.4.7}$$

where

$$k^2 = \frac{4\sigma\sigma_0}{\hat{x}^2 + (\sigma + \sigma_0)^2} \tag{2.4.8}$$

To compute $I_{mn}$ we expand the numerator of the last integrand in (2.4.7) in a polynomial with respect to $\cos\omega$, and express the resulting individual integrals in terms of complete elliptic integrals of the first and second kind (Gradshteyn & Ryshik 1980, section 2.58). Substituting the resulting expressions into (2.4.5) we find

$$
\begin{aligned}
M_{xx}(\mathbf{x}_0, \mathbf{x}) &= 2k\left(\frac{\sigma}{\sigma_0}\right)^{1/2}\left(F + \frac{\hat{x}^2}{\hat{r}^2}E\right) \\
M_{x\sigma}(\mathbf{x}_0, \mathbf{x}) &= k\frac{\hat{x}}{(\sigma_0\sigma)^{1/2}}\left[F - (\sigma_0^2 - \sigma^2 + \hat{x}^2)\frac{E}{\hat{r}^2}\right] \\
M_{\sigma x}(\mathbf{x}_0, \mathbf{x}) &= -k\frac{\hat{x}}{\sigma_0}\left(\frac{\sigma}{\sigma_0}\right)^{1/2}\left[F + (\sigma_0^2 - \sigma^2 - \hat{x}^2)\frac{E}{\hat{r}^2}\right] \\
M_{\sigma\sigma}(\mathbf{x}_0, \mathbf{x}) &= \frac{k}{\sigma_0\sigma}\left(\frac{\sigma}{\sigma_0}\right)^{1/2}\Big\{(\sigma_0^2 + \sigma^2 + 2\hat{x}^2)F \\
&\qquad - [2\hat{x}^4 + 3\hat{x}^2(\sigma_0^2 + \sigma^2) + (\sigma_0^2 - \sigma^2)^2]\frac{E}{\hat{r}^2}\Big\}
\end{aligned}
\tag{2.4.9}
$$

where $\hat{r}^2 = \hat{x}^2 + (\sigma - \sigma_0)^2$, and $F$ and $E$ are the complete elliptic integrals of the first and second kind with argument $k$, defined as

$$
F(k) = \int_0^{\pi/2} \frac{d\omega}{(1 - k^2\cos^2\omega)^{1/2}} \qquad E(k) = \int_0^{\pi/2} (1 - k^2\cos^2\omega)^{1/2}\, d\omega \tag{2.4.10}
$$

One efficient method for computing $F$ and $E$ is provided by the recursive formulae

$$
F = \frac{\pi}{2}(1 + K_1)(1 + K_2)(1 + K_3)\cdots \qquad E = F\left(1 - \frac{k^2}{2}P\right) \tag{2.4.11}
$$

where

$$
K_0 = k \qquad K_p = \frac{1 - C}{1 + C} \qquad C = \sqrt{1 - K_{p-1}^2}
$$

and

$$
P = 1 + \frac{K_1}{2}\left(1 + \frac{K_2}{2}\left(1 + \frac{K_3}{2}\left(\cdots\right)\cdots\right)\right)
$$

(Davis 1962, p. 139). Another method is to use the polynomial approximations compiled by Abramowitz & Stegun (1972, section 17).

It will be noted that in the limit as $\mathbf{x}$ approaches $\mathbf{x}_0$, $k$ tends to unity, and $F$ assumes an infinite value. To assess the corresponding behaviour of the matrix $\mathbf{M}$ given in (2.4.9), we use the asymptotic expansions

$$
F \approx \ln\frac{4}{(1 - k^2)^{1/2}} + \cdots \approx -\ln\hat{r} + \cdots \qquad E \approx 1 + \cdots \tag{2.4.12}
$$

and find that the off-diagonal components of $\mathbf{M}$ tend to finite values, whereas the diagonal components exhibit logarithmic singularities $M_{xx} \approx M_{\sigma\sigma} \approx -2\ln\hat{r} + \cdots$.

Proceeding, we relax the assumption of axisymmetric flow and consider three-dimensional flow in an axisymmetric domain. One example is shear flow over a plane wall that contains a spherical cavity or protrusion. Our goal will be to reduce the boundary integral equation to a system of one-dimensional equations over the contour of the boundaries in an azimuthal plane. Similar reductions of the analogous integral equations of linear elastostatics have been discussed by several authors including Mayr, Drexler & Kuhn (1980) Bakr (1985), and Wang & Banerjee (1989).

To make our discussion more specific, we address the particular case of three-dimensional infinite flow past the axisymmetric body depicted in Figure 2.4.1. Thus, we consider the simplified boundary integral representation (2.3.30), and express the Cartesian components of all boundary variables in terms of their cylindrical polar components, writing, for instance,

$$f_y = f_\sigma \cos\phi - f_\phi \sin\phi \qquad f_z = f_\sigma \sin\phi + f_\phi \cos\phi \tag{2.4.13}$$

Substituting (2.4.13) into the single-layer integral $\mathbf{I}^S$ on the right-hand side of (2.3.30) we obtain

$$I_k^S(\mathbf{x}_0) = \int_D [G_{xk}(\mathbf{x},\mathbf{x}_0) f_x(\mathbf{x}) + [G_{yk}(\mathbf{x},\mathbf{x}_0)\cos\phi + G_{zk}(\mathbf{x},\mathbf{x}_0)\sin\phi] f_\sigma(\mathbf{x})$$
$$+ [-G_{yk}(\mathbf{x},\mathbf{x}_0)\sin\phi + G_{zk}(\mathbf{x},\mathbf{x}_0)\cos\phi] f_\phi(\mathbf{x})]\, dS(\mathbf{x}) \tag{2.4.14}$$

The radial and azimuthal components of the single-layer potential are given by

$$I_\sigma^S(\mathbf{x}_0) = I_y^S(\mathbf{x}_0)\cos\phi_0 + I_z^S(\mathbf{x}_0)\sin\phi_0$$
$$I_\phi^S(\mathbf{x}_0) = -I_y^S(\mathbf{x}_0)\sin\phi_0 + I_z^S(\mathbf{x}_0)\cos\phi_0 \tag{2.4.15}$$

Substituting (2.4.14) into (2.4.15) we obtain

$$I_\alpha^S(\mathbf{x}_0) = \int_D N_{\alpha\beta}(\mathbf{x}_0,\mathbf{x}) f_\beta(\mathbf{x})\, dS(\mathbf{x}) \tag{2.4.16}$$

where Greek indices run through $(x, \sigma, \phi)$. Identifying $\mathbf{G}$ with the Stokeslet yields

$$N_{xx} = G_{xx} = \frac{1}{r} + \frac{\hat{x}^2}{r^3}$$

$$N_{\sigma x} = G_{xy}\cos\phi_0 + G_{xz}\sin\phi_0 = \frac{\hat{x}}{r^3}(\sigma\cos\hat{\phi} - \sigma_0)$$

$$N_{\phi x} = -G_{xy}\sin\phi_0 + G_{xz}\cos\phi_0 = \frac{\hat{x}\sigma}{r^3}\sin\hat{\phi}$$

$$N_{x\sigma} = G_{yx}\cos\phi + G_{zx}\sin\phi = \frac{\hat{x}}{r^3}(\sigma - \sigma_0\cos\hat{\phi})$$

$$N_{\sigma\sigma} = \cos\phi_0 (G_{yy}\cos\phi + G_{zy}\sin\phi) + \sin\phi_0 (G_{yz}\cos\phi + G_{zz}\sin\phi)$$
$$= \left(\frac{1}{r} + \frac{\sigma^2 + \sigma_0^2}{r^3}\right)\cos\hat{\phi} - \frac{\sigma_0\sigma}{r^3}(\cos^2\hat{\phi} + 1)$$

$$N_{\phi\sigma} = -\sin\phi_0(G_{yy}\cos\phi + G_{zy}\sin\phi) + \cos\phi_0(G_{yz}\cos\phi + G_{zz}\sin\phi)$$
$$= \left[\left(\frac{1}{r} + \frac{\sigma^2}{r^3}\right) - \frac{\sigma\sigma_0}{r^3}\cos\hat{\phi}\right]\sin\hat{\phi}$$

$$N_{x\phi} = -G_{yx}\sin\phi + G_{zx}\cos\phi = -\frac{\hat{x}\sigma_0}{r^3}\sin\hat{\phi}$$

$$N_{\sigma\phi} = \cos\phi_0(-G_{yy}\sin\phi + G_{zy}\cos\phi) + \sin\phi_0(-G_{yz}\sin\phi + G_{zz}\cos\phi)$$
$$= \left[-\left(\frac{1}{r} + \frac{\sigma_0^2}{r^3}\right) + \frac{\sigma\sigma_0}{r^3}\cos\hat{\phi}\right]\sin\hat{\phi}$$

$$N_{\phi\phi} = -\sin\phi_0(-G_{yy}\sin\phi + G_{zy}\cos\phi) + \cos\phi_0(-G_{yz}\sin\phi + G_{zz}\cos\phi)$$
$$= \frac{1}{r}\cos\hat{\phi} + \frac{\sigma\sigma_0}{r^3}\sin^2\hat{\phi} \tag{2.4.17}$$

where $\hat{x} = x - x_0$ and $\hat{\phi} = \phi - \phi_0$. It will be important to note that the matrix $\mathbf{N}$ is a function of $\hat{x}$, $\sigma, \sigma_0$, and $\hat{\phi}$.

Next, we expand the cylindrical polar components of the boundary surface force in a complex Fourier series with respect to the azimuthal angle $\phi$ as

$$f_\beta(l,\phi) = \sum_{n=-\infty}^{\infty} \mathscr{f}_{\beta n} \mathrm{e}^{\mathrm{i}n\phi} \tag{2.4.18}$$

Writing $\mathrm{d}S = \sigma \mathrm{d}\phi \mathrm{d}l$, and substituting (2.4.18) into (2.4.16), we express the single-layer potential in the form of a complex Fourier series as

$$I_\alpha^{\mathrm{S}}(\mathbf{x}_0) = \sum_{n=-\infty}^{\infty} \mathrm{e}^{\mathrm{i}n\phi_0} \int_C Q_{\alpha\beta n}(\hat{x}, \sigma, \sigma_0) \mathscr{f}_{\beta n} \sigma \mathrm{d}l \tag{2.4.19}$$

where

$$Q_{\alpha\beta n}(\hat{x}, \sigma, \sigma_0) = \int_0^{2\pi} N_{\alpha\beta}(\hat{x}, \sigma, \sigma_0, \hat{\phi}) \mathrm{e}^{\mathrm{i}n\hat{\phi}} \mathrm{d}\hat{\phi} \tag{2.4.20}$$

are the complex Fourier coefficients of $\mathbf{N}$. Substituting (2.4.17) into (2.4.20) and evaluating the integrals we obtain the specific expressions

$$Q_{\alpha\beta 0} = \begin{bmatrix} I_{10} + \hat{x}^2 I_{30} & \hat{x}(\sigma I_{30} - \sigma_0 I_{31}) & 0 \\ \hat{x}(\sigma I_{31} - \sigma_0 I_{30}) & I_{10} + (\sigma_0^2 + \sigma^2) I_{31} - \sigma\sigma_0(I_{32} + I_{30}) & 0 \\ 0 & 0 & I_{31} + \sigma\sigma_0(I_{30} - I_{32}) = 2I_{11} \end{bmatrix} \tag{2.4.21}$$

and

$$Q_{\alpha\beta 1} = Q_{\alpha\beta 1}^{\mathrm{R}} + \mathrm{i}Q_{\alpha\beta 1}^{\mathrm{I}} \tag{2.4.22}$$

where the superscripts R and I indicate the real and imaginary parts respectively and the integrals $I_{mn}$ were defined in (2.4.7). $Q_{\alpha\beta 1}^{\mathrm{R}}$ may be derived from $Q_{\alpha\beta 0}$ by augmenting the second index of $I_{mn}$ by a one, whereas

$$Q^{\mathrm{I}}_{\alpha\beta 1} = \begin{bmatrix} 0 & 0 & -\hat{x}\sigma_0(I_{32}-I_{30}) \\ 0 & 0 & -1+I_{12}-\sigma_0^2(I_{30}-I_{32}) \\ & & +\sigma\sigma_0(I_{31}-I_{33}) \\ \hat{x}\sigma(I_{32}-I_{30}) & 1-I_{12}+\sigma^2(I_{30}-I_{32}) & 0 \\ & -\sigma\sigma_0(I_{31}-I_{33}) & \end{bmatrix} \tag{2.4.23}$$

Now, we expand the incident velocity over the boundary in a complex Fourier series as

$$u^{\infty}_{\alpha}(l, \phi_0) = \sum_{n=-\infty}^{\infty} u_{\alpha n}(l) \mathrm{e}^{in\phi_0} \tag{2.4.24}$$

In order to ensure that the velocity is real, we require $u_{\alpha n} = u^*_{\alpha(-n)}$, where an asterisk indicates the complex conjugate. Applying (2.3.30) at the boundary, using the Fourier expansions (2.4.19) and (2.4.24), enforcing the boundary condition $\mathbf{u} = 0$, and then requiring that the sum of corresponding Fourier coefficients is equal to zero, we finally obtain

$$\int_C Q_{\alpha\beta n}(\hat{x}, \sigma, \sigma_0) f_{\beta n} \sigma \, \mathrm{d}l = 8\pi\mu u_{\alpha n} \tag{2.4.25}$$

Equation (2.4.25) provides us with a system of Fredholm integral equations of the first kind for the Fourier coefficients $f_{\beta n}$.

To present a particular application of (2.4.25), we consider unidirectional shearing flow $\mathbf{u}^{\infty}(\mathbf{x}) = kx\mathbf{j}$ past an axisymmetric body, where $\mathbf{j}$ is the unit vector in the direction of the $y$-axis. In this case, the only non-zero components of the Fourier coefficients $u$ are

$$u_{\sigma 1} = u_{\sigma(-1)} = kx/2, \qquad u_{\phi 1} = u^*_{\phi(-1)} = \mathrm{i}kx/2. \tag{2.4.26}$$

Inspecting (2.4.25) it becomes evident that all Fourier coefficients of $f$ except for the first must be equal to zero. Symmetry considerations dictate

$$f_{x1} = f^{\mathrm{R}}_{x1}, f_{\sigma 1} = f^{\mathrm{R}}_{\sigma 1}, f_{\phi 1} = \mathrm{i} f^{\mathrm{I}}_{\phi 1} \tag{2.4.27}$$

Substituting (2.4.26) and (2.4.27) into (2.4.25), we obtain a system of three scalar integral equations, namely,

$$\int_C \begin{bmatrix} Q^{\mathrm{R}}_{xx1} & Q^{\mathrm{R}}_{x\sigma 1} & -Q^{\mathrm{I}}_{x\phi 1} \\ Q^{\mathrm{R}}_{\sigma x1} & Q^{\mathrm{R}}_{\sigma\sigma 1} & -Q^{\mathrm{I}}_{\sigma\phi 1} \\ Q^{\mathrm{I}}_{\phi x1} & Q^{\mathrm{I}}_{\phi\sigma 1} & Q^{\mathrm{R}}_{\phi\phi 1} \end{bmatrix} \cdot \begin{bmatrix} f^{\mathrm{R}}_{x1} \\ f^{\mathrm{R}}_{\sigma 1} \\ f^{\mathrm{I}}_{\phi 1} \end{bmatrix} \sigma \, \mathrm{d}l = 4\pi\mu kx \begin{bmatrix} 0 \\ 1 \\ 1 \end{bmatrix} \tag{2.4.28}$$

**Problems**

2.4.1 Assess the leading order asymptotic behavior of the axisymmetric kernel **M** defined in (2.4.9) in the limit as $\mathbf{x}_0$ moves far away from $\mathbf{x}$.

2.4.2 Using (2.4.25), show that the swirling flow produced by the axial rotation of an axisymmetric body is described by the boundary integral equation

$$u_\phi(\mathbf{x}_0) = -\frac{1}{4\pi\mu}\int_C I_{11}(\mathbf{x}_0, \mathbf{x}) f_\phi(\mathbf{x}) \sigma \, dl(\mathbf{x})$$

where $I_{11}$ is defined in (2.4.7).

2.4.3 Using (2.4.25) verify that an axisymmetric flow past a rigid body is described by the boundary integral equation

$$u_\alpha(\mathbf{x}_0) = u_\alpha^\infty(\mathbf{x}_0) - \frac{1}{8\pi\mu}\int_C M_{\alpha\beta}(\mathbf{x}_0, \mathbf{x}) f_\beta(\mathbf{x}) \, dl(\mathbf{x})$$

where the Greek indices take the values $x$ and $\sigma$, and the matrix $\mathbf{M}$ is defined in (2.4.9).

2.4.4 Derive a system of one-dimensional integral equations describing the flow due to the translation or rotation of an axisymmetric rigid body.

2.4.5 Derive a system of one-dimensional integral equations describing shear flow over a plane wall that contains a depression or projection in the shape of a section of a sphere. Explore the possibility of simplifying the equations by using an appropriate Green's function.

## 2.5 Applications to particulate flows

In this section we discuss several applications of the boundary integral equation in the field of the low Reynolds number hydrodynamics of particulate flows.

### *The far flow due to an immersed particle*

First, we consider the asymptotic behaviour of the disturbance flow $\mathbf{u}^\mathrm{D}$ due to a rigid particle that is immersed in an ambient incident flow $\mathbf{u}^\infty$. Using (2.3.30), and interchanging the arguments and the subscripts of the Green's function, we express $\mathbf{u}^D$ in terms of a single-layer potential over the surface of the particle $S_P$, namely

$$u_j^\mathrm{D}(\mathbf{x}_0) = -\frac{1}{8\pi\mu}\int_{S_P} G_{ji}(\mathbf{x}_0, \mathbf{x}) f_i(\mathbf{x}) \, dS(\mathbf{x}) \tag{2.5.1}$$

To study the asymptotic behaviour of the flow far from $S_P$, we expand $\mathbf{G}$ in a Taylor series with respect to $\mathbf{x}$ about the point $\mathbf{x}_c$ which is presumed to be in the vicinity or in the interior of the particle. Thus, we obtain an asymptotic expansion for the disturbance velocity in the form

$$u_j^\mathrm{D}(\mathbf{x}_0) = -\frac{1}{8\pi\mu}\left[ G_{ji}(\mathbf{x}_0, \mathbf{x}_c)\int_{S_P} f_i(\mathbf{x}) \, dS(\mathbf{x}) + \frac{\partial G_{ji}}{\partial x_{c,k}}(\mathbf{x}_0, \mathbf{x}_c)\int_{S_P} (x_k - x_{c,k}) f_i(\mathbf{x}) \, dS(\mathbf{x}) + \cdots \right] \tag{2.5.2}$$

known as the *multipole expansion*. The first integral on the right-hand side of (2.5.2) is simply the force $\mathbf{F}$. The second integral is the first moment of the surface force, whereas subsequent integrals express higher moments of the surface force exerted on the particle. The first term on the right-hand side of (2.5.2) represents the flow due to a point force; the second term represents the flow due to a point force doublet; subsequent terms represent the flow due to point force quadruples and higher order multipoles. Equation (2.5.2) states that far from the particle the disturbance flow produced by the particle is similar to that due to a point force with strength $-\mathbf{F}$. When $\mathbf{F}=0$, i.e. for a force-free particle, the far flow is similar to that produced by a point force doublet. It will be noted that in the case of unbounded flow, the far disturbance flow decays as $1/r$ unless $\mathbf{F}=0$ in which case the far flow decays at least as $1/r^2$.

Proceeding, we split the coefficient of the point force doublet in (2.5.2) into a symmetric component $\mathscr{S}$, an isotropic component, and an antisymmetric component $\mathscr{R}$, setting

$$\int_{S_P} \hat{x}_k f_i(\mathbf{x})\,dS(\mathbf{x}) = \mathscr{S}_{ki} + \frac{1}{3}\delta_{ki}\int_{S_P} \hat{x}_l f_l(\mathbf{x})\,dS(\mathbf{x}) + \mathscr{R}_{ki} \tag{2.5.3}$$

where $\hat{\mathbf{x}} = \mathbf{x} - \mathbf{x}_c$,

$$\mathscr{S}_{ki} = \frac{1}{2}\int_{S_P} [\hat{x}_k f_i(\mathbf{x}) + \hat{x}_i f_k(\mathbf{x}) - \tfrac{2}{3}\delta_{ik}\hat{x}_l f_l(\mathbf{x})]\,dS(\mathbf{x}) \tag{2.5.4}$$

and

$$\mathscr{R}_{ki} = \frac{1}{2}\int_{S_P} [\hat{x}_k f_i(\mathbf{x}) - \hat{x}_i f_k(\mathbf{x})]\,dS(\mathbf{x}) \tag{2.5.5}$$

It will be noted that the antisymmetric component $\mathscr{R}$ is related to the torque $\mathbf{L}$ exerted on the particle with respect to the point $\mathbf{x}_c$, through the equation

$$\mathscr{R}_{ki} = \tfrac{1}{2}\varepsilon_{kil} L_l \tag{2.5.6}$$

where

$$\mathbf{L} \equiv \int_{S_P} \hat{\mathbf{x}} \times \mathbf{f}\,dS \tag{2.5.7}$$

Correspondingly, we split the point force dipole into a symmetric and an antisymmetric component with respect to the indices $k$ and $i$. Using (2.5.3), (2.1.3) and (2.5.7) we obtain

$$\frac{\partial G_{ji}}{\partial x_{c,k}}(\mathbf{x}_0, \mathbf{x}_c)\int_{S_P} \hat{x}_k f_i(\mathbf{x})\,dS(\mathbf{x}) = \frac{1}{2}\left(\frac{\partial G_{ji}}{\partial x_{c,k}} + \frac{\partial G_{jk}}{\partial x_{c,i}}\right)(\mathbf{x}_0, \mathbf{x}_c)\mathscr{S}_{ki}$$
$$+ \frac{1}{4}\varepsilon_{kil}\left(\frac{\partial G_{ji}}{\partial x_{c,k}} - \frac{\partial G_{jk}}{\partial x_{c,i}}\right)(\mathbf{x}_0, \mathbf{x}_c)L_l \tag{2.5.8}$$

The first term on the right-hand side of (2.5.8) represents the flow due to a symmetric pair of point force doublets. In section 7.2 we shall see that this term may be decomposed further into two derivative singularities, namely the stresslet and the point source. This decomposition allows us to call $\mathscr{S}$ the *coefficient of the stresslet.* The second term on the right-hand side of (2.5.8) represents the flow due to a fundamental singularity called the couplet or rotlet (see (7.2.22)). It will be noted that for a force-free particle the values of $\mathscr{S}$ and $\mathbf{L}$ are independent of the choice of $\mathbf{x}_c$. Equation (2.5.8) implies that if the torque $\mathbf{L}$ vanishes, the far flow is similar to that produced by a symmetric pair of point force dipoles.

Extending the above analysis, we consider the disturbance flow produced by a fluid particle that is immersed in an ambient incident flow. The simplified boundary integral representation (2.3.30) is no longer effective, and we must work with the full boundary integral equation. Considering equation (1) of problem 2.3.7 and repeating the above procedure we obtain an expression of the disturbance far flow in terms of fundamental singularities, namely

$$\begin{aligned} u_j^{\mathrm{D}}(\mathbf{x}_0) = & -\frac{1}{8\pi\mu_1}\left[ G_{ji}(\mathbf{x}_0, \mathbf{x}_c)\int_{S_P^+} f_i(\mathbf{x})\,dS(\mathbf{x}) \right. \\ & \left. + \frac{\partial G_{ji}}{\partial x_{c,k}}(\mathbf{x}_0, \mathbf{x}_c)\int_{S_P^+} \hat{x}_k f_i(\mathbf{x})\,dS(\mathbf{x}) + \cdots \right] \\ & + \frac{1}{8\pi}\left[ T_{ijl}(\mathbf{x}_c, \mathbf{x}_0)\int_{S_P} u_i(\mathbf{x}) n_l(\mathbf{x})\,dS(\mathbf{x}) \right. \\ & \left. + \frac{\partial T_{ijl}}{\partial x_{c,k}}(\mathbf{x}_c, \mathbf{x}_0)\int_{S_P} \hat{x}_k u_i(\mathbf{x}) n_l(\mathbf{x})\,dS(\mathbf{x}) + \cdots \right] \end{aligned} \tag{2.5.9}$$

where $\hat{\mathbf{x}} = \mathbf{x} - \mathbf{x}_c$, $\mu_1$ is the viscosity of the ambient fluid, and the superscript $+$ indicates that the surface force is evaluated on the external surface of the particle.

Maintaining only the point force and point force doublets in the first series, and the stresslet $\mathbf{T}$ in the second series on the right-hand side of (2.5.9), using (2.1.8), and exploiting the symmetry property (2.3.14), we obtain the simplified expansion

$$\begin{aligned} u_j^{\mathrm{D}}(\mathbf{x}_0) = & -\frac{1}{8\pi\mu_1}\left\{ G_{ji}(\mathbf{x}_0, \mathbf{x}_c)\int_{S_P^+} f_i(\mathbf{x})\,dS(\mathbf{x}) \right. \\ & \left. + \frac{\partial G_{ji}}{\partial x_{c,k}}(\mathbf{x}_0, \mathbf{x}_c)\int_{S_P^+} [\hat{x}_k f_i(\mathbf{x}) - \mu_1(u_k n_i + u_i n_k)(\mathbf{x})]\,dS(\mathbf{x}) + \cdots \right\} \\ & - \frac{1}{8\pi} p_j(\mathbf{x}_c, \mathbf{x}_0)\int_{S_P} u_l(\mathbf{x}) n_l(\mathbf{x})\,dS(\mathbf{x}) + \cdots \end{aligned} \tag{2.5.10}$$

The first series on the right-hand side of (2.5.10) contains multipoles of the point force, whereas the second series contains multipoles of the pressure. The coefficient of the pressure in the second series is proportional to the flow rate $Q$ through the surface of the particle, namely

$$Q = \int_{S_P} \mathbf{u} \cdot \mathbf{n} \, dS \tag{2.5.11}$$

Clearly, unless $Q$ is required to vanish (as is the case for an internal flow), $\mathbf{p}(\mathbf{x}_c, \mathbf{x}_0)$ represents the velocity field at the point $\mathbf{x}_0$ due to a point source of strength $-8\pi$ with pole at $\mathbf{x}_c$ (see also section 3.3).

Now, using the Stokes equation and the symmetry property (2.3.14) we find

$$\frac{\partial p_j}{\partial x_{c,k}}(\mathbf{x}_c, \mathbf{x}_0) = \nabla_c^2 G_{jk}(\mathbf{x}_0, \mathbf{x}_c) \tag{2.5.12}$$

where the subscript c of $\nabla_c$ indicates differentiation with respect to $\mathbf{x}_c$. Equation (2.5.12) suggests that all derivatives of the pressure may be expressed in terms of the Laplacian of the Green's function. In turn, this implies that all terms except for the first one in the second series on the right-hand side of (2.5.10) may be incorporated into the first series, thereby ensuring that the far disturbance flow may be represented in terms of an expansion of multipoles of the point force enhanced by a point source.

Proceeding, we decompose the coefficient of the point force doublet in (2.5.10) into a symmetric, an isotropic, and an antisymmetric component. The symmetric component is the coefficient of the stresslet, given by

$$\mathscr{S}_{ki} = \frac{1}{2} \int_{S_P^+} [\hat{x}_k f_i(\mathbf{x}) + \hat{x}_i f_k(\mathbf{x}) - \delta_{ik} \tfrac{2}{3} \hat{x}_l f_l(\mathbf{x}) - 2\mu_1 (u_k n_i + u_i n_k)(\mathbf{x})] \, dS(\mathbf{x}) \tag{2.5.13}$$

whereas the antisymmetric component $\mathscr{R}$ was given in (2.5.5) and (2.5.7). The values of $\mathscr{S}$ and $\mathscr{R}$ remain unchanged when the domain of integration $D$ is replaced by an arbitrary surface that encloses $D$; in addition, for a force-free particle, they are independent of the choice of $\mathbf{x}_c$.

Splitting the point force dipole into a symmetric and antisymmetric component and following a procedure similar to the one that led us to (2.5.8), we obtain a representation of the far flow in terms of a stresslet, a couplet, and a point source.

### *The viscosity of a suspension of force-free particles*

The coefficient of the point force dipole is useful not only for describing the far disturbance flow due to an immersed particle but also for expressing the global rheological properties of a suspension of force-free solid or liquid particles, viewed as a homogeneous medium (Batchelor 1970a). To establish this important connection, let us consider the flow of a suspension of liquid particles and compute the average value of the stress tensor over a volume $\mathscr{V}$ that encloses both ambient fluid and suspended particles, namely

$$\langle \boldsymbol{\sigma} \rangle \equiv \frac{1}{\mathscr{V}} \int_{\mathscr{V}} \boldsymbol{\sigma} \, \mathrm{d}V = \frac{1}{\mathscr{V}} \int_{V} \boldsymbol{\sigma} \, \mathrm{d}V + \frac{1}{\mathscr{V}} \int_{V_{\mathrm{P}}^{+}} \boldsymbol{\sigma} \, \mathrm{d}V \qquad (2.5.14)$$

where $V$ is the volume occupied by the ambient fluid, and $V_{\mathrm{P}}^{+}$ is the volume occupied by the particles. The superscript + in $V_{\mathrm{P}}^{+}$ indicates that the volume of the particles is enhanced with a thin layer of ambient fluid lining the external surface of the particles. In the absence of a body force, the average value of the stress over $V_{\mathrm{P}}^{+}$ is

$$\int_{V_{\mathrm{P}}^{+}} \sigma_{ij} \, \mathrm{d}V = \int_{S_{\mathrm{P}}^{+}} x_i f_j \, \mathrm{d}S \qquad (2.5.15)$$

where $\mathbf{f} = \boldsymbol{\sigma} \cdot \mathbf{n}$, and $S_{\mathrm{P}}^{+}$ is the external surface of the particles (problem 1.2.1). Decomposing the stress tensor into its pressure and viscous constituents, we obtain

$$\int_{V} \sigma_{ij} \, \mathrm{d}V = -\delta_{ij} \int_{V} P \, \mathrm{d}V + 2\mu_1 \int_{\mathscr{V}} e_{ij} \, \mathrm{d}V - 2\mu_1 \int_{V_{\mathrm{P}}^{+}} e_{ij} \, \mathrm{d}V \qquad (2.5.16)$$

where $\mu_1$ is the viscosity of the ambient fluid. Using the divergence theorem, and requiring that the velocity is continuous across the surface of the particles, we convert the last volume integral on the right-hand side of (2.5.16) into a surface integral over the external surface of the particles. Substituting the resulting expression, along with (2.5.15), into (2.5.14) we obtain

$$\langle \sigma_{ij} \rangle = -\delta_{ij} \frac{1}{\mathscr{V}} \int_{V} P \, \mathrm{d}V + 2\mu_1 \langle e_{ij} \rangle + \frac{1}{\mathscr{V}} \int_{S_{\mathrm{P}}^{+}} [f_i x_j - \mu_1 (u_i n_j + u_j n_i)] \, \mathrm{d}S \qquad (2.5.17)$$

where

$$\langle \mathbf{e} \rangle \equiv \frac{1}{\mathscr{V}} \int_{\mathscr{V}} \mathbf{e} \, \mathrm{d}V \qquad (2.5.18)$$

is the average value of the rate of deformation tensor over the selected volume of the suspension. Consulting with (2.5.10), we recognize the second integral on the right-hand side of (2.5.17) as the coefficient of the point force dipole in the far field expansion. Decomposing this term further into a symmetric, an isotropic, and an antisymmetric component we obtain

$$\langle\sigma_{ij}\rangle = \delta_{ij}\frac{1}{\mathscr{V}}\left(-\int_V P\,\mathrm{d}V + \frac{1}{3}\int_{S_P^+} x_k f_k\,\mathrm{d}V\right) + 2\mu_1\langle e_{ij}\rangle$$
$$+ \frac{1}{\mathscr{V}}\sum(\mathscr{S}_{ij} + \tfrac{1}{2}\varepsilon_{ijm}L_m) \tag{2.5.19}$$

where $\mathscr{S}$ is the coefficient of the stresslet, $\mathbf{L}$ is the torque exerted on a particle, and the summation is over all particles. The first bracket on the right-hand side of (2.5.19) represents an isotropic stress due to an effective pressure. The second term on the right-hand side of (2.5.19) is what we would expect if the suspension behaved similarly to a homogeneous Newtonian fluid with viscosity equal to that of the ambient fluid. The third term is the particle stress tensor

$$\Sigma_{ij} \equiv \frac{1}{\mathscr{V}}\sum(\mathscr{S}_{ij} + \tfrac{1}{2}\varepsilon_{ijm}L_m) \tag{2.5.20}$$

representing the effect of the suspended particles. The computation of $\boldsymbol{\Sigma}$ is the ultimate objective in quantifying the rheology of a suspension.

The above analysis applies for a suspension of liquid particles. The case of solid particles will arise as a special case, if we assume that the velocity at the surface of the particles expresses rigid body motion.

Now, as a specific application, we consider the particle stress tensor for a dilute suspension of force-free and torque-free solid spheres. In the limit of infinite dilution each sphere finds itself in an isolated environment, and responds as if it were immersed in an infinite linear ambient flow. The rotational component of this flow will cause each sphere to spin about the local axis of vorticity without introducing a disturbance flow. Thus, for the purposes of our analysis, it will be consistent to assume that each sphere is immersed in a purely straining incident flow $\mathbf{u} = \mathbf{E}\cdot\mathbf{x}$, where $\mathbf{E}$ is a symmetric traceless matrix. A detailed calculation shows that the coefficient of the stresslet on each sphere is $\mathscr{S} = (20\pi\mu a^3/3)\mathbf{E}$, where $a$ is the radius of a sphere and $\mu$ is the viscosity of the suspending fluid (see (7.3.24)). Using (2.5.20) we find that the particle stress tensor is given by

$$\boldsymbol{\Sigma} = \frac{20\pi}{3}\mu a^3\frac{N}{\mathscr{V}}\mathbf{E} = 5\mu c\mathbf{E} \tag{2.5.21}$$

where $N$ is the number of spheres per unit volume of the suspension, and $c$ is the volumetric concentration of the spheres in the suspension. Adding to (2.5.21) the stress tensor of the incident flow, namely $\boldsymbol{\sigma} = 2\mu\mathbf{E}$, we obtain the total stress tensor of the suspension

$$\boldsymbol{\sigma} + \boldsymbol{\Sigma} = \mu(1 + \tfrac{5}{2}c)2\mathbf{E} \tag{2.5.22}$$

Noting the linear relationship between the stress tensor and the rate of deformation tensor, we conclude that the suspension behaves like a Newtonian fluid with an effective viscosity equal to $\mu[1 + (5/2)c]$ (Einstein 1906, 1956).

### *Generalized Faxen relations*

In order to compute the linear and angular velocities of a suspended particle, as well as the contribution of the particle to the particle stress tensor, we must know the force **F**, the torque **L** exerted on the particle, and the coefficient of the stresslet $\mathscr{S}$ arising when the particle is held stationary in an incident ambient flow. The generalized Faxen relations (named after Faxen (1924) who developed relations for the force and torque on a solid spherical particle), provide expressions for **F**, **L**, and $\mathscr{S}$ simply in terms of the value of the incident velocity or its derivatives at specific locations in the interior or on the surface of the particle.

First, we consider the Faxen relations for a solid particle. Specifically, we consider the force **F** exerted on a solid particle when it is held stationary in an ambient flow $\mathbf{u}^\infty$. Following Kim (1985) and Kim & Lu (1987), we assume that the flow due to the translation of the particle may be represented in terms of a discrete or continuous distribution of Green's functions as

$$u_i(\mathbf{x}) = V_k \mathscr{L}^0_{kj} \langle G_{ij}(\mathbf{x}, \mathbf{x}_0) \rangle \tag{2.5.23}$$

where **V** is the velocity of translation, and $\mathscr{L}^0$ is an integral, differential, or integro-differential operator acting with respect to $\mathbf{x}_0$ so that the poles of the Green's function and its derivatives are always in the interior of the particle. The surface force on the particle is

$$f^{\mathrm{T}}_i(\mathbf{x}) = \mu V_k \mathscr{L}^0_{kj} \langle T_{ijl}(\mathbf{x}, \mathbf{x}_0) \rangle n_l(\mathbf{x}) \tag{2.5.24}$$

where the superscript T indicates translation. In order to satisfy the boundary condition $\mathbf{u} = \mathbf{V}$ on the surface of the particle, we require

$$\mathscr{L}^0_{kj} \langle G_{ij}(\mathbf{x}, \mathbf{x}_0) \rangle = \delta_{ki} \tag{2.5.25}$$

when **x** is on the surface of the particle. Comparing (2.5.24) with (1.4.6), and using (1.4.10) we find

$$F_k = -\mu \int_{S_P} u^\infty_i(\mathbf{x}) \mathscr{L}^0_{kj} \langle T_{ijl}(\mathbf{x}, \mathbf{x}_0) \rangle n_l(\mathbf{x}) \, dS(\mathbf{x}) \tag{2.5.26}$$

where $S_P$ is the surface of the particle.

Next, we write the boundary integral equation (2.3.11) for the ambient flow within the volume occupied by the particle, and operate on both sides by $\mathscr{L}^0$ obtaining

$$\mathscr{L}^0_{kj}\langle u^\infty_j(\mathbf{x}_0)\rangle = \frac{1}{8\pi\mu}\int_{S_P} f^\infty_i(\mathbf{x})\,\mathscr{L}^0_{kj}\langle G_{ij}(\mathbf{x},\mathbf{x}_0)\rangle\,\mathrm{d}S(\mathbf{x}) - \frac{1}{8\pi}\int_{S_P} u^\infty_i(\mathbf{x})\,\mathscr{L}^0_{kj}\langle T_{ijl}(\mathbf{x},\mathbf{x}_0)\rangle n_l(\mathbf{x})\,\mathrm{d}S(\mathbf{x}) \tag{2.5.27}$$

where the normal vector $\mathbf{n}$ is directed outside the particle. Incorporating the boundary condition (2.5.25) into the first integral on the right-hand side of (2.5.27), noting that the force due to the ambient flow on the volume occupied by the particle is equal to zero, and combining the resulting expression with (2.5.26) we obtain the generalized Faxen relation for the force

$$F_k = 8\pi\mu\,\mathscr{L}^0_{kj}\langle u^\infty_j(\mathbf{x}_0)\rangle \tag{2.5.28}$$

Considering next the torque $\mathbf{L}$ acting on the particle, we assume that the flow produced by the rotation of a particle may be represented in terms of a discrete or continuous distribution of Green's function as

$$u_i(\mathbf{x}) = \omega_k\,\mathscr{L}^0_{kj}\langle G_{ij}(\mathbf{x},\mathbf{x}_0)\rangle \tag{2.5.29}$$

where $\boldsymbol{\omega}$ is the angular velocity of rotation. Repeating the above procedure, we find that the torque exerted on the particle when it is held stationary in an ambient flow $\mathbf{u}^\infty$ is given by the generalized Faxen relation

$$L_k = 8\pi\mu\,\mathscr{L}^0_{kj}\langle u^\infty_j(\mathbf{x}_0)\rangle \tag{2.5.30}$$

Similarly, if we assume that the disturbance flow produced when the particle is held stationary in the purely straining flow $\mathbf{u}^\infty = \mathbf{E}\cdot\mathbf{x}$ may be represented as

$$u^D_i(\mathbf{x}) = E_{lk}\,\mathscr{L}^0_{lkj}\langle G_{ij}(\mathbf{x},\mathbf{x}_0)\rangle \tag{2.5.31}$$

where $\mathbf{E}$ is a symmetric traceless matrix, we will find that the coefficient of the stresslet arising when the particle is held still in an ambient flow $\mathbf{u}^\infty$ is given by the generalized Faxen relation

$$\mathscr{S}_{lk} = -4\pi\mu[\mathscr{L}^0_{lkj}\langle u^\infty_j(\mathbf{x}_0)\rangle + \mathscr{L}^0_{klj}\langle u^\infty_j(\mathbf{x}_0)\rangle] \tag{2.5.32}$$

As a specific application, we note that the flow due to the translation or rotation of a solid spherical particle in an unbounded fluid may be presented as

$$u_i(\mathbf{x}) = V_k\delta_{kj}(\tfrac{3}{4}a + \tfrac{1}{8}a^3\nabla_0^2)\mathscr{S}_{ij}(\mathbf{x},\mathbf{x}_0) \tag{2.5.33}$$

$$u_i(\mathbf{x}) = a^3\omega_k\tfrac{1}{2}\varepsilon_{klj}\frac{\partial\mathscr{S}_{ij}}{\partial x_{0,l}}(\mathbf{x},\mathbf{x}_0) \tag{2.5.34}$$

where $\mathscr{S}$ is the Stokeslet, $a$ is the radius of the particle, and $\mathbf{x}_0$ is the centre of the particle (see (7.3.10) and (7.3.30)). Furthermore, the disturbance flow produced when a solid spherical particle is immersed in a purely straining ambient flow may be represented as

$$u_i^D(\mathbf{x}) = -\tfrac{1}{6}E_{jl}\frac{\partial}{\partial x_{0,l}}(5a^3 + \tfrac{1}{2}a^5\nabla_0^2)\mathscr{S}_{ij}(\mathbf{x}, \mathbf{x}_0) \tag{2.5.35}$$

(see (7.3.26)). Using (2.5.28), (2.5.30), and (2.5.32) we obtain the force, torque, and coefficient of the stresslet as

$$\mathbf{F} = 6\pi\mu a\mathbf{u}^\infty(\mathbf{x}_0) + \pi\mu a^3\nabla_0^2\mathbf{u}^\infty(\mathbf{x}_0) \tag{2.5.36}$$

$$\mathbf{L} = 4\pi\mu a^3\nabla_0 \times \mathbf{u}^\infty(\mathbf{x}_0) \tag{2.5.37}$$

$$\mathscr{S}_{ij} = \tfrac{10}{3}\pi\mu a^3\left[\frac{\partial u_j^\infty}{\partial x_i} + \frac{\partial u_i^\infty}{\partial x_j} + \frac{1}{10}a^2\nabla_0^2\left(\frac{\partial u_j^\infty}{\partial x_i} + \frac{\partial u_i^\infty}{\partial x_j}\right)\right](\mathbf{x}_0) \tag{2.5.38}$$

in agreement with Faxen (1924) and Batchelor & Green (1972). The subscript 0 indicates evaluation at the centre of the particle. Now, comparing (2.5.36) and (2.5.37) with (1.4.15), we derive the identities

$$\frac{1}{4\pi a^2}\int_{\text{sphere}} \mathbf{u}\,dS = \mathbf{u}(\mathbf{x}_0) + \tfrac{1}{6}a^2\nabla^2\mathbf{u}(\mathbf{x}_0) \tag{2.5.39}$$

$$\int_{\text{sphere}} (\mathbf{x} - \mathbf{x}_0) \times \mathbf{u}\,dS = \frac{4\pi}{3}a^4\nabla \times \mathbf{u}(\mathbf{x}_0) \tag{2.5.40}$$

(see problem 1.4.3). Equation (2.5.39) relates the mean value of the velocity over the surface of a sphere to the values of the velocity and its Laplacian at the center of the sphere. Thus this equation is the counterpart to the mean value theorem for potential flow, which states that the mean value of a potential function over the surface of a sphere is equal to the value of the potential function at the center of the sphere. Equation (2.5.40) relates the mean value of the angular momentum over the surface of a sphere to the value of the vorticity at the center of the sphere.

Proceeding, we set as our next objective the derivation of Faxen relations for a fluid particle that is held stationary in an ambient flow $\mathbf{u}^\infty$. As in the case of the solid particle, it will be convenient to decompose the total flow $\mathbf{u}$ as $\mathbf{u} = \mathbf{u}^\infty + \mathbf{u}^\text{D}$, where $\mathbf{u}^\text{D}$ is the disturbance flow due to the particle. We require that the normal component of the total velocity $\mathbf{u}$ vanishes over the surface of the particle, whereas the tangential component of the total velocity and surface force are continuous across the surface of the particle. We allow the normal component of the total surface force to be discontinuous across the surface of the particle, assuming that the discontinuity of the surface force is compensated by surface tension or by a difference in the densities of the drop and the ambient fluid (see discussion in section 5.5).

First, we consider the Faxen relation for the force. Tracing our previous steps for a solid particle, we represent the exterior flow due to the translation of the fluid particle as in (2.5.23). Applying the reciprocal theorem for the disturbance flow $\mathbf{u}^D$ and the flow produced by the translation of the particle, we obtain

$$\int_{S_P^+} \mathbf{u}^T \cdot \mathbf{f}^D \, dS = \int_{S_P^+} \mathbf{u}^D \cdot \mathbf{f}^T \, dS \tag{2.5.41}$$

where the superscript T stands for translation, and the superscript + indicates the external surface of the particle. Writing $\mathbf{u}^D = \mathbf{u} - \mathbf{u}^\infty$ we restate (2.5.41) in the equivalent form

$$\int_{S_P^+} (\mathbf{u}^T - \mathbf{V}) \cdot \mathbf{f}^D \, dS + \mathbf{V} \cdot \mathbf{F}^D = \int_{S_P^+} \mathbf{u} \cdot \mathbf{f}^T \, dS - \int_{S_P^+} \mathbf{u}^\infty \cdot \mathbf{f}^T \, dS \tag{2.5.42}$$

where $\mathbf{F}^D$ is the disturbance force exerted on the particle. Multiplying (2.5.27) by $\mathbf{V}$ we obtain

$$V_k \mathscr{L}_{ki}^0 \langle u_i^\infty(\mathbf{x}_0) \rangle = \frac{1}{8\pi\mu} \int_{S_P^+} f_i^\infty u_i^T \, dS - \frac{1}{8\pi\mu} \int_{S_P^+} u_i^\infty f_i^T \, dS \tag{2.5.43}$$

where $\mu$ is the viscosity of the ambient fluid. Combining (2.5.43) with (2.5.42) and rearranging we obtain

$$\int_{S_P^+} (\mathbf{u}^T - \mathbf{V}) \cdot \mathbf{f} \, dS + \mathbf{V} \cdot \mathbf{F}^D - 8\pi\mu \mathbf{V} \cdot \mathscr{L}^0 \langle \mathbf{u}^\infty(\mathbf{x}_0) \rangle = \int_{S_P^+} \mathbf{u} \cdot \mathbf{f}^T \, dS \tag{2.5.44}$$

Since the velocities $\mathbf{u}^T - \mathbf{V}$ and $\mathbf{u}$ are tangential to the surface of the particle and the tangential components of $\mathbf{f}^T$ and $\mathbf{f}$ are continuous across the surface of the particle, we may switch the domain of integration of both integrals in (2.5.44) from the external to the internal side of $S_P$. Furthermore, applying the reciprocal theorem for the flows $\mathbf{u}^T - \mathbf{V}$ and $\mathbf{u}$ over the interior of the particle, we deduce that the two integrals in (2.5.44) are identical. In this manner, we obtain the generalized Faxen relation for the force expressed by (2.5.28). Following a similar line of argument, we find that the generalized Faxen relations for the torque and coefficient of the stresslet expressed by (2.5.30) and (2.5.32) are also valid for liquid particles (Kim & Lu 1987).

To present a specific application of (2.5.28), we note that the external flow produced by the translation of a spherical drop may be represented as

$$u_i^{\text{Ext}}(\mathbf{x}) = \frac{1}{8} U_j a \left[ 2\left( \frac{3\lambda + 2}{\lambda + 1} \right) + a^2 \left( \frac{\lambda}{\lambda + 1} \right) \nabla_0^2 \right] \mathscr{S}_{ij}(\mathbf{x}, \mathbf{x}_0) \tag{2.5.45}$$

where $\lambda$ is the ratio of the viscosities of the ambient fluid and the drop, $\mathscr{S}$ is the Stokeslet, $a$ is the radius of the drop, and $\mathbf{x}_0$ is the centre of the drop (see (7.3.42)). Using (2.5.45) and (2.5.28) we obtain the Faxen relation for the force:

$$\mathbf{F} = \pi\mu a\left[2\left(\frac{3\lambda+2}{\lambda+1}\right) + a^2\left(\frac{\lambda}{\lambda+1}\right)\nabla_0^2\right]\mathbf{u}^\infty(\mathbf{x}_0) \tag{2.5.46}$$

in agreement with Hetsroni & Haber (1970) and Rallison (1978).

The reader will note that in order to derive the Faxen relations for a specific particle, we must know the singularity representations of the flow due to the translation and rotation of the particle, as well as the disturbance flow produced when the particle is placed in an ambient straining flow. Unfortunately, exact singularity representations are known only for a limited number of boundary configurations and particle shapes (see discussion in section 7.3, and problems 2.5.3, 2.5.4). Approximate singularity representations, however, may be constructed using numerical methods, as will be discussed in section 7.4.

### *Boundary effects on the motion of a particle*

Solid boundaries have a strong influence on the force and torque exerted on a stationary particle as well as on the translational and rotational velocities of a freely suspended particle. To compute the precise effect of a boundary, one must solve the equations of Stokes flow subject to the condition that the velocity over the boundary is equal to zero. This results in a rather complicated problem that has been tackled only for simple particle shapes and boundary configurations (Happel & Brenner 1973, Chapter 7; Kim & Karrila 1991, Chapter 12). Fortunately, when the particle is located far away from a boundary, the effect of the boundary may be estimated in a remarkably simple and general manner by means of an asymptotic analysis of the boundary integral equation (Williams 1966a).

Let us consider a rigid particle translating with velocity $\mathbf{V}$ and rotating with angular velocity $\mathbf{\Omega}$ in the presence of a remote boundary. The flow due to the motion of the particle may be described by the boundary integral equation (2.3.27), where $\mathbf{G}$ is a Green's function that vanishes over the solid boundary. It will be convenient to decompose the Green's function $\mathbf{G}$ into the Stokeslet $\mathcal{S}$ and a complementary matrix $\mathbf{J}$, $\mathbf{G} = \mathcal{S} + \mathbf{J}$, where $\mathbf{J}$ is regular (non-singular) throughout the domain of flow. We select a point $\mathbf{x}_c$ within the particle, and expand $\mathbf{J}$ as a double Taylor series with respect to $\mathbf{x}$ and $\mathbf{x}_0$ about the point $\mathbf{x}_c$ obtaining

$$\mathbf{J}(\mathbf{x},\mathbf{x}_0) = \mathbf{J}(\mathbf{x}_c,\mathbf{x}_c) + (\mathbf{x}-\mathbf{x}_c)\cdot\nabla\mathbf{J}(\mathbf{x}_c,\mathbf{x}_c) + (\mathbf{x}_0-\mathbf{x}_c)\cdot\nabla_0\mathbf{J}(\mathbf{x}_c,\mathbf{x}_c) + \cdots \tag{2.5.47}$$

It is important to note at the outset that the matrix $\mathbf{J}(\mathbf{x}_c,\mathbf{x}_c)$ and its derivatives depend only on the boundary configuration, and are independent of the geometry of the particle. Provided that the distance of the particle from the boundary is much larger than the characteristic size of the particle, the second, third, and subsequent terms on the right-hand side of (2.5.47) are smaller than the first term, and may be neglected. Equation (2.3.27) then simplifies to

$$\mathbf{u}_j(\mathbf{x}_0) + \frac{1}{8\pi\mu}\mathbf{F}\cdot\mathbf{J}(\mathbf{x}_c, \mathbf{x}_c) = -\frac{1}{8\pi\mu}\int_{S_P} f_i(\mathbf{x})\mathscr{S}_{ij}(\mathbf{x}, \mathbf{x}_0)\,dS(\mathbf{x}) \tag{2.5.48}$$

where $\mathbf{F}$ is the force exerted on the particle. Applying (2.5.48) on the surface of the particle we obtain a Fredholm integral equation of the first kind for the surface force $\mathbf{f}$.

Recalling now that the velocity on the surface of the particle is $\mathbf{u} = \mathbf{V} + \boldsymbol{\Omega} \times \mathbf{x}$, we realize that (2.5.48) describes the flow produced when the particle translates with velocity $\mathbf{V} + \mathbf{F}\cdot\mathbf{J}(\mathbf{x}_c, \mathbf{x}_c)/8\pi\mu$ and rotates with angular velocity $\boldsymbol{\Omega}$ in infinite fluid. This observation allows us to express the solution of (2.5.48) in the standard form

$$\mathbf{f} = -\mu\mathscr{G}^{\mathrm{T}}\cdot[\mathbf{V} + \mathbf{F}\cdot\mathbf{J}(\mathbf{x}_c, \mathbf{x}_c)/8\pi\mu] - \mu\mathscr{G}^{\mathrm{R}}\cdot\boldsymbol{\Omega} \tag{2.5.49}$$

where $\mathscr{G}^{\mathrm{T}}$ and $\mathscr{G}^{\mathrm{R}}$ are the translational and rotary surface force resistance matrices for unbounded flow introduced in (1.4.6) and (1.4.11). The force and torque exerted on the particle are given by

$$\begin{bmatrix}\mathbf{F}\\ \mathbf{L}\end{bmatrix} = -\mu\begin{bmatrix}\mathbf{X}\\ \mathbf{P}'\end{bmatrix}\cdot\left(\mathbf{V} + \frac{1}{8\pi\mu}\mathbf{F}\cdot\mathbf{J}(\mathbf{x}_c, \mathbf{x}_c)\right) - \mu\begin{bmatrix}\mathbf{P}\\ \mathbf{Y}\end{bmatrix}\cdot\boldsymbol{\Omega} \tag{2.5.50}$$

where $\mathbf{X}, \mathbf{P}', \mathbf{P}, \mathbf{Y}$ are the resistance matrices defined in (1.4.18) and (1.4.19). Solving the first equation in (2.5.50) for $\mathbf{F}$ we obtain

$$\mathbf{F} = -\mu(\mathbf{X}\cdot\mathbf{V} + \mathbf{P}\cdot\boldsymbol{\Omega})\cdot\mathbf{J}^{-1}\cdot\left(\mathbf{J}^{-1} + \frac{1}{8\pi}\mathbf{X}^{\mathrm{T}}\right)^{-1} \tag{2.5.51}$$

Substituting (2.5.51) into the second equation in (2.5.50) we obtain an explicit expression for the torque merely in terms of $\mathbf{V}$ and $\boldsymbol{\Omega}$.

In summary, equations (2.5.49) and (2.5.50) provide us with approximate expressions for the traction, force, and torque exerted on the particle in terms of the resistance matrices for unbounded flow and the boundary dependent matrix $\mathbf{J}(\mathbf{x}_c, \mathbf{x}_c)$.

### Problems

2.5.1 Show that the values of $\mathscr{S}$ and $\mathscr{R}$ defined in (2.5.5) and (2.5.13) remain unchanged when the domain of integration is replaced by an arbitrary surface that encloses $S_P$.

2.5.2 Prove the Faxen relation for the stresslet given in (2.5.32).

2.5.3 Chwang & Wu (1975) derived a singularity representation of the flow due to the translation of a prolate spheroid in terms of distributions of Stokeslets and potential dipoles over the focal length of the spheroid. Their results may be cast in the form

$$u_i(\mathbf{x}) = V_k a_{kj} \int_{-c}^{c} \left[ 1 + \frac{1-e^2}{4e^2}(c^2 - x_0^2)\nabla_0^2 \right] \mathcal{S}_{ij}(\mathbf{x}, \mathbf{x}_0)\, \mathrm{d}x_0$$

where $c$ is the focal length of the spheroid defined by $c^2 = a^2 - b^2, e = c/a$ is the eccentricity of the spheroid $(0 < e < 1)$, $a$ and $b$ are the major and minor semi-axes of the spheroid, and $\mathbf{a}$ is a diagonal matrix given by

$$a_{11} = \frac{e^2}{-2e + (1+e^2)\ln\left(\dfrac{1+e}{1-e}\right)} \qquad a_{22} = a_{33} = \frac{-2e^2}{-2e + (1-3e^2)\ln\left(\dfrac{1+e}{1-e}\right)}$$

The 1-axis is aligned with the major axis of the spheroid. Using the above expressions we find that the Faxen relation for the force on a prolate spheroid is given by

$$F_k = 8\pi\mu a_{kj} \int_{-c}^{c} \left[ 1 + \left(\frac{1-e^2}{4e^2}\right)(c^2 - x_0^2)\nabla_0^2 \right] u_j^\infty(\mathbf{x}_0)\, \mathrm{d}x_0$$

Verify that as $e \to 0$, i.e. the spheroid tends to become a sphere $a_{11}, a_{22}$ and $a_{33} \to (3/8)e$, recovering the Faxen relation for the sphere. Compute the force acting on a spheroid embedded in a uniform ambient flow $\mathbf{u}^\infty = \mathbf{U}$ or a paraboloidal ambient flow $\mathbf{u}^\infty = U[(y^2 + z^2)/a^2, 0, 0)] + V[0, (x^2 + z^2)/a^2, 0]$.

2.5.4 Chwang & Wu (1974) derived a singularity representation of the flow produced by the axial rotation of a prolate spheroid. Their results may be cast in the form

$$u_i(\mathbf{x}) = \omega_1 \frac{1}{\left(\dfrac{2e}{1-e^2}\right) - \ln\left(\dfrac{1+e}{1-e}\right)} \int_{-c}^{c} (c^2 - x_0^2)\tfrac{1}{2}\varepsilon_{1lj} \frac{\partial}{\partial x_{0,l}} \mathcal{S}_{ij}(\mathbf{x}, \mathbf{x}_0)\, \mathrm{d}x_0$$

Using this expression deduce that the Faxen relation for the axial component of the torque on a prolate spheroid is

$$L_1 = \frac{4\pi\mu}{\left(\dfrac{2e}{1-e^2}\right) - \ln\left(\dfrac{1+e}{1-e}\right)} \int_{-c}^{c} (c^2 - x_0^2)[\nabla_0 \times \mathbf{u}^\infty(\mathbf{x}_0)]_1\, \mathrm{d}x_0$$

Verify that as $e \to 0$, we recover the Faxen relation for the sphere. Compute the torque on a spheroid embedded in the linear flow $\mathbf{u}^\infty = (0, kz, 0)$.

2.5.5 Using the Faxen relations compute the force and torque exerted on a solid sphere due to a nearby point force. The answers are given in (3.3.26) and (3.3.27).

2.5.6 Perform an asymptotic analysis of the boundary integral equation to derive an approximate expression for the force and torque exerted on each of two remote particles that move in an infinite ambient fluid. Your results should be in terms of the resistance coefficients of the individual particles for unbounded flow (Williams 1966a). Extend your analysis to the general case of $N$ particles.

2.5.7 Show that the matrix $\mathbf{J}(\mathbf{x}_c, \mathbf{x}_c)$ defined in (2.5.47) is symmetric.

## 2.6 Two-dimensional Stokes flow

The Green's functions of two-dimensional Stokes flow represent solutions of the continuity equation and the two-dimensional version of the singularity forced Stokes equation (2.1.1); $\delta$ is now the two-dimensional delta function and $\nabla^2$ is the two-dimensional Laplacian operator. Introducing the Green's function $\mathbf{G}$ we write the solution for the velocity in the form

$$u_i(\mathbf{x}) = \frac{1}{4\pi\mu} G_{ij}(\mathbf{x}, \mathbf{x}_0) g_j \tag{2.6.1}$$

Physically, equation (2.6.1) expresses the flow due to a two-dimensional point force, and may be identified with the flow produced by the slow motion of a slender infinite cylinder with an infinitesimal cross-sectional area, parallel to its generators.

As in the case of three-dimensional flow, it is convenient to distinguish between the Green's function for infinite unbounded flow (i.e. the free-space Green's function), and the Green's functions for infinite bounded, and internal flow. The Green's functions for infinite flow are required either to vanish at infinity or to increase, at most, at a logarithmic rate. The Green's functions for infinite bounded or internal flow are required to vanish over the boundaries of the flow. All Green's functions exhibit a common logarithmic singular behaviour at the pole.

The continuity equation requires that $\mathbf{G}$ satisfies (2.1.3). Integrating (2.1.3) over an area $A$ enclosed by the curve $\mathscr{C}$, and using the divergence theorem to convert the area integral into a line integral, we find

$$\int_{\mathscr{C}} G_{ij}(\mathbf{x}, \mathbf{x}_0) n_i(\mathbf{x}) \, \mathrm{d}l(\mathbf{x}) = 0 \tag{2.6.2}$$

where the unit normal vector $\mathbf{n}$ is directed outside $\mathscr{C}$. Equation (2.6.2) is valid independently of whether the pole $\mathbf{x}_0$ is inside, on, or outside $\mathscr{C}$.

The vorticity, pressure, and stress fields corresponding to the flow (2.6.1) may be written in the standard forms

$$\omega_i(\mathbf{x}) = \frac{1}{4\pi\mu}\Omega_{ij}(\mathbf{x}, \mathbf{x}_0)g_j \tag{2.6.3}$$

$$P(\mathbf{x}) = \frac{1}{4\pi}p_j(\mathbf{x}, \mathbf{x}_0)g_j \tag{2.6.4}$$

$$\sigma_{ik}(\mathbf{x}) = \frac{1}{4\pi}T_{ijk}(\mathbf{x}, \mathbf{x}_0)g_j \tag{2.6.5}$$

where the stress tensor $\mathbf{T}$ is defined in (2.1.8). When the Green's function corresponds to an infinite domain of flow, all $\mathbf{\Omega}$, $\mathbf{p}$, and $\mathbf{T}$ are required to decay to zero far from the pole. Substituting (2.6.1), (2.6.4), and (2.6.5) into (2.1.1) we derive the equations

$$-\frac{\partial p_j}{\partial x_k}(\mathbf{x}, \mathbf{x}_0) + \nabla^2 G_{kj}(\mathbf{x}, \mathbf{x}_0) = -4\pi\delta_{kj}\delta(\mathbf{x} - \mathbf{x}_0) \tag{2.6.6}$$

and

$$\frac{\partial T_{ijk}}{\partial x_i}(\mathbf{x}, \mathbf{x}_0) = \frac{\partial T_{kji}}{\partial x_i}(\mathbf{x}, \mathbf{x}_0) = -4\pi\delta_{kj}\delta(\mathbf{x} - \mathbf{x}_0) \tag{2.6.7}$$

Furthermore, using (2.6.7) we find that

$$\frac{\partial}{\partial x_k}[\varepsilon_{ilm}x_l T_{mjk}(\mathbf{x}, \mathbf{x}_0)] = -4\pi\varepsilon_{ilj}x_l\delta(\mathbf{x} - \mathbf{x}_0) \tag{2.6.8}$$

Integrating (2.6.7) and (2.6.8) over the area $A$ enclosed by the smooth curve $\mathscr{C}$, and using the divergence theorem to convert the area integral into a line integral, we obtain the identities

$$\int_{\mathscr{C}} T_{ijk}(\mathbf{x}, \mathbf{x}_0)n_i(\mathbf{x})\,\mathrm{d}l(\mathbf{x}) = \int_{\mathscr{C}} T_{kji}(\mathbf{x}, \mathbf{x}_0)n_i(\mathbf{x})\,\mathrm{d}l(\mathbf{x}) = -\begin{bmatrix} 4\pi \\ 2\pi \\ 0 \end{bmatrix}\delta_{jk} \tag{2.6.9}$$

and

$$\int_{\mathscr{C}} \varepsilon_{ilm}x_l T_{mjk}(\mathbf{x}, \mathbf{x}_0)n_k(\mathbf{x})\,\mathrm{d}l(\mathbf{x}) = -\begin{bmatrix} 4\pi \\ 2\pi \\ 0 \end{bmatrix}\varepsilon_{ilj}x_{0,l} \tag{2.6.10}$$

where the unit normal vector $\mathbf{n}$ in both (2.6.9) and (2.6.10) is directed outside $A$. The values $4\pi, 2\pi$, and 0 apply when the point $\mathbf{x}_0$ is located inside, on, or outside $\mathscr{C}$ respectively. When $\mathbf{x}_0$ is on $\mathscr{C}$, the integrals on the right-hand sides of (2.6.9) and (2.6.10) must be interpreted in the principal value sense.

The pressure vector $\mathbf{p}$ and stress tensor $\mathbf{T}$ associated with a Green's function for infinite unbounded or bounded flow, constitute two fundamental solutions of Stokes flow (see section 3.2). Specifically, $\mathbf{p}(\mathbf{x}, \mathbf{x}_0)$ represents the velocity at the point $\mathbf{x}_0$ due to a point source of strength

$-4\pi$ with pole at $\mathbf{x}$, whereas

$$u_j(\mathbf{x}_0) = T_{ijk}(\mathbf{x}, \mathbf{x}_0) q_{ik} \tag{2.6.11}$$

where $\mathbf{q}$ is a constant matrix, represents the velocity field due to a two-dimensional stresslet with pole at $\mathbf{x}$. The pressure field corresponding to (2.6.11) may be expressed in terms of a new pressure tensor $\mathbf{\Pi}$ as

$$P(\mathbf{x}_0) = \mu \Pi_{ik}(\mathbf{x}_0, \mathbf{x}) q_{ik} \tag{2.6.12}$$

### *The two-dimensional Stokeslet*

To compute the two-dimensional free-space Green's function, we follow a procedure similar to that outlined in section 2.2. Thus, we replace the delta function on the right-hand side of (2.1.1) with the equivalent expression

$$\delta(\mathbf{x} - \mathbf{x}_0) = \frac{1}{2\pi} \nabla^2 \ln r \tag{2.6.13}$$

Recalling that the pressure is a harmonic function, and balancing the dimensions of the pressure gradient and the delta function in (2.1.1), we set

$$P = \frac{1}{2\pi} (\nabla \ln r) . \mathbf{g} \tag{2.614}$$

Furthermore, we introduce the function $H$ as in (2.2.4), and note that the continuity equation is satisfied for any choice of $H$. Substituting (2.6.13), (2.6.14), and (2.2.4) into (2.1.1), we derive a Poisson equation for $H$, namely

$$\nabla^2 H = \frac{1}{2\pi} \ln r \tag{2.6.15}$$

Combining (2.6.15) and (2.6.13) we find that $H$ is the Green's function of the biharmonic equation, i.e. $\nabla^4 H = \delta(\mathbf{x} - \mathbf{x}_0)$ (see (2.6.28)). Setting

$$H = \frac{1}{8\pi} r^2 (\ln r - 1) \tag{2.6.16}$$

and substituting into (2.2.4) we obtain the velocity field in the standard form (2.6.1) where $\mathbf{G}$ is the free-space Green's function or *two-dimensional Stokeslet*, namely,

$$G_{ij} = \mathscr{S}_{ij} = -\delta_{ij} \ln r + \frac{\hat{x}_i \hat{x}_j}{r^2} \tag{2.6.17}$$

and $r = |\hat{\mathbf{x}}|$, $\hat{\mathbf{x}} = \mathbf{x} - \mathbf{x}_0$. The associated vorticity, pressure, and stress fields are given by (2.6.3), (2.6.4), and (2.6.5) with

$$\Omega_{ij} = 2\varepsilon_{ijk} \frac{\hat{x}_k}{r^2} \tag{2.6.18}$$

$$p_i = 2\frac{\hat{x}_i}{r^2} \tag{2.6.19}$$

$$T_{ijk} = -4\frac{\hat{x}_i \hat{x}_j \hat{x}_k}{r^4} \tag{2.6.20}$$

The pressure tensor $\mathbf{\Pi}$ defined in (2.6.12) is given by

$$\Pi_{ik}(\mathbf{x}_0, \mathbf{x}) = 2\left(-\frac{\delta_{ik}}{r^2} + 2\frac{\hat{x}_i \hat{x}_k}{r^4}\right) \tag{2.6.21}$$

The stream functions corresponding to a Stokeslet that is oriented in the $x$ or $y$ direction are given by

$$\Psi = -y(\ln r - 1) \qquad \Psi = x(\ln r - 1) \tag{2.6.22}$$

respectively.

### *The boundary integral equation*

To derive the boundary integral equation for two-dimensional flow, we follow a procedure identical to that outlined in section 2.3. In this manner, we find that for a point $\mathbf{x}_0$ that is located outside a selected area of flow

$$\int_{\mathscr{C}} f_i(\mathbf{x}) G_{ij}(\mathbf{x}, \mathbf{x}_0)\, \mathrm{d}l(\mathbf{x}) - \mu \int_{\mathscr{C}} u_i(\mathbf{x}) T_{ijk}(\mathbf{x}, \mathbf{x}_0) n_k(\mathbf{x})\, \mathrm{d}l(\mathbf{x}) = 0 \tag{2.6.23}$$

where the curve $\mathscr{C}$ is the boundary of the selected area of flow. Furthermore, we find that for a point $\mathbf{x}_0$ that is located inside a selected area of flow

$$u_j(\mathbf{x}_0) = -\frac{1}{4\pi\mu}\int_{\mathscr{C}} f_i(\mathbf{x}) G_{ij}(\mathbf{x}, \mathbf{x}_0)\, \mathrm{d}l(\mathbf{x}) + \frac{1}{4\pi}\int_{\mathscr{C}} u_i(\mathbf{x}) T_{ijk}(\mathbf{x}, \mathbf{x}_0) n_k(\mathbf{x})\, \mathrm{d}l(\mathbf{x}) \tag{2.6.24}$$

The normal vector $\mathbf{n}$ in both (2.6.23) and (2.6.24) is oriented into the area of flow.

As the point $\mathbf{x}_0$ approaches the boundary $\mathscr{C}$ from either side, the double-layer potential on the right-hand side of (2.6.24) tends to two different values. Specifically, if $\mathscr{C}$ is a Lyapunov line, i.e. it has a continuously varying normal vector, and the velocity $\mathbf{u}$ varies over $\mathscr{C}$ in a continuous manner,

$$\lim_{\mathbf{x}_0 \to \mathscr{C}} \int_{\mathscr{C}} u_i(\mathbf{x}) T_{ijk}(\mathbf{x}, \mathbf{x}_0) n_k(\mathbf{x})\, \mathrm{d}l(\mathbf{x})$$

$$= \pm 2\pi u_j(\mathbf{x}_0) + \int_{\mathscr{C}}^{\mathscr{PV}} u_i(\mathbf{x}) T_{ijk}(\mathbf{x}, \mathbf{x}_0) n_k(\mathbf{x})\, \mathrm{d}l(\mathbf{x}) \tag{2.6.25}$$

where the plus sign applies for the internal side, in the direction of the normal vector, and the minus sign for the external side. The principal value of the double-layer integral is defined as the value of the improper double-layer integral corresponding to the case where $\mathbf{x}_0$ is located on $\mathscr{C}$. Clearly, the double-layer potential undergoes a discontinuity $4\pi\mathbf{u}$ across $\mathscr{C}$. Combining (2.6.25) with (2.6.24) we find that for a point $\mathbf{x}_0$ that lies on $\mathscr{C}$

$$u_j(\mathbf{x}_0) = -\frac{1}{2\pi\mu}\int_{\mathscr{C}} f_i(\mathbf{x})G_{ij}(\mathbf{x},\mathbf{x}_0)\,\mathrm{d}l(\mathbf{x}) + \frac{1}{2\pi}\int_{\mathscr{C}}^{\mathscr{PV}} u_i(\mathbf{x})T_{ijk}(\mathbf{x},\mathbf{x}_0)n_k(\mathbf{x})\,\mathrm{d}l(\mathbf{x}) \tag{2.6.26}$$

The boundary integral equation for a point $\mathbf{x}_0$ that is located at the corner of a boundary will be discussed in problem 2.6.4.

Now, inspecting (2.6.24) suggests an expression for the pressure in terms of two distributions corresponding to the single-layer and double-layer potential, namely

$$P(\mathbf{x}_0) = -\frac{1}{4\pi}\int_{\mathscr{C}} p_i(\mathbf{x}_0,\mathbf{x})f_i(\mathbf{x})\,\mathrm{d}l(\mathbf{x}) + \frac{\mu}{4\pi}\int_{\mathscr{C}} u_i(\mathbf{x})\Pi_{ik}(\mathbf{x}_0,\mathbf{x})n_k(\mathbf{x})\,\mathrm{d}l(\mathbf{x}) \tag{2.6.27}$$

where we recall that $\mathbf{\Pi}$ is the pressure field associated with $\mathbf{T}$ (see (2.6.12)).

All of the properties, interpretations, and simplifications of the boundary integral equation discussed previously in sections 2.3 and 2.4 for three-dimensional flow apply for two-dimensional flow as well. Two exceptions are the results for infinite flow past a body, and flow produced by a body that moves in an infinite fluid. In both cases, if the force acting on the body is not equal to zero, the disturbance velocity far from the body increases at a logarithmic rate and the boundary integrals at infinity may not be overlooked.

To study two-dimensional infinite flow, we must divide the domain of flow into two regions: one on the vicinity of the boundaries (inner flow), and the second far from the boundaries (outer flow). To obtain a consistent mathematical formulation, we must reconcile the inner flow with the outer flow using the method of matched asymptotic expansions (Van Dyke 1975). It will be noted that in the cases of incident streaming flow, or flow due to the translation of a body, the outer flow is governed by Oseen's equation, and may be formulated in terms of a boundary integral equation (Oseen 1927; Williams 1965).

### *Semi-direct and indirect boundary integral methods*

The boundary integral equations discussed previously in this section provide us with a representation of a flow in terms of primary variables, i.e. the velocity, the pressure, and the stress. In practice, it may be more convenient to use an alternative representation in terms of a secondary variable such as the stream function or the Airy stress function, or a tertiary variable such as the density of a hydrodynamic potential. These representations suggest a new class of boundary integral methods called semi-direct or indirect methods, corresponding to formulations in terms of secondary and tertiary variables, respectively.

First, we discuss a semi-direct formulation in terms of the stream function $\psi$. In section 1.2 we saw that by taking the Laplacian of the Stokes equation and noting that the pressure is a harmonic function, we find that $\psi$ satisfies the biharmonic equation $\nabla^4\psi = 0$. To derive a boundary integral equation for the stream function, we introduce the Green's function $G$ of the biharmonic equation, defined as a solution of the singularly forced biharmonic equation $\nabla^4 G = \delta(\mathbf{x} - \mathbf{x}_0)$, where $\delta$ is the two-dimensional delta function. If the flow is bounded by a solid boundary $\mathscr{C}$ we require that $G$, as well as its normal derivative, vanishes along $\mathscr{C}$, i.e. $G(\mathbf{x}, \mathbf{x}_0) = 0$ and $\nabla G(\mathbf{x}, \mathbf{x}_0)\cdot\mathbf{n} = 0$ when $\mathbf{x}$ is on $\mathscr{C}$. The free-space Green's function is given by

$$G(\mathbf{x}, \mathbf{x}_0) = \frac{1}{8\pi} r^2(\ln r - 1) \tag{2.6.28}$$

where $r = |\mathbf{x} - \mathbf{x}_0|$. The Green's function for semi-infinite flow bounded by a plane wall is given by

$$G(\mathbf{x}, \mathbf{x}_0) = \frac{1}{8\pi} r^2 \ln \frac{r}{R} + \frac{1}{4\pi}(y - w)(y_0 - w) \tag{2.6.29}$$

where the wall is located at $y = w$, $\mathbf{x}_0^{IM} = (x_0, 2w - y_0)$ is the image of the pole with respect to the wall, and $R = |\mathbf{x} - \mathbf{x}_0^{IM}|$ (Hansen 1987). The Green's function for flow in the interior of a circle of radius $a$ is given by

$$G(\mathbf{x}, \mathbf{x}_0) = \frac{1}{8\pi} r^2 \ln \frac{ra}{R|\mathbf{x}_0|} + \frac{1}{16\pi a^2}(|\mathbf{x}_0|^2 - a^2)(|\mathbf{x}|^2 - a^2) \tag{2.6.30}$$

where $R = |\mathbf{x} - \mathbf{x}_0^{IM}|$, and $\mathbf{x}_0^{IM}$ is the image of the pole with respect to the circle located at the same polar angle as $\mathbf{x}_0$ and at a radial position $|\mathbf{x}_0^{IM}| = a^2/|\mathbf{x}_0|$ (Garabedian 1964, p. 272).

Next, we select an area of flow $D$ that is enclosed by the contour $\mathscr{C}$, and use the second Green's identity to obtain

$$\int_D (\psi\nabla^4\chi - \chi\nabla^4\psi)\,\mathrm{d}A = \int_{\mathscr{C}} (\psi\nabla\varpi - \varpi\nabla\psi + \omega\nabla\chi - \chi\nabla\omega)\cdot\mathbf{n}\,\mathrm{d}l \tag{2.6.31}$$

where $\varpi = -\nabla^2\chi$, $\omega = -\nabla^2\psi$, and $\mathbf{n}$ is the unit normal vector pointing into $D$. Note that $\omega$ is the magnitude of the vorticity associated with $\psi$. Identifying $\chi$ with the Green's function $G$, and using the properties of the delta function to manipulate the left-hand side of (2.6.31), we obtain

$$\psi(\mathbf{x}_0) = \int_{\mathscr{C}} [\psi(\mathbf{x})\nabla\varpi(\mathbf{x}, \mathbf{x}_0) - \varpi(\mathbf{x}, \mathbf{x}_0)\nabla\psi(\mathbf{x}) \\ + \omega(\mathbf{x})\nabla G(\mathbf{x}, \mathbf{x}_0) - G(\mathbf{x}, \mathbf{x}_0)\nabla\omega(\mathbf{x})]\cdot\mathbf{n}(\mathbf{x})\,\mathrm{d}l(\mathbf{x}) \qquad (2.6.32)$$

where $\varpi = -\nabla^2 G$. Equation (2.6.32) provides us with a representation of the stream function in terms of the boundary values of the stream function $\psi$, the vorticity $\omega$, the tangential component of the boundary velocity $\nabla\psi\cdot\mathbf{n}$, and the normal derivative of the boundary vorticity $\nabla\omega\cdot\mathbf{n}$, which is proportional to the tangential derivative of the pressure, $\nabla\omega\cdot\mathbf{n} = -(1/\mu)\nabla P\cdot\mathbf{t}$ (problem 2.6.8). If we use a Green's function that satisfies the conditions $G = 0$ and $\nabla G\cdot\mathbf{n} = 0$ over $\mathscr{C}$, we will find that the last two terms of the integral in (2.6.32) disappear yielding a simplified representation that involves only the boundary values of the stream function and its normal derivatives, namely, the boundary velocity.

It will be instructive to consider the specific form of (2.6.32) for the particular case of the free-space Green's function. Substituting (2.6.28) into (2.6.32) we obtain

$$\psi(\mathbf{x}_0) = \frac{1}{2\pi}\int_{\mathscr{C}} \{-\psi(\mathbf{x})\nabla\ln r + \ln r\nabla\psi(\mathbf{x}) + \tfrac{1}{4}\omega(\mathbf{x})\nabla[r^2(\ln r - 1)] \\ - \tfrac{1}{4}[r^2(\ln r - 1)]\nabla\omega(\mathbf{x})\}\cdot\mathbf{n}(\mathbf{x})\,\mathrm{d}l(\mathbf{x}) \qquad (2.6.33)$$

The first three terms within the integral on the right-hand side of (2.6.33) represent distributions of two-dimensional potential dipoles, point vortices, and point forces.

Next, we recall that the vorticity is a harmonic function, and use the third Green's identity to express the vorticity in terms of the boundary values of the vorticity and its normal derivative obtaining

$$\omega(\mathbf{x}_0) = -\int_{\mathscr{C}} [\omega(\mathbf{x})\nabla G^{\mathrm{L}}(\mathbf{x}, \mathbf{x}_0) - G^{\mathrm{L}}(\mathbf{x}, \mathbf{x}_0)\nabla\omega(\mathbf{x})]\cdot\mathbf{n}(\mathbf{x})\,\mathrm{d}l(\mathbf{x}) \qquad (2.6.34)$$

where the unit normal vector $\mathbf{n}$ points into the interior of $\mathscr{C}$, $G^{\mathrm{L}}$ is the Green's function of the two-dimensional Laplace equation $\nabla^2 G^{\mathrm{L}} = \delta(\mathbf{x} - \mathbf{x}_0)$, and the point $\mathbf{x}_0$ is presumed to be located within the domain of flow (Kellogg 1954, p. 215). The free-space Green's function is given by $G^{\mathrm{L}} = (1/2\pi)\ln r$.

Equations (2.6.31) and (2.6.34) are valid at a point $\mathbf{x}_0$ that lies within the domain of flow. To apply these equations at a point $\mathbf{x}_0$ on the boundary $\mathscr{C}$, we must multiply the right-hand side of each equation by a factor of two and, in addition, when necessary, we must interpret the integrals in the principal value sense. Applying (2.6.31) and (2.6.34) over $\mathscr{C}$, we obtain a system of two scalar Fredholm integral equations for the boundary distributions of $\psi, \omega, \nabla\psi \cdot \mathbf{n}$, and $\nabla\omega \cdot \mathbf{n}$. Specifying two or a combination of two of these distributions brings the boundary integral formulation of the problem to completion (Kelmanson 1983a, b, c). For a theoretical analysis of the properties of the pertinent integral equations, the reader is referred to the monograph of Mikhlin (1957, Chapter V). It should be noted that Bézine and Bonneau (1981) presented an alternative boundary integral representation for the stream function in terms of boundary distributions of the velocity, shear stress, and normal derivative of the vorticity. The reader is directed to their paper for further details.

Now, maintaining selected terms in the integrand on the right-hand side of (2.6.33), we obtain indirect boundary integral methods. Two useful and popular methods arise by expressing the stream function in one of the forms

$$\psi(\mathbf{x}) = |\mathbf{x} - \mathbf{x}_1|^2 \phi_1(\mathbf{x}) + \phi_2(\mathbf{x})$$
$$\psi(\mathbf{x}) = (y - y_1)\phi_1(\mathbf{x}) + \phi_2(\mathbf{x}) \tag{2.6.35}$$

where $\phi_1$ and $\phi_2$ are two harmonic functions and $\mathbf{x}_1$ is an arbitary point. To effect the boundary integral representation, we express $\phi_1$ and $\phi_2$ in terms of a single-layer or double-layer harmonic potential with unknown density functions (see Jaswon & Symm 1977, Chapters 9 and 15, and Bhattacharya & Symm 1984). Applying (2.6.35) at the boundary and enforcing the required boundary conditions yields a system of integral equations of the first or second kind for the densities of the distributions.

Coleman (1980) has developed an alternative semi-direct boundary integral representation in complex variables based on the stream function and the Airy stress function (see section 1.2 and also Mikhlin 1957, Chapter V). The starting point in his formulation is equation (1.2.19). Applying the integral representation at the boundary and enforcing the required boundary conditions, he obtained two real Fredholm integal equations for the unknown boundary functions.

**Problems**

2.6.1 Derive the two-dimensional Stokeslet using the method of Fourier transforms.

2.6.2 Apply the boundary integral equations (2.6.23), (2.6.24), and (2.6.26) for flow expressing rigid body motion to derive the identities (2.6.9) and (2.6.10).

2.6.3 Using (2.6.9) prove (2.6.25). (Hint: see problem 4.3.1.)

2.6.4 Show that the boundary integral equation (2.6.26) is also valid at a point that is located at the corner of a boundary provided that the coefficient $1/(2\pi)$

in front of the single-layer and double-layer potentials is replaced by $1/(2\alpha)$, where $\alpha$ is the angle subtended by the corner.

2.6.5 Derive the fundamental solution to the biharmonic equation (2.6.28).

2.6.6 Using the second Green's identity derive the boundary integral equation (2.6.32).

2.6.7 Verify that the Green's functions (2.6.29) and (2.6.30) satisfy the boundary conditions $G = 0$ and $\nabla G \cdot \mathbf{n} = 0$ over the corresponding bounding surfaces.

2.6.8 Show that $\nabla\omega \cdot \mathbf{n} = -(1/\mu)\nabla P \cdot \mathbf{t}$, where $\mathbf{n}$ and $\mathbf{t}$ are the unit normal and unit tangent vectors with respect to a line in a flow.

2.6.9 Verify that (2.6.35) provides us with two solutions of the biharmonic equation.

## 2.7 Unsteady Stokes flow

Turning now our attention to unsteady Stokes flow, we consider the time-transformed non-dimensional unsteady Stokes equation (1.6.9), and repeat our previous analysis for steady flow. Recall that, for convenience, we dropped the primes indicating non-dimensional variables and incorporated the effect of the body force into the modified pressure. Under these simplifications, we find that the Green's functions of unsteady Stokes flow represent solutions of the continuity equation and the singularly forced unsteady Stokes equation

$$(\nabla^2 - \lambda^2)\mathbf{u} = \nabla P - \mathbf{g}\delta(\mathbf{x} - \mathbf{x}_0) \tag{2.7.1}$$

where $\mathbf{g}$ is a constant vector.

### *The unsteady Stokeslet*

To compute the free-space Green's function, we follow a procedure similar to that outlined in section 2.2. Thus, we substitute (2.2.1), (2.2.2), and (2.2.4) into (2.7.1) to obtain

$$(\nabla^2 - \lambda^2)H = -\frac{1}{4\pi r} \tag{2.7.2}$$

where $r = |\hat{\mathbf{x}}|$, $\hat{\mathbf{x}} = \mathbf{x} - \mathbf{x}_0$. Using (2.2.1) we find that $H$ is the fundamental solution of the modified biharmonic equation $\nabla^2(\nabla^2 - \lambda^2)H = \delta(\mathbf{x} - \mathbf{x}_0)$:

$$H = \frac{1}{4\pi\lambda R}(1 - e^{-R}) \tag{2.7.3}$$

where $R = \lambda r$. Substituting (2.7.3) into (2.2.4) we derive the velocity field due to an unsteady point force in the standard form

$$u_i(\mathbf{x}, \mathbf{x}_0) = \frac{1}{8\pi}\mathscr{S}_{ij}(\hat{\mathbf{x}})g_j \tag{2.7.4}$$

where

$$\mathscr{S}_{ij}(\hat{\mathbf{x}}) = A(R)\frac{\delta_{ij}}{r} + B(R)\frac{\hat{x}_i\hat{x}_j}{r^3} \tag{2.7.5}$$

is the *unsteady Stokeslet*. The functions $A(R)$ and $B(R)$ are defined as

$$A = 2\mathrm{e}^{-R}\left(1 + \frac{1}{R} + \frac{1}{R^2}\right) - \frac{2}{R^2} \qquad B = -2\mathrm{e}^{-R}\left(1 + \frac{3}{R} + \frac{3}{R^2}\right) + \frac{6}{R^2} \tag{2.7.6}$$

It will be noted that $A(0) = B(0) = 1$, suggesting that at small frequencies or close to the pole, the unsteady Stokeslet reduces to the regular Stokeslet for steady flow. To examine the asymptotic limit of small frequencies in more detail, we expand $\mathscr{S}$ in a Taylor series for small $\lambda$ obtaining

$$\mathscr{S} = \mathscr{S}^0 + \lambda\mathscr{S}^1 + \lambda^2\mathscr{S}^2 + \lambda^3\mathscr{S}^3 \cdots \tag{2.7.7}$$

where $\mathscr{S}^0$ is the steady Stokeslet, and

$$\mathscr{S}^1_{ij} = -\tfrac{4}{3}\delta_{ij} \qquad \mathscr{S}^2_{ij} = \frac{1}{4}\left(3r\delta_{ij} - \frac{\hat{x}_i\hat{x}_j}{r}\right) \qquad \mathscr{S}^3_{ij} = \tfrac{2}{15}(2r^2\delta_{ij} - \hat{x}_i\hat{x}_j) \tag{2.7.8}$$

It will be useful to note that $\mathscr{S}^1$ represents streaming flow.

To examine the behaviour of the Green's function at high values of $\lambda$ or far away from the pole, we expand $\mathscr{S}$ in an asymptotic series for large $R$ obtaining

$$\mathscr{S}_{ij} = \frac{2}{\lambda^2}\left(-\frac{\delta_{ij}}{r^3} + \frac{\hat{x}_i\hat{x}_j}{r^5}\right) + 2\mathrm{e}^{-R}\left(\frac{\delta_{ij}}{r} - \frac{\hat{x}_i\hat{x}_j}{r^3}\right) + \cdots \tag{2.7.9}$$

In order to ensure that $\mathscr{S}$ vanishes at infinity, we require that the real part of $\lambda$ is positive. We note that the expression in the first parenthesis on the right-hand side of (2.7.9) is the steady potential dipole; this suggests that at high frequencies or large distances, the unsteady Stokeslet produces irrotational flow (see section 7.2).

The vorticity, pressure, and stress fields associated with the unsteady point force are given by

$$\omega_i = \frac{1}{8\pi}\Omega_{ij}g_j \qquad p = \frac{1}{8\pi}p_j g_j \qquad \sigma_{ik} = \frac{1}{8\pi}T_{ijk}g_j \tag{2.7.10}$$

where

$$\Omega_{ij} = 2\varepsilon_{ijl}\frac{\hat{x}_l}{r^3}\mathrm{e}^{-R}(R+1) \tag{2.7.11}$$

$$p_i = 2\frac{\hat{x}_i}{r^3} \tag{2.7.12}$$

and

$$T_{ijk} = -\frac{2}{r^3}(\delta_{ij}\hat{x}_k + \delta_{kj}\hat{x}_i)[e^{-R}(R+1) - B] - \frac{2}{r^3}\delta_{ik}\hat{x}_j(1-B)$$

$$-2\frac{\hat{x}_i\hat{x}_j\hat{x}_k}{r^5}[5B - 2e^{-R}(R+1)] \qquad (2.7.13)$$

When the domain of flow is infinite, all $\mathbf{\Omega}$, $\mathbf{p}$, and $\mathbf{T}$ are required to decay to zero as the observation point moves far away from the pole.

The pressure vector $\mathbf{p}$ and stress tensor $\mathbf{T}$ are two acceptable unsteady Stokes flows representing a point source and a stresslet respectively (see section 2.2 and also (2.1.14) and (2.1.15)). The pressure matrix $\mathbf{\Pi}$ corresponding to the stresslet will be discussed in problem 2.7.6.

It will be instructive to consider the surface force on a spherical surface that is centered at the unsteady point force. After some algebra we find

$$f_i = \sigma_{ij}n_j = \frac{1}{8\pi}\left[\frac{\delta_{ij}}{r^2}K(R) + \frac{\hat{x}_i\hat{x}_j}{r^4}L(R)\right]g_j \qquad (2.7.14)$$

where the functions $K(R)$ and $L(R)$ are defined as

$$K = 2[B - e^{-R}(R+1)] \qquad L = 2[e^{-R}(R+1) - 1 - 3B] \quad (2.7.15)$$

and the function $B$ was given in (2.7.6). One may show that $K(0) = 0$ and $L(0) = -6$, consistent with out previous results for steady Stokes flow. Using (2.7.14) we find that the force exerted on the spherical surface is given by

$$\mathbf{F} = \tfrac{1}{6}(3K + L)\mathbf{g} = -\tfrac{1}{3}[2e^{-R}(R+1) + 1]\mathbf{g} \qquad (2.7.16)$$

We note that the force acting on a small sphere of infinitesimal radius is equal to $-\mathbf{g}$, whereas the force on a sphere of infinitely large radius is equal to $-\frac{1}{3}\mathbf{g}$. The difference between these two values is equal to the rate of change of momentum of the fluid that surrounds the point force.

### *The boundary integral equation*

To derive a boundary integral equation for unsteady Stokes flow, we follow the procedure outlined in section 2.3. In this manner, we derive (2.3.4), (2.3.11), and (2.3.13) for points in the exterior, interior, and on the boundary of the flow. The properties of these equations are similar to those of the corresponding equations for steady flow discussed in the preceding sections. It is worth noting, in particular, that the boundary integral equation for unsteady flow may be simplified by eliminating either the single-layer or the double-layer potential as discussed in section 2.3 (see problem 2.7.4). The pressure field is given by the boundary integral representation (2.3.17). The pressure matrix $\mathbf{\Pi}$ of the free-space Green's function will be discussed in problem 2.7.6.

It will prove useful to examine the asymptotic behaviour of the boundary integral equation at small values of the frequency parameter $\lambda$. Following Williams (1966b), we expand the Green's function $\mathbf{G}$ (which for simplicity we identify with the unsteady Stokeslet $\mathscr{S}$) and its associated stress tensor $\mathbf{T}$ in a Taylor series with respect to $\lambda$, as indicated in (2.7.7). Furthermore, we expand the boundary surface force $\mathbf{f}$ and the boundary velocity $\mathbf{u}$ in Taylor series with respect to $\lambda$, as $\mathbf{f} = \mathbf{f}^0 + \lambda\mathbf{f}^1 + \dots$, $\mathbf{u} = \mathbf{u}^0 + \lambda\mathbf{u}^1 + \lambda\mathbf{u}^2 + \dots$. Substituting these expansions into (2.3.13) and collecting terms of zero, first, and second order in $\lambda$ we obtain

$$u_j^0(\mathbf{x}_0) = -\frac{1}{4\pi}\int_D f_i^0(\mathbf{x})\mathscr{S}_{ij}^0(\mathbf{x}, \mathbf{x}_0)\,\mathrm{d}S(\mathbf{x}) + \frac{1}{4\pi}\int_D^{\mathscr{PV}} u_i^0(\mathbf{x})T_{ijk}^0(\mathbf{x}, \mathbf{x}_0)n_k(\mathbf{x})\,\mathrm{d}S(\mathbf{x}) \tag{2.7.17}$$

$$u_j^1(\mathbf{x}_0) - \frac{1}{3\pi}F_j^0 = -\frac{1}{4\pi}\int_D f_i^1(\mathbf{x})\mathscr{S}_{ij}^0(\mathbf{x}, \mathbf{x}_0)\,\mathrm{d}S(\mathbf{x}) + \frac{1}{4\pi}\int_D^{\mathscr{PV}} u_i^1(\mathbf{x})T_{ijk}^0(\mathbf{x}, \mathbf{x}_0)n_k(\mathbf{x})\,\mathrm{d}S(\mathbf{x}) \tag{2.7.18}$$

and

$$u_j^2(\mathbf{x}_0) - \frac{1}{3\pi}F_j^1 = -\frac{1}{4\pi}\int_D f_i^2(\mathbf{x})\mathscr{S}_{ij}^0(\mathbf{x}, \mathbf{x}_0)\,\mathrm{d}S(\mathbf{x}) - \frac{1}{4\pi}\int_D f_i^0(\mathbf{x})\mathscr{S}_{ij}^2(\mathbf{x}, \mathbf{x}_0)\,\mathrm{d}S(\mathbf{x}) + \frac{1}{4\pi}\int_D^{\mathscr{PV}} u_i^2(\mathbf{x})T_{ijk}^0(\mathbf{x}, \mathbf{x}_0)n_k(\mathbf{x})\,\mathrm{d}S(\mathbf{x}) + \frac{1}{4\pi}\int_D u_i^0(\mathbf{x})T_{ijk}^2(\mathbf{x}, \mathbf{x}_0)n_k(\mathbf{x})\,\mathrm{d}S(\mathbf{x}) \tag{2.7.19}$$

where

$$\mathbf{F}^\alpha = \int_D \mathbf{f}^\alpha\,\mathrm{d}S \qquad \alpha = 0, 1, 2\dots \tag{2.7.20}$$

is the force acting on the boundary $D$. Prescribing the boundary velocity renders (2.7.17), (2.7.18), and (2.7.19) a system of Fredholm integral equations of the first kind for the boundary surface forces $\mathbf{f}^0$, $\mathbf{f}^1$, and $\mathbf{f}^2$.

As a specific application of the above equations, we consider the flow due to the translational or rotational vibrations of a rigid particle in an infinite fluid. Neglecting the integrals at infinity and imposing the boundary conditions $\mathbf{u}^0 = \mathbf{V} + \mathbf{\Omega} \times \mathbf{x}$ and $\mathbf{u}^1 = \mathbf{u}^2 = \dots = 0$ on the surface of the particle, where $\mathbf{V}$ and $\mathbf{\Omega}$ are the translational and rotational velocities of oscillation, we obtain two integral equations for $\mathbf{f}^0$ and $\mathbf{f}^1$, namely

$$V_j + \varepsilon_{jkl}\Omega_k x_l = -\frac{1}{8\pi}\int_{S_P} f_i^0(\mathbf{x})\mathscr{S}_{ij}^0(\mathbf{x}, \mathbf{x}_0)\,\mathrm{d}S(\mathbf{x}) \tag{2.7.21}$$

and

$$-\frac{1}{6\pi}F_j^0 = -\frac{1}{8\pi}\int_{S_P} f_i^1(\mathbf{x})\mathcal{S}_{ij}^0(\mathbf{x}, \mathbf{x}_0)\,dS(\mathbf{x}) \qquad (2.7.22)$$

where $S_P$ is the surface of the particle. It will be convenient to express the solution of (2.7.21) in the form

$$\mathbf{f}^0 = -\mathcal{G}^T\cdot\mathbf{V} - \mathcal{G}^R\cdot\mathbf{\Omega} \qquad (2.7.23)$$

where $\mathcal{G}^T$ and $\mathcal{G}^R$ are respectively the steady translational and steady rotary surface force resistance matrices introduced in (1.4.6) and (1.4.11). Using (1.4.18) and (1.4.19) we find that the steady force and steady torque exerted on the particle are given by

$$\mathbf{F}^0 = -\mathbf{X}\cdot\mathbf{V} - \mathbf{P}\cdot\mathbf{\Omega}, \qquad \mathbf{L}^0 = -\mathbf{P}'\cdot\mathbf{V} - \mathbf{Y}\cdot\mathbf{\Omega} \qquad (2.7.24)$$

where $\mathbf{X}$, $\mathbf{P}$, $\mathbf{P}'$, and $\mathbf{Y}$ are the steady force and steady torque resistance matrices. Comparing the last four equations it becomes evident that the first corrections to the surface force, force, and torque are given by

$$\mathbf{f}^1 = \frac{1}{6\pi}\mathcal{G}^T\cdot\mathbf{F}^0 = -\frac{1}{6\pi}\mathcal{G}^T\cdot(\mathbf{X}\cdot\mathbf{V} + \mathbf{P}\cdot\mathbf{\Omega}) \qquad (2.7.25)$$

$$\mathbf{F}^1 = \frac{1}{6\pi}\mathbf{X}\cdot\mathbf{F}^0 = -\frac{1}{6\pi}\mathbf{X}\cdot(\mathbf{X}\cdot\mathbf{V} + \mathbf{P}\cdot\mathbf{\Omega}) \qquad (2.7.26)$$

and

$$\mathbf{L}^1 = \frac{1}{6\pi}\mathbf{P}'\cdot\mathbf{F}^0 = -\frac{1}{6\pi}\mathbf{P}'\cdot(\mathbf{X}\cdot\mathbf{V} + \mathbf{P}\cdot\mathbf{\Omega}) \qquad (2.7.27)$$

Remarkably, we find that the first-order corrections may be computed directly from the resistance matrices for steady flow.

### Problems

2.7.1 Verify that as $R$ tends to zero, the vorticity and stress tensors $\mathbf{\Omega}$ and $\mathbf{T}$ defined in (2.7.11) and (2.7.13) reduce to those for steady flow given in (2.2.9) and (2.2.11).

2.7.2 The Laplace transform of the velocity field due to the impulsive point force $\mathbf{g}\delta(\mathbf{x} - \mathbf{x}_0)\delta(t - t_0)$ where $\mathbf{g}$ is a constant, is given by

$$\hat{u}_i(\hat{\mathbf{x}}; s) = \frac{1}{8\pi}\mathcal{S}_{ij}(\hat{\mathbf{x}})g_j \qquad (1)$$

where $\hat{\mathbf{x}} = \mathbf{x} - \mathbf{x}_0$ and $\mathcal{S}$ is the unsteady Stokeslet with $\lambda^2 = s$. To compute the long-time behaviour of the flow we use the expansion (2.7.7) finding

$$\hat{u}_i(\hat{\mathbf{x}}; s) = \frac{1}{8\pi}[\mathcal{S}_{ij}^0(\hat{\mathbf{x}}) + s^{1/2}\mathcal{S}_{ij}^1(\hat{\mathbf{x}}) + s\mathcal{S}_{ij}^2(\hat{\mathbf{x}}) + s^{3/2}\mathcal{S}_{ij}^3(\hat{\mathbf{x}}) + \cdots]g_j \qquad (2)$$

Inverting (2) show that the long-time behaviour of the flow is described by

$$u_i(\hat{\mathbf{x}};t) = \frac{1}{(4\pi t)^{3/2}}\left[-\tfrac{1}{2}\mathcal{S}^1_{ij}(\hat{\mathbf{x}}) + \frac{3}{4t}\mathcal{S}^3_{ij}(\hat{\mathbf{x}}) + O\left(\frac{r^4}{t^2}\right)\right]g_j \tag{3}$$

Note that equation (3) finds application in the computation of the long-time decay of the angular velocity autocorrelation function of a rigid Brownian particle (Hocquart & Hinch 1983).

2.7.3 Show that the generalized Faxen relations discussed in section 2.5 apply also for unsteady Stokes flow (Pozrikidis 1989a).

2.7.4 Show that an unsteady Stokes flow with a prescribed constant velocity $\mathbf{u} = \mathbf{U}$ on the surface $S_P$ of a body may be represented in terms of a single-layer potential as

$$u_j(\mathbf{x}_0) = -\frac{1}{8\pi}\int_{S_P}[f_i(\mathbf{x}) - \lambda^2 \mathbf{V}\cdot\mathbf{x}n_i(\mathbf{x})]\,G_{ij}(\mathbf{x},\mathbf{x}_0)\,\mathrm{d}S(\mathbf{x}) \tag{1}$$

2.7.5 The Green's function for two-dimensional unsteady flow may be derived in a procedure analogous to that described in section 2.6. The results may be expressed in terms of the fundamental solution of the two-dimensional Helmholtz equation (Stakgold 1968, Vol. II, p. 265). Derive the two-dimensional unsteady Stokeslet and show that in the limit of small $\lambda$, it reduces to the steady Stokeslet.

2.7.6 Using the results of section 7.5 show that the equivalent of (2.2.13) for unsteady flow is

$$\Pi_{ik}(\mathbf{x}_0,\mathbf{x}) = 2\frac{\delta_{ik}}{r^3}(R^2 - 2) + 12\frac{\hat{x}_i\hat{x}_k}{r^5}$$

## 2.8 Swirling flow

In section 2.4 we developed a simplified boundary integral representation for flow in an axisymmetric domain. In this section we wish to develop an even more simplified representation for swirling flow produced by the axial rotation of an axisymmetric body. An example is the flow produced by the axial rotation of a spheroid whose major axis is aligned with the centre line of a cylindrical tube. Admittedly, a swirling flow may be treated within the general framework of the boundary integral equation for flow in an axisymmetric domain. It will be useful and instructive, however, to pursue an independent derivation based on the simplified equations of flow (see section 1.2).

First, we consider steady swirling flow. In section 1.2 we saw that the swirl $\Omega = \sigma u_\phi$ satisfies the equation $\mathrm{E}^2\Omega = 0$, where $u_\phi$ is the azimuthal component of the velocity and the operator $\mathrm{E}^2$ was defined in (1.2.25). The boundary conditions require that on the surface of the body $u_\phi = W\sigma$ or $\Omega = W\sigma^2$, where $W$ is the angular velocity of rotation.

The equation $E^2\Omega = 0$ would seem to be the natural starting point for developing a boundary integral representation. It will prove more convenient, however, to revert to Cartesian coordinates, and to note that the pressure is constant throughout the flow whereas the $y$ and $z$ components of the velocity $u_y$ and $u_z$ are harmonic functions, where $u_y = -u_\phi \sin\phi$, $u_z = u_\phi \cos\phi$. Thus, using the third Green's identity, we obtain the boundary integral representation for $u_z$, i.e.

$$u_z(\mathbf{x}_0) = -\int_D [u_z(\mathbf{x})\nabla G^L(\mathbf{x}, \mathbf{x}_0) - G^L(\mathbf{x}, \mathbf{x}_0)\nabla u_z(\mathbf{x})]\cdot\mathbf{n}\,dS(\mathbf{x}) \quad (2.8.1)$$

where the point $\mathbf{x}_0$ is presumed to be within the domain of flow, $G^L$ is the Green's function of Laplace's equation, i.e. $\nabla^2 G^L = \delta(\mathbf{x} - \mathbf{x}_0)$, $D$ is the boundary of the flow, and the normal vector $\mathbf{n}$ points into the flow. The integrals of the two terms on the right-hand side of (2.8.1) represent the single-layer and double-layer harmonic potentials respectively.

Now, to simplify (2.8.1), we wish to eliminate the double-layer potential. For this purpose, we introduce a complementary swirling flow $u_z^{Int}$ in the interior of D such that $u_z^{Int} = W\sigma\cos\phi$ on $D$. Following a standard procedure of potential flow, we transform (2.8.1) into the simplified form

$$u_z(\mathbf{x}_0) = \int_D G^L(\mathbf{x}, \mathbf{x}_0)\nabla(u_z - u_z^{Int})\cdot\mathbf{n}(\mathbf{x})\,dS(\mathbf{x}) \quad (2.8.2)$$

(Kellogg 1954, p. 220). Clearly, rigid body rotation, $u_z^{Int} = W\sigma\cos\phi$, is an acceptable choice for $u_z^{Int}$. Then

$$\begin{aligned}\nabla(u_z - u_z^{Int})\cdot\mathbf{n} &= \frac{\partial u_z}{\partial x}n_x + \frac{\partial}{\partial\sigma}(u_z - u_z^{Int})n_\sigma \\ &= \frac{\partial u_z}{\partial x}n_x + \left[\frac{\partial u_z}{\partial\sigma} - W\frac{\partial}{\partial\sigma}(\sigma\cos\phi)\right]n_\sigma \\ &= \frac{\partial u_z}{\partial x}n_x + \sigma\frac{\partial}{\partial\sigma}\left(\frac{u_z}{\sigma}\right)n_\sigma = \frac{1}{\mu}\cos\phi\, f_\phi \quad (2.8.3)\end{aligned}$$

where $f_\phi$ is the azimuthal component of the surface force. Substituting (2.8.3) into (2.8.2) and recalling that $u_z = u_\phi\cos\phi$ we obtain

$$u_\phi(\mathbf{x}_0) = \frac{1}{\mu\cos\phi_0}\int_D G^L(\mathbf{x}, \mathbf{x}_0)\cos\phi\, f_\phi(\mathbf{x})\,dS(\mathbf{x}) \quad (2.8.4)$$

Furthermore, realizing that $f_\phi$ is independent of the azimuthal angle $\phi$, we write

$$u_\phi(\mathbf{x}_0) = \frac{1}{\mu\cos\phi_0}\int_C\int_0^{2\pi} G^L(\mathbf{x}, \mathbf{x}_0)\cos\phi\,d\phi\, f_\phi\sigma\,dl \quad (2.8.5)$$

where $C$ is the trace of the boundary $D$ in an azimuthal plane. It will be noted that the inner integral with respect to $\phi$ on the right-hand side of (2.8.5) is simply the coefficient of the first term of the cosine Fourier series of the Green's function.

At this point, in order to make our analysis more specific, we restrict our attention to the free-space problem, setting

$$G^L(\mathbf{x}, \mathbf{x}_0) = -\frac{1}{4\pi r} \tag{2.8.6}$$

where $r = |\mathbf{x} - \mathbf{x}_0|$. Substituting (2.8.6) into (2.8.5) we obtain

$$u_\phi(\mathbf{x}_0) = -\frac{1}{4\pi\mu \cos\phi_0} \int_C \int_0^{2\pi} \frac{\cos\phi}{r} \mathrm{d}\phi \, f_\phi \sigma \, \mathrm{d}l(\mathbf{x}) \tag{2.8.7}$$

or equivalently

$$u_\phi(\mathbf{x}_0) = -\frac{1}{4\pi\mu} \int_C I_{11}(\hat{x}, \sigma, \sigma_0) f_\phi \sigma \, \mathrm{d}l(\mathbf{x}) \tag{2.8.8}$$

where $\hat{x} = x - x_0$ and $I_{11}$ was defined in (2.4.7). Equation (2.8.8) is in perfect agreement with and may be derived from the more general boundary integral equation (2.4.25) for flow in an axisymmetric domain (see problem 2.4.2).

Now, applying (2.8.8) on $C$ and enforcing the boundary condition $u_\phi = W\sigma$ we obtain a Fredholm integral equation of the first kind for the boundary surface force $f_\phi$. In the particular case of flow due to the rotation of a circular disk of infinitesimal thickness, this equation may be conveniently recast in terms of a Fredholm integral equation of the second kind which in turn may be solved using an iterative method (Collins 1962, Williams 1962). The solution is

$$f_\phi = -\mu W a \frac{4}{\pi} \frac{\sigma}{(a^2 - \sigma^2)^{1/2}} \tag{2.8.9}$$

where $a$ is the radius of the disk. It will be noted that once (2.8.8) is solved, the torque $\mathbf{L}$ exerted on the body may be computed using the equation

$$\mathbf{L} = 2\pi \int_C f_\phi \sigma^2 \, \mathrm{d}l \, \hat{\mathbf{i}} \tag{2.8.10}$$

Proceeding next to unsteady flow, we consider swirling flow produced by the axial rotary oscillation of an axisymmetric body. On the surface of the body the amplitude of the azimuthal velocity is $u_\phi = W\sigma$. Nondimensionalizing all variables as indicated in section 1.1, using as characteristic velocity $U = WL$, and noting that the pressure is uniform throughout the flow, we find that the $y$ and $z$ components of the velocity satisfy the

Helmholtz equation

$$(\nabla^2 - \lambda^2)u_y = (\nabla^2 - \lambda^2)u_z = 0 \tag{2.8.11}$$

where $\lambda^2 = -\mathrm{i}\omega L^2/\nu$ and $\omega$ is the angular frequency of the oscillations (see section 1.6). Introducing the Green's function of the Helmholtz equation $G^{\mathrm{H}}$ and applying the second Green's identity for the Helmholtz equation, we obtain the boundary integral representations (2.8.1) and (2.8.2). The free-space Green's function is given by

$$G^{\mathrm{H}}(\mathbf{x}, \mathbf{x}_0) = -\frac{1}{4\pi r}\exp(-\lambda r) \tag{2.8.12}$$

where $r = |\mathbf{x} - \mathbf{x}_0|$.

Considering (2.8.2) in particular, we recall that $u_z^{\mathrm{Int}}$ represents an internal flow that satisfies the boundary condition $u_z = \sigma\cos\phi$ on $D$. Unfortunately, it is not possible to find $u_z^{\mathrm{Int}}$ explicitly for a body of arbitrary shape. Anyway, setting, $\nabla(u_z - u_z^{\mathrm{Int}})\cdot\mathbf{n} = f\cos\phi$ and integrating (2.8.2) in the azimuthal direction we obtain

$$u_\phi(\mathbf{x}_0) = -\frac{1}{4\pi}\int_C K(\hat{x}, \sigma, \sigma_0) f \sigma \, \mathrm{d}l(\mathbf{x}) \tag{2.8.13}$$

where the kernel $K$ is given by

$$K(\hat{x}, \sigma, \sigma_0) = \int_0^{2\pi} \cos\phi \frac{\exp\{-\lambda[(x - x_0)^2 + \sigma^2 + \sigma_0^2 - 2\sigma\sigma_0\cos\phi]^{1/2}\}}{[(x - x_0)^2 + \sigma^2 + \sigma_0^2 - 2\sigma\sigma_0\cos\phi]^{1/2}} \mathrm{d}\phi \tag{2.8.14}$$

Applying (2.8.12) over $C$ and imposing the boundary condition $u_\phi = \sigma$ we obtain a Fredholm integral equation of the first kind for the unknown function $f$. Once this equation is solved, the velocity field, surface force, and torque exerted on $D$ may be computed using the integral representation (2.8.13).

In the particular case of swirling flow due to the axial rotary oscillations of a circular disk of infinitesimal thickness, the integral equation (2.8.13) may be conveniently converted into a Fredholm integral equation of the second kind, which in turn may be solved using an iterative method (Collins 1962; Williams 1962). At small frequencies the solution is given by the asymptotic series

$$f = -\frac{4}{\pi}\frac{\sigma}{(1 - \sigma^2)^{1/2}} - \lambda^2 \frac{2}{3\pi}\frac{(2 - \sigma^2)}{(1 - \sigma^2)^{1/2}} + \cdots \tag{2.8.15}$$

where the characteristic length $L$ is the radius of the disk.

**Problems**

2.8.1 Following the perturbation analysis of section 2.5 show that the torque exerted on an axisymmetric body that rotates steadily around its axis inside a large coaxial axisymmetric tube is given by the approximate expression

$$L = L^{\infty} \Big/ \left(1 + \frac{\alpha}{8\pi^2 \mu W} L^{\infty}\right)$$

where $W$ is the angular velocity, $L^{\infty}$ is the torque experienced by the body in the absence of the tube and $\alpha$ is a small coefficient whose value depends exclusively on the shape of the tube (Williams 1964; Shail & Townsend 1966; Kanwal 1971, p. 261). (Hint: consider the equivalent of (2.8.8) for flow bounded by a tube, where the kernel vanishes over the tube; decompose the kernel into that for the free-space problem and a regular complementary part; using regularity arguments, express the complementary part in the form $\sigma\sigma_0 H(\hat{x}, \sigma, \sigma_0)$ where $H$ is a regular function).

2.8.2 Show that the torque exerted on a circular disk of infinitesimal thickness that rotates steadily around its axis is $L = -(32/3)\mu W a^3$, where $W$ is the angular velocity and $a$ is the radius of the disk.

2.8.3 Perform an asymptotic analysis for low frequencies of the boundary integral equation for unsteady rotary flow due to the slow rotary axial oscillations of an axisymmetric body and thus derive an integral equation for the first correction to the density function $f$ (see Kanwal 1971, p. 264).

# 3

# *Green's functions*

In section 2.3 we saw that the domain of integration in the boundary integral equation may be conveniently reduced by using a Green's function that vanishes over selected boundaries of the flow. This reduction not only facilitates the theoretical analysis, but also allows accurate and efficient numerical solution of the integral equations arising from boundary integral representations. Motivated by these simplifications, we dedicate this chapter to the study of Green's functions in domains of flow that are bounded internally or externally by a solid surface $S_B$.

## 3.1 Properties of Green's functions

First, we recall that the velocity field associated with a Green's function is required to vanish over a selected boundary $S_B$. This implies that

$$\mathbf{G}(\mathbf{x}, \mathbf{x}_0) = 0 \qquad \text{when } \mathbf{x} \text{ lies on } S_B \tag{3.1.1}$$

When the pole $\mathbf{x}_0$ is moved onto $S_B$, as the field point $\mathbf{x}$ approaches the pole $\mathbf{x}_0$, $\mathbf{G}$ must exhibit singular behaviour, and yet, must vanish in order to satisfy the defining constraint (3.1.1). This is possible only when $\mathbf{G}$ is identically equal to zero for any choice of $\mathbf{x}$, implying that

$$\mathbf{G}(\mathbf{x}, \mathbf{x}_0) = \mathbf{p}(\mathbf{x}, \mathbf{x}_0) = \mathbf{T}(\mathbf{x}, \mathbf{x}_0) = 0 \qquad \text{when } \mathbf{x}_0 \text{ lies on } S_B \tag{3.1.2}$$

Thus, the flow due to the Green's function and the associated pressure and stress fields vanish when the pole of the Green's function is placed right on the boundary $S_B$.

The Green's functions satisfy the symmetry property

$$G_{ij}(\mathbf{x}, \mathbf{x}_0) = G_{ji}(\mathbf{x}_0, \mathbf{x}) \tag{3.1.3}$$

Physically, equation (3.1.3) provides us with a relation between the flow due to a point force with pole at $\mathbf{x}_0$ and the flow due to another point force with pole at $\mathbf{x}$. It will be interesting to note that symmetry properties similar to (3.1.3) exist for the Green's functions of potential flow and linear elastostatics (see Stakgold 1968, Vol. I, p. 131, and Jaswon & Symm 1977, p. 80, respectively).

To demonstrate the validity of (3.1.3), we consider the reciprocal identity (1.4.3) and select

$$u_i(\mathbf{x}) = \frac{1}{8\pi\mu} G_{ik}(\mathbf{x}, \mathbf{x}_1) a_k \qquad u'_i(\mathbf{x}) = \frac{1}{8\pi\mu} G_{ik}(\mathbf{x}, \mathbf{x}_2) b_k \tag{3.1.4}$$

where $\mathbf{a}$ and $\mathbf{b}$ are two arbitrary constant vectors, and $\mathbf{x}_1$ and $\mathbf{x}_2$ are two arbitrary points in the domain of flow. Correspondingly, we have

$$\frac{\partial \sigma_{ij}}{\partial x_j} = -a_i \delta(\mathbf{x} - \mathbf{x}_1) \qquad \frac{\partial \sigma'_{ij}}{\partial x_j} = -b_i \delta(\mathbf{x} - \mathbf{x}_2) \tag{3.1.5}$$

where $\delta$ is the three-dimensional delta function. Substituting (3.1.4) and (3.1.5) into the right-hand side of (1.4.3) we obtain

$$\frac{\partial}{\partial x_j}(u'_i \sigma_{ij} - u_i \sigma'_{ij}) = -\frac{1}{8\pi\mu}[G_{ik}(\mathbf{x}, \mathbf{x}_2) b_k a_i \delta(\mathbf{x} - \mathbf{x}_1) - G_{ik}(\mathbf{x}, \mathbf{x}_1) a_k b_i \delta(\mathbf{x} - \mathbf{x}_2)] \tag{3.1.6}$$

Next, we integrate (3.1.6) over a control volume that is confined by the solid boundary $S_B$, two spherical surfaces of infinitesimal radii enclosing $\mathbf{x}_1$ and $\mathbf{x}_2$, and in the case of infinite flow, a large surface that encloses $S_B$ as well as the points $\mathbf{x}_1$ and $\mathbf{x}_2$. Using the divergence theorem, we convert the volume integral on the left-hand side of the resulting equation into a surface integral over all surfaces enclosing the control volume. The surface integral over $S_B$ vanishes, because the Green's function and hence the velocities $\mathbf{u}$ and $\mathbf{u}'$ vanish over $S_B$. Letting the radius of the large surface tend to infinity and the radii of the small spheres enclosing $\mathbf{x}_1$ and $\mathbf{x}_2$ tend to zero, we find that the corresponding surface integrals make insignificant contributions. Hence, the whole of the integral of the left-hand side of (3.1.6) is equal to zero. Using the properties of the delta function to manipulate the right-hand side, we arrive at the equation

$$a_k b_i [G_{ki}(\mathbf{x}_1, \mathbf{x}_2) - G_{ik}(\mathbf{x}_2, \mathbf{x}_1)] = 0 \tag{3.1.7}$$

The validity of (3.1.3) becomes evident by noting that $\mathbf{a}$ and $\mathbf{b}$ are arbitrary constant vectors.

It will be instructive to present an alternative proof of (3.1.3) using concepts from the theory of generalized boundary integral representations, to be discussed in detail in Chapter 4. Thus, we consider the flow due to the translation of a rigid particle in the proximity of the surface $S_B$. Applying the boundary integral equation (2.3.27) and noting that the Green's function vanishes over $S_B$ we obtain

$$u_j(\mathbf{x}_0) = -\frac{1}{8\pi\mu} \int_{S_P} f_i^+(\mathbf{x}) G_{ij}(\mathbf{x}, \mathbf{x}_0) \, dS(\mathbf{x}) \tag{3.1.8}$$

where $S_P$ is the surface of the particle, and the superscript + emphasizes that the surface force is evaluated over the external surface of the particle. Alternatively, we may represent the flow in terms of a boundary distribution of Green's functions as

$$u_j(\mathbf{x}_0) = \frac{1}{8\pi\mu} \int_{S_P} G_{ji}(\mathbf{x}_0, \mathbf{x}) q_i(\mathbf{x}) \, dS(\mathbf{x}) \tag{3.1.9}$$

where $\mathbf{q}$ is the density of the distribution. Recalling that the velocity must satisfy the boundary condition $\mathbf{u} = \mathbf{U}$ on $S_P$, where $\mathbf{U}$ is the velocity of translation, we deduce that in the interior of the particle (3.1.9) must express rigid body translation. This implies that $\mathbf{u} = \mathbf{U}$ throughout the volume of the particle, and $\mathbf{f}^- = -P\mathbf{n}$ over the internal side of $S_P$, where $P$ is the constant internal pressure; the superscript minus indicates the internal side of $S_P$. In section 4.1 we shall see that the surface force corresponding to the flow (3.1.9) undergoes a discontinuity across the distribution domain $S_P$. Specifically, we shall see that $\mathbf{f}^+ - \mathbf{f}^- = -\mathbf{q}$. Recalling that $\mathbf{f}^- = -P\mathbf{n}$ we find $\mathbf{f}^+ = -P\mathbf{n} - \mathbf{q}$. Substituting this expression into (3.1.8) and using (2.1.4) we obtain

$$u_j(\mathbf{x}_0) = \frac{1}{8\pi\mu} \int_{S_P} q_i(\mathbf{x}) G_{ij}(\mathbf{x}, \mathbf{x}_0) \, dS(\mathbf{x}) \tag{3.1.10}$$

Subtracting (3.1.10) from (3.1.9) and noting that $S_P$ is an arbitrary surface yields (3.1.3).

Next, we wish to demonstrate that the force and torque exerted on $S_B$ due to the flow associated with the Green's function, i.e. the flow due to a point force, is related to the value of the velocity at the pole of the Green's function produced when $S_B$ moves as a rigid body. For this purpose, we use the reciprocal identity (1.4.3), and identify $\mathbf{u}$ with the velocity produced when $S_B$ translates with velocity $\mathbf{U}$ and $\mathbf{u}'$ with the velocity due to a point force with pole at $\mathbf{x}_0$. Thus, we set

$$u_i'(\mathbf{x}) = \frac{1}{8\pi\mu} G_{ik}(\mathbf{x}, \mathbf{x}_0) g_k \qquad \frac{\partial \sigma_{ij}'}{\partial x_j} = -g_i \delta(\mathbf{x} - \mathbf{x}_0) \tag{3.1.11}$$

where $\mathbf{g}$ is a constant vector. Furthermore, exploiting the linearity of the equations of Stokes flow, we express the corresponding force exerted on $S_B$ in the form

$$F_i' = X_{ij}(\mathbf{x}_0) g_j \tag{3.1.12}$$

where $\mathbf{X}$ is a resistance matrix. Now, we substitute (3.1.11) into (1.4.3) and integrate the resulting equation over a volume defined by the solid boundary $S_B$, a spherical surface of infinitesimal radius centered at $\mathbf{x}_0$, and in the case of infinite flow, a large surface enclosing $S_B$ as well as $\mathbf{x}_0$. Using the divergence theorem, we convert the volume integral on the left-hand side of the resulting equation into a surface integral over all surfaces enclosing the control volume. Because $\mathbf{G}$ vanishes over $S_B$, the surface integral over $S_B$ involving $\mathbf{u}'$ is equal to zero. Sending the radius of the large surface to infinity and the radius of the small spherical surface to zero, we find that the corresponding surface integrals make vanishing contributions. Finally, noting that $\mathbf{u} = \mathbf{U}$ over $S_B$, introducing the definition (3.1.12), and using the properties of the delta function to manipulate the right-hand side of the resulting simplified equation, we arrive at the expression

$$u_j(\mathbf{x}_0) = U_i X_{ij}(\mathbf{x}_0) \tag{3.1.13}$$

Equation (3.1.13) states that knowledge of the flow produced when $S_B$ translates as a rigid body, is sufficient for computing the resistance matrix $\mathbf{X}$, or equivalently, the force exerted on $S_B$ due to the flow produced by a point force. It will be noted that in the case of internal flow $\mathbf{X}$ is the identity matrix and (3.1.13) yields $\mathbf{u} = \mathbf{U}$.

Working as above, we may show that if the torque exerted on $S_B$ due to the flow (3.1.11) is given by

$$L_i' = Y_{ij}(\mathbf{x}_0) g_j \tag{3.1.14}$$

then the velocity $\mathbf{u}$ produced when $S_B$ rotates as a rigid body with angular velocity $\mathbf{\Omega}$ will satisfy the equation

$$u_j(\mathbf{x}_0) = \Omega_i Y_{ij}(\mathbf{x}_0) \tag{3.1.15}$$

In addition, we may show that knowledge of the disturbance velocity produced when $S_B$ is immersed in a purely straining flow is sufficient for computing the coefficient of the stresslet generated by $S_B$ due to the flow produced by a point force (problem 3.1.5).

## Problems

3.1.1 Prove the following identities

$$\frac{\partial G_{ij}}{\partial x_j}(\mathbf{x}_0, \mathbf{x}) = 0 \qquad \int_D G_{ij}(\mathbf{x}_0, \mathbf{x}) n_j(\mathbf{x}) \, dS(\mathbf{x}) = 0$$

where $D$ is a closed surface.

3.1.2 Multiplying (2.3.30) by the surface force exerted on a translating or rotating particle and using the symmetry property (3.1.3) derive (1.4.10) and (1.4.12).

3.1.3 Comment on the implications of (3.1.15) for internal flow.

3.1.4 Using (7.3.5) and (7.3.28) for the flow due to the translation or rotation of a sphere, compute the matrices $\mathbf{X}$ and $\mathbf{Y}$ for an infinite flow bounded internally by a solid sphere. Verify that the force and torque exerted on a sphere due to a point force are given by (3.3.26) and (3.3.27).

3.1.5 Establish a relation between the coefficient of the stresslet generated by a solid boundary $S_B$ due to a nearby point force and the disturbance velocity produced when $S_B$ is immersed in a purely straining ambient flow.

## 3.2 Properties of the pressure and stress

In this section we proceed to investigate the properties of the pressure vector **p** and the stress tensor **T** associated with the Green's functions. First, we shall show that, for Green's functions corresponding to an infinite (unbounded or bounded) domain of flow,

$$\mathbf{u}(\mathbf{x}_0) = \beta \mathbf{p}(\mathbf{x}, \mathbf{x}_0) \tag{3.2.1}$$

and

$$u_j(\mathbf{x}_0) = T_{ijk}(\mathbf{x}, \mathbf{x}_0) q_{ik} \tag{3.2.2}$$

where $\beta$ and **q** are arbitrary constants, are acceptable solutions to the equations of Stokes flow. For this purpose, we write the boundary integral equation for a flow **u** in a domain that is confined by an arbitrary surface $D$, the solid boundary $S_B$, and a large surface extending to infinity. Exploiting (3.1.1), noting that **u** vanishes over $S_B$, and stipulating that the flow vanishes at infinity, we obtain the simplified expression

$$u_j(\mathbf{x}_0) = -\frac{1}{8\pi\mu}\int_D G_{ji}(\mathbf{x}_0, \mathbf{x}) f_i(\mathbf{x})\, dS(\mathbf{x}) + \frac{1}{8\pi}\int_D u_i(\mathbf{x}) T_{ijk}(\mathbf{x}, \mathbf{x}_0) n_k(\mathbf{x})\, dS(\mathbf{x}) \tag{3.2.3}$$

The first integral on the right-hand side of (3.2.3) represents the flow due to a surface distribution of point forces of density $-\mathbf{f}$. The second integral represents the flow due to a double-layer potential of density **u**. We note that $D$ is an arbitrary surface, and that **u** may be prescribed over $D$ in an arbitrary manner without any constraints; this suggests that the flow (3.2.2) is an acceptable solution to the equations of Stokes flow.

It will be important to note that if the domain of the flow were not infinite, i.e. **G** vanished over all external boundaries of the flow, **u** would have to satisfy the constraint of vanishing flow rate across $D$. This constraint will disqualify (3.2.2) from representing a valid solution to the equations of Stokes flow.

Next, invoking the definition (2.1.8), and using the symmetry property (3.1.3), we write

$$T_{ijk}(\mathbf{x}, \mathbf{x}_0) = -p_j(\mathbf{x}, \mathbf{x}_0)\delta_{ik} + \frac{\partial G_{ji}(\mathbf{x}_0, \mathbf{x})}{\partial x_k} + \frac{\partial G_{jk}(\mathbf{x}_0, \mathbf{x})}{\partial x_i} \tag{3.2.4}$$

The differential terms on the right-hand side of (3.2.4) represent point force dipoles and therefore they are legitimate solutions to the equations of Stokes flow. As a result, the pressure term, and therefore (3.2.1), must also be a legitimate solution to the equations of Stokes flow.

Now, we recall that $\mathbf{p}$ is composed of a regular part and a singular part; the latter is the pressure associated with the Stokeslet. Noting that the singular part represents the flow due to a point source, and taking into consideration (3.1.2), we deduce that the velocity field (3.2.1) expresses the flow due to a point source of strength $-8\pi\beta$ with pole at $\mathbf{x}$, in the presence of the solid boundary $S_B$. Conservation of mass requires

$$\int_S p_j(\mathbf{x}, \mathbf{x}_0)\, n_j(\mathbf{x}_0)\, \mathrm{d}S(\mathbf{x}_0) = \begin{cases} -8\pi & \text{if } \mathbf{x} \text{ is inside } S \\ -4\pi & \text{if } \mathbf{x} \text{ is on } S \\ 0 & \text{if } \mathbf{x} \text{ is outside } S \end{cases} \tag{3.2.5}$$

where $S$ is an arbitrary closed surface, and the normal vector is directed outward from $S$. Equation (3.2.5) states that when $S$ does not enclose the pole $\mathbf{x}$, the volume enclosed by $S$ is free of singularities, and the flow rate through $S$ must vanish. On the other hand, when $S$ encloses $\mathbf{x}$, the flow rate is equal to that due to a point source with strength $-8\pi$. When $\mathbf{x}$ is right on $S$, the integral on the left-hand side of (3.2.5) is a principal value integral that may be shown to be equal to the average of the values on either side of $S$, namely $-4\pi$.

Considering next the flow (3.2.2) and using (3.2.4) we find

$$\int_S T_{ijk}(\mathbf{x}, \mathbf{x}_0) n_j(\mathbf{x}_0)\, \mathrm{d}S(\mathbf{x}_0)$$
$$= -\delta_{ik} \int_S p_j(\mathbf{x}, \mathbf{x}_0) n_j(\mathbf{x}_0)\, \mathrm{d}S(\mathbf{x}_0) + \int_S \left[ \frac{\partial G_{ji}(\mathbf{x}_0, \mathbf{x})}{\partial x_k} + \frac{\partial G_{jk}(\mathbf{x}_0, \mathbf{x})}{\partial x_i} \right] n_j(\mathbf{x}_0)\, \mathrm{d}S(\mathbf{x}_0) \tag{3.2.6}$$

The continuity equation requires that the last integral on the right-hand side of (3.2.6) is equal to zero. Using (3.2.5) we find

$$\int_S T_{ijk}(\mathbf{x}, \mathbf{x}_0) n_j(\mathbf{x}_0)\, \mathrm{d}S(\mathbf{x}_0) = \delta_{ik} \begin{cases} 8\pi & \text{if } \mathbf{x} \text{ is inside } S \\ 4\pi & \text{if } \mathbf{x} \text{ is on } S \\ 0 & \text{if } \mathbf{x} \text{ is outside } S \end{cases} \tag{3.2.7}$$

When $\mathbf{x}$ lies on $S_B$, the integral on the right-hand side of (3.2.7) is to be interpreted in the principal value sense. In a sense, the identity (3.2.7) is complementary to (2.1.12).

Finally, we turn our attention to the pressure fields associated with the flows (3.2.1) and (3.2.2). Consulting with (3.2.4) we find that if the pressure field associated with the flow (3.2.1) is given by $\beta\mu\mathscr{P}(\mathbf{x}_0, \mathbf{x})$, the pressure field associated with the flow (3.2.2) will be given by

$$P(\mathbf{x}_0, \mathbf{x}) = \mu \Pi_{ik}(\mathbf{x}_0, \mathbf{x}) q_{ik} \tag{3.2.8}$$

where

$$\Pi_{ik}(\mathbf{x}_0, \mathbf{x}) = -\delta_{ik} \mathscr{P}(\mathbf{x}_0, \mathbf{x}) + \frac{\partial p_i(\mathbf{x}_0, \mathbf{x})}{\partial x_k} + \frac{\partial p_k(\mathbf{x}_0, \mathbf{x})}{\partial x_i} \tag{3.2.9}$$

and $\mathbf{p}$ is the pressure vector associated with the Green's function.

**Problem**

3.2.1 Using the Stokes equation show that

$$u_i(\mathbf{x}_0) = \frac{\partial p_i}{\partial x_j}(\mathbf{x}, \mathbf{x}_0)\, d_j$$

where $\mathbf{d}$ is a constant vector, represents the flow due to a point source dipole in a domain bounded by $S_B$.

## 3.3 Computation of Green's functions

Having established several properties of Green's functions and their associated pressure and stress, we proceed to address the important issue of their computation. The general theory of boundary value problems in partial differential equations provides us with several methods for computing Green's functions including the integral equation method and the methods of images, of eigenfunction expansions and of integral transforms (Stakgold 1968, Vol. II, p. 146). The effectiveness of each method depends upon the particular geometry of the boundary $S_B$.

As a general rule, a Green's function may be computed most effectively after it has been decomposed into the free-space Green's function (the Stokeslet), a collection of image singularities with poles outside the domain of flow, and a regular complementary component that is required to satisfy proper boundary conditions on $S_B$. The details of this decomposition will be exemplified in sections 3.4 and 3.5.

Before proceeding to discuss Green's functions for specific domains of flow, we wish to demonstrate that the Green's function and its associated stress tensor may be computed by solving a Fredholm integral equation of the first kind over the solid boundary $S_B$. Thus, we consider the reciprocal identity (1.4.3), and select

$$u_i(\mathbf{x}) = \frac{1}{8\pi\mu} \mathscr{S}_{ik}(\mathbf{x}, \mathbf{x}_1) g_k \quad u_i'(\mathbf{x}) = \frac{1}{8\pi\mu} G_{ik}(\mathbf{x}, \mathbf{x}_2) g_k \tag{3.3.1}$$

where $\mathscr{S}$ is the Stokeslet, $\mathbf{g}$ is an arbitrary constant vector, and $\mathbf{x}_1$ and

$\mathbf{x}_2$ are two arbitrary points in the domain of flow. The corresponding stress tensors satisfy the equations

$$\frac{\partial \sigma_{ij}}{\partial x_j} = -g_i \delta(\mathbf{x} - \mathbf{x}_1) \qquad \frac{\partial \sigma'_{ij}}{\partial x_j} = -g_i \delta(\mathbf{x} - \mathbf{x}_2) \tag{3.3.2}$$

where $\delta$ is the three-dimensional delta function. Substituting (3.3.1) and (3.3.2) into (1.4.3) we obtain

$$\frac{1}{8\pi} g_k g_l \frac{\partial}{\partial x_j} [G_{ik}(\mathbf{x}, \mathbf{x}_2) T^{\mathrm{ST}}_{ilj}(\mathbf{x}, \mathbf{x}_1) - \mathscr{S}_{ik}(\mathbf{x}, \mathbf{x}_1) T_{ilj}(\mathbf{x}, \mathbf{x}_2)]$$
$$= -g_k g_i [G_{ik}(\mathbf{x}, \mathbf{x}_2) \delta(\mathbf{x} - \mathbf{x}_1) - \mathscr{S}_{ik}(\mathbf{x}, \mathbf{x}_1) \delta(\mathbf{x} - \mathbf{x}_2)] \tag{3.3.3}$$

where $\mathbf{T}$ is the stress tensor associated with the Green's function, and $\mathbf{T}^{\mathrm{ST}}$ the stress tensor associated with the Stokeslet, given in (2.2.11). Next, we integrate (3.3.3) over a control volume that is confined by the solid boundary $S_{\mathrm{B}}$, a large surface enclosing $S_{\mathrm{B}}$ as well as the poles $\mathbf{x}_1$ and $\mathbf{x}_2$, and two spherical surfaces of infinitesimal radii enclosing $\mathbf{x}_1$ and $\mathbf{x}_2$. Using the divergence theorem, we convert the volume integral on the left-hand side into a surface integral over all surfaces enclosing the control volume, and note that the surface integral over $S_{\mathrm{B}}$ involving $\mathbf{G}$ is equal to zero. Letting the radius of the large surface tend to infinity and the radii of the small spherical surfaces tend to zero, we find that the corresponding surface integrals make insignificant contributions. Changing the dummy index $i$ to $l$ on the right-hand side of the resulting equation, discarding the arbitrary constant vector $\mathbf{g}$, using the properties of the delta function to manipulate the right-hand side, and interchanging the arguments of the Stokeslet, we finally arrive at the equation

$$G_{lk}(\mathbf{x}_1, \mathbf{x}_2) = \mathscr{S}_{lk}(\mathbf{x}_1, \mathbf{x}_2) + \frac{1}{8\pi} \int_{S_{\mathrm{B}}} \mathscr{S}_{ik}(\mathbf{x}, \mathbf{x}_1) T_{ilj}(\mathbf{x}, \mathbf{x}_2) n_j(\mathbf{x}) \, \mathrm{d}S(\mathbf{x}) \tag{3.3.4}$$

It will be noted that (3.3.4) provides us with a relation between the Green's function $\mathbf{G}$ and the associated surface force $T_{ilj} n_j$. Moving $\mathbf{x}_1$ onto $S_{\mathrm{B}}$ and requiring that $\mathbf{G}$ vanishes over $S_{\mathrm{B}}$, we obtain a Fredholm integral equation of the first kind for $T_{ilj} n_j$;

$$\int_{S_{\mathrm{B}}} \mathscr{S}_{ik}(\mathbf{x}, \mathbf{x}_1) T_{ilj}(\mathbf{x}, \mathbf{x}_2) n_j(\mathbf{x}) \mathrm{d}S(\mathbf{x}) = -8\pi \mathscr{S}_{lk}(\mathbf{x}_1, \mathbf{x}_2) \tag{3.3.5}$$

Once this equation is solved, $\mathbf{G}$ may be computed by simple integration using (3.3.4).

At this point, we proceed to discuss Green's functions for specific domains of flow. At the outset, we wish to present a list of all known Green's functions placed in chronological order of derivation. These are Green's functions for flow bounded by a solid plane wall (discussed below), flow bounded internally by a solid sphere (also discussed below), flow bounded by two parallel plane walls (Liron & Mochon 1976), flow bounded by two intersecting planes (Sano & Hasimoto 1977, 1978), flow bounded externally by a circular pipe (Liron & Shahar 1978; Blake 1979), flow in the interior of a circular cone (Kim 1979), flow bounded by a semi-infinite plane wall (Hasimoto, Kim & Miyazaki 1983), and flow bounded by a plane wall with a circular hole (Miyazaki & Hasimoto 1984). Hasimoto (1976) and Hasimoto & Sano (1980) provide guidelines for the computation of Green's functions in domains that are bounded by cylindrical walls.

In the remaining part of this section we discuss two particular Green's functions, the first for semi-infinite flow bounded by a plane wall, and the second for infinite flow bounded internally by a solid sphere.

### *Flow bounded by an infinite plane wall*

The simplest and most frequently used Green's function is that for flow bounded by a plane wall (Lorentz 1907; see Happel & Brenner 1973). Assuming that the wall is located at $x = w$ we require

$$\mathbf{G}^{\mathrm{W}}(x = w, y, z; \mathbf{x}_0) = 0 \tag{3.3.6}$$

where the superscript W indicates the plane wall. Blake (1971) showed that $\mathbf{G}$ may be constructed from a Stokeslet and a few image singularities including a Stokeslet, a potential dipole, and a Stokeslet doublet (see section 7.2). Specifically, he suggested the decomposition

$$\mathbf{G}^{\mathrm{W}}(\mathbf{x}, \mathbf{x}_0) = \mathscr{S}(\hat{\mathbf{x}}) - \mathscr{S}(\hat{\mathbf{X}}) + 2h_0^2\mathbf{G}^{\mathrm{D}}(\hat{\mathbf{X}}) - 2h_0\mathbf{G}^{\mathrm{SD}}(\hat{\mathbf{X}}) \tag{3.3.7}$$

where $\mathscr{S}$ is the Stokeslet, $h_0 = x_0 - w$, $\hat{\mathbf{x}} = \mathbf{x} - \mathbf{x}_0, \hat{\mathbf{X}} = \mathbf{x} - \mathbf{x}_0^{\mathrm{IM}}$, and $\mathbf{x}_0^{\mathrm{IM}} = (2w - x_0, y_0, z_0)$ is the image of $\mathbf{x}_0$ with respect to the wall. The matrices $\mathbf{G}^{\mathrm{D}}$ and $\mathbf{G}^{\mathrm{SD}}$ contain respectively potential dipoles and Stokeslet doublets and are given by

$$G_{ij}^{\mathrm{D}}(\mathbf{x}) = \pm\frac{\partial}{\partial x_j}\left(\frac{x_i}{|\mathbf{x}|^3}\right) = \pm\left(\frac{\delta_{ij}}{|\mathbf{x}|^3} - 3\frac{x_i x_j}{|\mathbf{x}|^5}\right) \tag{3.3.8}$$

$$G_{ij}^{\mathrm{SD}}(\mathbf{x}) = \pm\frac{\partial S_{i1}}{\partial x_j} = x_1 G_{ij}^{\mathrm{D}}(\mathbf{x}) \pm \frac{\delta_{j1}x_i - \delta_{i1}x_j}{|\mathbf{x}|^3} \tag{3.3.9}$$

with a minus sign for $j = 1$, corresponding to the $x$ direction, and a plus sign for $j = 2, 3$, corresponding to the $y$ and $z$ directions. By analogy with (3.3.7), we express the pressure associated with the Green's function in the form

$$p_i^{\mathrm{W}}(\mathbf{x}, \mathbf{x}_0) = p_i^{\mathrm{ST}}(\hat{\mathbf{x}}) - p_i^{\mathrm{ST}}(\hat{\mathbf{X}}) - 2h_0 p_i^{\mathrm{SD}}(\hat{\mathbf{X}}) \tag{3.3.10}$$

where

$$p_i^{\mathrm{ST}}(\mathbf{x}) = 2\frac{x_i}{|\mathbf{x}|^3}, \tag{3.3.11}$$

and

$$p_i^{\mathrm{SD}}(\mathbf{x}) = \pm 2\frac{\partial}{\partial x_i}\left(\frac{x_1}{|\mathbf{x}|^3}\right) = \pm 2\left(\frac{\delta_{i1}}{|\mathbf{x}|^3} - 3\frac{x_1 x_i}{|\mathbf{x}|^5}\right) \tag{3.3.12}$$

It will be noted that because the potential dipoles are irrotational singularities they make a vanishing contribution to the pressure (see section 7.2).

In section 3.2 we argued that $\mathbf{p}$ represents the flow due to a point source of strength $-8\pi$ located at $\mathbf{x}$. To demonstrate this explicitly, we substitute (3.3.11) and (3.3.12) into (3.3.10) to obtain

$$p_i^{\mathrm{W}}(\mathbf{x}, \mathbf{x}_0) = -2\left[\frac{X_i}{|\mathbf{X}|^3} + \frac{Y_i}{|\mathbf{Y}|^3} + 2Y_i\left(-\frac{1}{|\mathbf{Y}|^3} + 3\frac{Y_1^2}{|\mathbf{Y}|^5}\right) - 2h\left(-\frac{\delta_{i1}}{|\mathbf{Y}|^3} + 3\frac{Y_1 Y_i}{|\mathbf{Y}|^5}\right)\right] \tag{3.3.13}$$

where $h = x - w$, $\mathbf{X} = \mathbf{x}_0 - \mathbf{x}$, $\mathbf{Y} = \mathbf{x}_0 - \mathbf{x}^{\mathrm{IM}}$, and $\mathbf{x}^{\mathrm{IM}} = (2w - x, y, z)$ is the image of $\mathbf{x}$ with respect to the wall. Inspecting (3.3.13), we recognize the first two terms on the right-hand side as two point sources of strength $-8\pi$ with poles at $\mathbf{x}$ and $\mathbf{x}^{\mathrm{IM}}$, respectively. The third term is a Stokeslet doublet whereas the fourth term is a potential dipole; each has a pole at $\mathbf{x}^{\mathrm{IM}}$. Blake & Chwang (1974) derived (3.3.13) in an alternative fashion using the method of Fourier transforms.

The pressure field corresponding to the velocity field (3.3.13) is due exclusively to the Stokeslet doublet, and is given by

$$\mathscr{P}(\mathbf{x}_0, \mathbf{x}) = 8\frac{\partial}{\partial Y_1}\left(\frac{Y_1}{|\mathbf{Y}|^3}\right) = \frac{8}{|\mathbf{Y}|^3} - 24\frac{Y_1^2}{|\mathbf{Y}|^5} \tag{3.3.14}$$

The stress tensor associated with the Green's function is given by

$$\mathbf{T}^{\mathrm{W}}(\mathbf{x}, \mathbf{x}_0) = \mathbf{T}^{\mathrm{ST}}(\hat{\mathbf{x}}) - \mathbf{T}^{\mathrm{ST}}(\hat{\mathbf{X}}) + 2h_0^2\mathbf{T}^{\mathrm{D}}(\hat{\mathbf{X}}) - 2h_0\mathbf{T}^{\mathrm{SD}}(\hat{\mathbf{X}}) \tag{3.3.15}$$

where

$$T_{ijk}^{\mathrm{ST}}(\mathbf{x}) = -6\frac{x_i x_j x_k}{|\mathbf{x}|^5} \tag{3.3.16}$$

$$T_{ijk}^{\mathrm{D}}(\mathbf{x}) = \pm 6\left(-\frac{\delta_{ik}x_j + \delta_{ij}x_k + \delta_{kj}x_i}{|\mathbf{x}|^5} + 5\frac{x_i x_j x_k}{|\mathbf{x}|^7}\right) \tag{3.3.17}$$

$$T_{ijk}^{\mathrm{SD}}(\mathbf{x}) = x_1 T_{ijk}^{\mathrm{D}}(\mathbf{x}) \pm 6\left(\frac{\delta_{ik}x_j x_1 - \delta_{j1}x_i x_k}{|\mathbf{x}|^5}\right) \tag{3.3.18}$$

One may confirm by direct evaluation that indeed all $\mathbf{G}$, $\mathbf{p}$, and $\mathbf{T}$ vanish when the pole of the Green's function is placed right on the wall, i.e. when $x_0 = w$.

At this point, we turn to assess the asymptotic behaviour of the Green's function in the limit as the distance $h_0$ between the pole of the Green's function and the wall becomes much smaller than the distance $|\mathbf{x} - \mathbf{x}_0|$ between the observation point and the pole. Expanding $\mathbf{x} - \mathbf{x}_0$ in a Taylor series about $\mathbf{x} - \mathbf{x}_0^W$, where $\mathbf{x}_0^W = (w, y_0, z_0)$ is a point on the wall, substituting the series into (3.3.7), expanding the resulting expression in a Taylor series with respect to $h_0$, and maintaining linear and quadratic terms, we obtain

$$\mathbf{G}^W(\mathbf{x}, \mathbf{x}_0) = \left[\mathcal{S} - h_0\frac{\partial \mathcal{S}}{\partial x_1} + \frac{h_0^2}{2}\frac{\partial \mathcal{S}}{\partial x_1}\right](\mathbf{Z}) - \left[\mathcal{S} + h_0\frac{\partial \mathcal{S}}{\partial x_1} + \frac{h_0^2}{2}\frac{\partial \mathcal{S}}{\partial x_1}\right](\mathbf{Z})$$
$$+ 2h_0^2\mathbf{G}^D(\mathbf{Z}) - 2h_0\left[\mathbf{G}^{SD} + h_0\frac{\partial \mathbf{G}^{SD}}{\partial x_1}\right](\mathbf{Z}) + O(h_0^3) \quad (3.3.19)$$

where $\mathbf{Z} = \mathbf{x} - \mathbf{x}_0^W$ (Blake & Chwang 1974). After some simplifications we obtain

$$\mathbf{G}^W(\mathbf{x}, \mathbf{x}_0) = -2h_0\left[\frac{\partial \mathcal{S}}{\partial x_1} + \mathbf{G}^{SD}\right](\mathbf{Z}) + 2h_0^2\left[\mathbf{G}^D - \frac{\partial \mathbf{G}^{SD}}{\partial x_1}\right](\mathbf{Z}) + O(h_0^3) \quad (3.3.20)$$

Substituting (3.3.8) and (3.3.9) into the right-hand side of (3.3.20) we find that to leading order in $h_0$

$$G_{i1}^W(\mathbf{x}, \mathbf{x}_0) = -2h_0^2\frac{\partial}{\partial x_1}\left[\frac{x_i}{|\mathbf{x}|^3} - \frac{\partial \mathcal{S}_{i1}}{\partial x_1}\right](\mathbf{Z})$$
$$= -6h_0^2 Z_1\left(2\frac{Z_i}{|\mathbf{Z}|^5} + \delta_{i1}\frac{Z_1}{|\mathbf{Z}|^5} - 5\frac{Z_1^2 Z_i}{|\mathbf{Z}|^7}\right) \quad (3.3.21)$$

and

$$G_{ik}^W(\mathbf{x}, \mathbf{x}_0) = -2h_0\left[\frac{\partial \mathcal{S}_{ik}}{\partial x_1} + \frac{\partial \mathcal{S}_{i1}}{\partial x_k}\right](\mathbf{Z}) = 12h_0\frac{Z_1 Z_i Z_k}{|\mathbf{Z}|^5} \quad (3.3.22)$$

where $k = 2, 3$. The right-hand sides of (3.3.21) and (3.3.22) express a stresslet doublet and a stresslet, respectively (see section 7.2).

Inspecting the right-hand sides of (3.3.21) and (3.3.22) we find that the flow induced by a point force oriented perpendicular to the wall decays as $|\mathbf{Z}|^{-3}$, whereas the flow induced by a point force oriented parallel to the wall decays as $|\mathbf{Z}|^{-2}$ far from the wall. These results find useful applications in studies of the flow due to cilia motion (Blake 1972), and in the computation of the free surface of a liquid film flowing down an inclined wall due to a small particle captured on the wall (Pozrikidis & Thoroddsen 1991).

***Flow bounded internally by a solid sphere***

Oseen (1927) derived the Green's function $\mathbf{G}^{\text{SPH}}$ for an infinite flow that is bounded internally by a solid sphere. His results are

$$G_{ij}^{\text{SPH}}(\mathbf{x}, \mathbf{x}_0) = \left(\frac{\delta_{ij}}{r} + \frac{\hat{x}_i\hat{x}_j}{r^3}\right) - \frac{1}{R}\frac{\delta_{ij}}{r^*} - \frac{1}{R^3}\frac{\hat{x}_i^*\hat{x}_j^*}{(r^*)^3}$$
$$- \frac{R^2-1}{R}\left[X_i^*X_j^*\left(\frac{1}{r^*} + 2X_k^*\frac{\hat{x}_k^*}{(r^*)^3}\right) - \frac{1}{R^2}\frac{X_j^*\hat{x}_i^* + X_i^*\hat{x}_j^*}{(r^*)^3}\right]$$
$$- \frac{(|\mathbf{x}|^2-1)(\mathrm{R}^2-1)}{2R^3}\frac{\partial\phi_j}{\partial x_i} \tag{3.3.23}$$

where

$$\frac{\partial\phi_j}{\partial x_i} = -3X_j\frac{\hat{x}_i^*}{(r^*)^3} + \frac{\delta_{ij}}{(r^*)^3} - 3\frac{\hat{x}_i^*\hat{x}_j^*}{(r^*)^5} - 2\frac{X_i^*X_j}{(r^*)^3} + 6X_jX_k^*\frac{\hat{x}_i^*\hat{x}_k^*}{(r^*)^5}$$
$$+3\frac{\hat{x}_i^*[X_j^*(r^*)^2 + \hat{x}_j^*(R^*)^2] + \delta_{ij}(r^*-R^*)(r^*)^2R^*}{(R^*)^2(r^*)^3[R^*r^* + x_kX_k^* - (R^*)^2]}$$
$$-3\frac{R^*\hat{x}_i + r^*X_i^*}{r^*R^* + x_kX_k^* - (R^*)^2}\left[\frac{X_j^*}{(R^*)^2} + \frac{\hat{x}_j^* - X_j^*}{r^*R^*} - \frac{\hat{x}_j}{(r^*)^2}\right]$$
$$-3\frac{x_iX_j^* + \delta_{ij}|\mathbf{x}|R^*}{|\mathbf{x}|(R^*)^2(|\mathbf{x}|R^* + x_kX_k^*)} + 3\frac{(R^*x_i + |\mathbf{x}|X_i^*)(R^*x_j + |\mathbf{x}|X_j^*)}{|\mathbf{x}|(R^*)^2(|\mathbf{x}|R^* + x_kX_k^*)^2} \tag{3.3.24}$$

and

$$\mathbf{X} = \mathbf{x}_0 \qquad R = |\mathbf{X}| \quad \mathbf{X}^* = \frac{\mathbf{x}_0}{|\mathbf{x}_0|^2} \qquad R^* = |\mathbf{X}^*|$$
$$\hat{\mathbf{x}} = \mathbf{x} - \mathbf{X} \qquad r = |\hat{\mathbf{x}}| \quad \hat{\mathbf{x}}^* = \mathbf{x} - \mathbf{X}^* \qquad r^* = |\hat{\mathbf{x}}^*| \tag{3.3.25}$$

For convenience, we have assumed that the origin is located at the center of the sphere and in addition, we reduced all distances with respect to the radius of the sphere. Thus, $\mathbf{G}^{\text{SPH}}(\mathbf{x}, \mathbf{x}_0) = 0$ when $|\mathbf{x}| = 1$.

Nigam & Srinivasan (1975) and Higdon (1979) observed that (3.3.23) may be interpreted as a line distribution of singularities with poles in the interior of the sphere. Specifically, they found that the radial component of the image system may be resolved into a Stokeslet, a potential dipole, and a stresslet with poles at the inverse point $\mathbf{X}^*$. The corresponding transverse component may be resolved into a line distribution of Stokeslets, potential dipoles, and Stokeslet doublets extending from the origin up to the inverse point. A thorough investigation of these distributions is presented by Kim & Karrila (1991, Chapter 10).

The force and torque acting on a sphere due to the flow produced by a point force of strength $\mathbf{g}$ may be obtained by examining the asymptotic behaviour of the Green's function far from the sphere. An alternative method is to use the Faxen relations (2.5.36) and (2.5.37), obtaining

$$\mathbf{F} = \frac{1}{2}\left(\frac{3}{R} - \frac{1}{R^3}\right)\mathbf{g}_r + \frac{1}{4}\left(\frac{3}{R} + \frac{1}{R^3}\right)\mathbf{g}_t \tag{3.3.26}$$

$$\mathbf{L} = \frac{1}{R^3}\mathbf{X} \times \mathbf{g} \tag{3.3.27}$$

where $\mathbf{g}_r = \mathbf{e}_r(\mathbf{g}\cdot\mathbf{e}_r)$ is the radial component of $\mathbf{g}$, $\mathbf{e}_r$ is the unit vector in the radial direction emanating from the center of the sphere, and $\mathbf{g}_t = \mathbf{g} - \mathbf{e}_r(\mathbf{g}\cdot\mathbf{e}_r)$ is the transverse component of $\mathbf{g}$ (see problem 2.5.5). It will be noted that as the point force approaches the sphere ($R \to 1$), $\mathbf{F} \to \mathbf{g}$ recovering the results for a plane wall (problem 3.3.1).

**Problems**

3.3.1 Show that the force exerted on an infinite plane wall due to a point force that is located above the wall is equal to the strength of the point force.

3.3.2 Consider a shearing flow over a plane wall that contains a solitary protrusion. Compute the disturbance force acting on the wall due to the protrusion.

3.3.3 Compute the force and torque on a fluid surface enclosing (a) a point force and a neighbouring solid sphere, (b) the solid sphere but not the point force, (c) the point force but not the solid sphere.

3.3.4 Using the results of problem 3.3.3 compute the leading order far field behaviour of the flow due to a point force in the presence of a solid sphere.

## 3.4 Axisymmetric flow

In section 2.4 we saw that in the case of axisymmetric flow, the domain of the boundary integral equation may conveniently be reduced into the trace $C$ of the boundary in a meridional plane. More specifically, we saw that the single-layer potential in the boundary integral equation may be written in the simplified form

$$I^{S}_{\alpha}(\mathbf{x}_0) = -\frac{1}{8\pi\mu}\int_C M_{\alpha\beta}(\mathbf{x}_0, \mathbf{x}) f_\beta(\mathbf{x})\, \mathrm{d}l(\mathbf{x}) \tag{3.4.1}$$

where $\alpha, \beta$ are either $x$ or $\sigma$, indicating the radial and axial components respectively. Probing the physical origin of (3.4.1), we found that

$$u_\alpha(\mathbf{x}) = \frac{1}{8\pi\mu} M_{\alpha\beta}(\mathbf{x}, \mathbf{x}_0) g_\beta \tag{3.4.2}$$

represents the velocity field at the point $\mathbf{x}$ due to a ring of point forces of density per unit length $\mathbf{g}$ passing through the point $\mathbf{x}_0$ (see (2.4.4)). In this light, it appears proper to call the kernel $\mathbf{M}(\mathbf{x}, \mathbf{x}_0)$ the *Green's function of axisymmetric flow.*

### *A ring of point forces*

To compute the free-space axisymmetric Green's function corresponding to a ring of point forces, $\mathbf{M}^{\mathrm{R}}$, we simply integrate the Stokeslet in the azimuthal direction as indicated in section 2.4, obtaining

$$M^{\mathrm{R}}_{xx} = 2k\sqrt{\frac{\sigma_0}{\sigma}}\left(F + \frac{\hat{x}^2}{r^2}E\right) \qquad M^{\mathrm{R}}_{x\sigma} = -k\frac{\hat{x}}{\sqrt{\sigma_0\sigma}}\left[F - (\sigma^2 - \sigma_0^2 + \hat{x}^2)\frac{E}{r^2}\right]$$

$$M^{\mathrm{R}}_{\sigma x} = k\frac{\hat{x}}{\sigma}\sqrt{\frac{\sigma_0}{\sigma}}\left[F + (\sigma^2 - \sigma_0^2 - \hat{x}^2)\frac{E}{r^2}\right] \tag{3.4.3}$$

$$M^{\mathrm{R}}_{\sigma\sigma} = \frac{k}{\sigma_0\sigma}\sqrt{\frac{\sigma_0}{\sigma}}\left\{(\sigma_0^2 + \sigma^2 + 2\hat{x}^2)F - [2\hat{x}^4 + 3\hat{x}^2(\sigma_0^2 + \sigma^2) + (\sigma^2 - \sigma_0^2)^2]\frac{E}{r^2}\right\}$$

where $\hat{x} = x - x_0$, $r^2 = \hat{x}^2 + (\sigma - \sigma_0)^2$, $F$ and $E$ are the complete elliptic integrals of the first and second kind with argument $k^2$ defined in (2.4.8) and (2.4.10), and the superscript R stands for ring. It is instructive to observe that (3.4.3) may be produced from (2.4.9) simply by switching the arguments $\mathbf{x}$ and $\mathbf{x}_0$.

### *A ring of point forces in a cylindrical tube*

Next, we consider the axisymmetric Green's function $\mathbf{M}^{\mathrm{T}}$ in a domain that is bounded externally by a cylindrical tube of radius $\sigma_c$. We require that $\mathbf{M}^{\mathrm{T}}$ and, hence, the associated velocity field given by (3.4.2) vanish over the surface of the tube at $\sigma = \sigma_c$. Physically, $\mathbf{M}^{\mathrm{T}}$ represents the flow produced by a ring of point forces whose axis is collinear with the centre line of the tube.

Following the general guidelines for computing Green's functions discussed in section 3.3, we decompose $\mathbf{M}^{\mathrm{T}}$ into a free-space and a complementary component, setting

$$\mathbf{M}^{\mathrm{T}} = \mathbf{M}^{\mathrm{R}} + \mathbf{M}^{\mathrm{C}} \tag{3.4.4}$$

where $\mathbf{M}^{\mathrm{R}}$ is given by (3.4.3), and the superscripts T and C stand for 'tube' and 'complementary' respectively. The complementary component $\mathbf{M}^{\mathrm{C}}$ was derived by Tözeren (1984) in terms of inverse Fourier integrals in the form

$$M^{\mathrm{C}}_{xx} = \sigma_0\int_0^\infty \left[\frac{tI_0(\omega)}{\omega I_1(\omega) + 2I_0(\omega)}\right]\cdot \mathbf{A}_x \cos(\hat{x}t)\,dt$$

$$M^{C}_{x\sigma} = \sigma_0 \int_0^\infty \begin{bmatrix} tI_0(\omega) \\ \omega I_1(\omega) + 2I_0(\omega) \end{bmatrix} \cdot \mathbf{A}_\sigma \sin(\hat{x}t)\,dt$$

$$M^{C}_{\sigma x} = \sigma_0 \int_0^\infty \begin{bmatrix} tI_1(\omega) \\ \omega I_0(\omega) \end{bmatrix} \cdot \mathbf{A}_x \sin(\hat{x}t)\,dt$$

$$M^{C}_{\sigma\sigma} = -\sigma_0 \int_0^\infty \begin{bmatrix} tI_1(\omega) \\ \omega I_0(\omega) \end{bmatrix} \cdot \mathbf{A}_\sigma \cos(\hat{x}t)\,dt \qquad (3.4.5)$$

where $\omega = t\sigma$. The vectors $\mathbf{A}_x$ and $\mathbf{A}_\sigma$ are functions of $\omega_0 = t\sigma_0, \omega_c = t\sigma_c$, and also explicitly of $t$ and arise as solutions to the linear system

$$\begin{bmatrix} tI_0(\omega_c) & \omega_c I_1(\omega_c) + 2I_0(\omega_c) \\ tI_1(\omega_c) & \omega_c I_0(\omega_c) \end{bmatrix} \cdot \mathbf{A}_\alpha = 4\mathbf{B}_\alpha \qquad (3.4.6)$$

where

$$\mathbf{B}_x(\omega_c, \omega_0) = \begin{bmatrix} B_{xx} \\ B_{x\sigma} \end{bmatrix} = \begin{bmatrix} -2K_0(\omega_c) + \omega_c K_1(\omega_c) & -K_0(\omega_c) \\ -\omega_c K_0(\omega_c) & K_1(\omega_c) \end{bmatrix} \cdot \begin{bmatrix} I_0(\omega_0) \\ \omega_0 I_1(\omega_0) \end{bmatrix}$$

$$\mathbf{B}_\sigma(\omega_c, \omega_0) = \begin{bmatrix} B_{\sigma x} \\ B_{\sigma\sigma} \end{bmatrix} = \begin{bmatrix} -\omega_c K_1(\omega_c) & K_0(\omega_c) \\ 2K_1(\omega_c) + \omega_c K_0(\omega_c) & -K_1(\omega_c) \end{bmatrix} \cdot \begin{bmatrix} I_1(\omega_0) \\ \omega_0 I_0(\omega_0) \end{bmatrix} \qquad (3.4.7)$$

and $I_0$, $K_0$, $I_1, K_1$ are the modified Bessel functions of zeroth and first order. For the purposes of computation, all $I_0$, $K_0$, and $I_1, K_1$ may be accurately approximated with polynomial expansions (Abramowitz & Stegun 1972, p. 378).

For convenience of notation, we write (3.4.5) in the equivalent symbolic form

$$\mathbf{M}^{C} = \sigma_0 \int_0^\infty \begin{bmatrix} F_{xx}\cos(\hat{x}t) & F_{x\sigma}\sin(\hat{x}t) \\ F_{\sigma x}\sin(\hat{x}t) & -F_{\sigma\sigma}\cos(\hat{x}t) \end{bmatrix} dt \qquad (3.4.8)$$

where $\mathbf{F}$ is a function of $(\omega, \omega_0, \omega_c, t)$. As $t$ tends to zero, all components of $\mathbf{F}$ vanish except for the $xx$ component for which $F_{xx} = 8\ln t + \cdots$. Because this is a singularity that is just logarithmic, the corresponding integrand in (3.4.8) is integrable at the origin. As $t$ tends to infinity, $\mathbf{F}$ decays at an exponential rate: $F_{ij} \sim \exp[(\sigma + \sigma_0 - 2\sigma_c)t]$. The exponential decay guarantees the existence of all integrals in (3.4.8) at any point within the cylinder, i.e. as long as $\sigma < \sigma_c$ and $\sigma_0 < \sigma_c$.

To obtain a regular integral representation of the $xx$ component of the complementary Green's function, it is imperative to subtract off the logarithmic singularity of the corresponding integrand in (3.4.8). For this purpose, we use the identity

$$\frac{\pi}{[\hat{x}^2 + (2\sigma_c - \sigma - \sigma_0)^2]^{1/2}} = 2\int_0^\infty K_0(2\omega_c - \omega - \omega_0)\cos(\hat{x}t)\,dt \qquad (3.4.9)$$

and write

$$M^{C}_{xx} = \sigma_0 \int_0^\infty [F_{xx} + 8K_0(2\omega_c - \omega - \omega_0)] \cos(\hat{x}t)\, dt$$
$$- \frac{4\pi\sigma_0}{[\hat{x}^2 + (2\sigma_c - \sigma - \sigma_0)^2]^{1/2}} \tag{3.4.10}$$

(Happel & Brenner 1973, p. 303). As $t$ tends to zero, $K_0(2\omega_c - \omega - \omega_0)$ behaves as $-\ln(2\omega_c - \omega - \omega_0)$, and the modified integrand

$$F'_{xx} \equiv F_{xx} + 8K_0(2\omega_c - \omega - \omega_0) \tag{3.4.11}$$

tends to a finite value. On the other hand, as $t$ tends to infinity, $F'_{xx}$ decays at the exponential rate $(\sigma + \sigma_0 - 2\sigma_c)t$. In retrospect, the argument of the Bessel function in (3.4.9) was chosen so that the rate of decay of $F'_{xx}$ is identical to that of the integrands in (3.4.8).

It will be instructive to note that the free-space axisymmetric Green's function $\mathbf{M}^R$ may also be expressed in terms of a Fourier series involving the vectors $\mathbf{B}_x$ and $\mathbf{B}_\sigma$, as

$$\mathbf{M}^R = -4\sigma_0 \int_0^\infty \begin{bmatrix} B_{xx}(\omega, \omega_0)\cos(\hat{x}t) & B_{x\sigma}(\omega, \omega_0)\sin(\hat{x}t) \\ B_{\sigma x}(\omega, \omega_0)\sin(\hat{x}t) & -B_{\sigma\sigma}(\omega, \omega_0)\cos(\hat{x}t) \end{bmatrix} dt \tag{3.4.12}$$

Examining the behaviour of the integrands in (3.4.12) for large values of $t$, we find that as $t$ tends to infinity $\mathbf{B}$ behaves as $B_{ij} \sim \exp(\sigma_0 - \sigma)t$, indicating that the Fourier integrals in (3.4.12) are proper only when $\sigma > \sigma_0$.

### *An array of rings of point forces in a cylindrical tube*

Next, we consider the flow due to an array of equidistant rings of point forces centered at the axis of the tube. Working as in the case of a single ring, we decompose the corresponding Green's function $\mathbf{M}^{TA}$ into a primary and a complementary component as

$$\mathbf{M}^{TA} = \mathbf{M}^{RA} + \mathbf{M}^{CA} \tag{3.4.13}$$

where $\mathbf{M}^{RA}$ represents the flow due to an array of rings of point forces in an infinite fluid and $\mathbf{M}^{CA}$ is the complementary Green's function. The superscript A stands for array. To compute $\mathbf{M}^{CA}$ we sum (3.4.8) over all rings, obtaining, for instance,

$$M^{CA}_{x\sigma} = \sigma_0 \int_0^\infty F_{x\sigma} \sum_{n=-N}^{N} \sin[(\hat{x} + nl)t]\, dt \tag{3.4.14}$$

where $l$ is the distance between two rings. The sum within the integral on the right-hand side of (3.4.14) may be computed in closed form using the identity (problem 3.4.2)

$$H(N) \equiv \sum_{n=-N}^{N} \exp[\mathrm{i}(\hat{x} + nl)t] = \exp(\mathrm{i}\hat{x}t) \frac{\cos(Nlt) - \cos[(N+1)lt]}{1 - \cos lt} \tag{3.4.15}$$

To obtain an infinite array of rings we pass to the limit as $N$ tends to infinity and use the asymptotic expression

$$H(\infty)=\frac{2\pi}{l}\exp(\mathrm{i}\hat{x}t)\sum_{m=-\infty}^{\infty}\delta\left(t-\frac{2\pi m}{l}\right) \tag{3.4.16}$$

where $\delta$ is the one-dimensional delta function (Lighthill 1958, p. 69). Substituting (3.4.16) into (3.4.14), recalling that $F_{x\sigma}(t=0)=0$, and adding the contribution from the rings we obtain the periodic Green's function

$$M_{x\sigma}^{\mathrm{TA}}=\sigma_0\frac{2\pi}{l}\sum_{m=1}^{\infty}F_{x\sigma}(t_m)\sin(t_m\hat{x})+\sum_{n=-\infty}^{\infty}M_{x\sigma}^{\mathrm{R}}(\hat{x}+nl) \tag{3.4.17}$$

where $t_m=2\pi m/l$. The two sums on the right-hand side of (3.4.17) converge at rates that are respectively exponential and algebraic.

Following a procedure similar to the above, and recalling that $F_{\sigma x}(t=0)=F_{\sigma\sigma}(t=0)=0$, we may derive expressions similar to (3.4.17) for the $\sigma x$ and the $\sigma\sigma$ components of $\mathbf{M}^{\mathrm{TA}}$. The computation of $M_{xx}^{\mathrm{TA}}$ is frustrated by the singular behaviour of $F_{xx}$ at the origin, i.e. $F_{xx}(t=0)=\infty$. To tame this singularity we write

$$M_{xx}^{\mathrm{CA}}=\sigma_0\int_0^{\infty}F'_{xx}\sum_{n=-N}^{N}\cos[(\hat{x}+nl)t]\,\mathrm{d}t$$
$$-4\pi\sigma_0\sum_{n=-N}^{N}\frac{1}{[(\hat{x}+nl)^2+(2\sigma_c-\sigma-\sigma_0)^2]^{1/2}} \tag{3.4.18}$$

where $F'_{xx}$ was defined in (3.4.11). Passing to the limit as $N$ tends to infinity and adding the contributions from the primary rings we obtain

$$M_{xx}^{\mathrm{TA}}=\sigma_0\frac{2\pi}{l}\left[\tfrac{1}{2}F'_{xx}(t=0)+\sum_{m=1}^{\infty}F'_{xx}(t_m)\cos(t_m\hat{x})\right]$$
$$+\sum_{n=-\infty}^{\infty}\left\{M_{xx}^{\mathrm{R}}(\hat{x}+nl)-\frac{4\pi\sigma_0}{[(\hat{x}+nl)^2+(2\sigma_c-\sigma-\sigma_0)^2]^{1/2}}\right\} \tag{3.4.19}$$

The first sum on the right-hand side of (3.4.19) decays at an exponential rate. To demonstrate that the second sum is convergent, we expand $M_{xx}^{\mathrm{R}}$ in an asymptotic series for large $nl$, obtaining

$$M_{xx}^{\mathrm{R}}(\hat{x}+nl)=\frac{2\pi\sigma_0}{[(\hat{x}+nl)^2+(\sigma+\sigma_0)^2]^{1/2}}\left[2-\frac{(\sigma-\sigma_0)^2}{(\hat{x}+nl)^2+(\sigma-\sigma_0)^2}\right]+\cdots \tag{3.4.20}$$

Comparing (3.4.20) with (3.4.19) we find that the terms in the second series on the right-hand side of (3.4.19) decay as $(nl)^{-3}$ and thus the corresponding sum is convergent.

**Problems**

3.4.1 Compute the axisymmetric Green's function corresponding to a ring of point forces above a solid plane wall (Liron & Blake 1981; Pozrikidis 1990b).

3.4.2 Prove the identity (3.4.15) and show that the zeros of the function $H(N)$ occur for $t = 2\pi m/[l(2N+1)]$ where $m$ is an integer. Also, show that additional zeros of the real part of $H(N)$ occur for $t = (m + 1/2)\pi/\hat{x}$ and of the imaginary part of $H(N)$ occur for $t = m\pi/\hat{x}$.

## 3.5 Two-dimensional flow

The Green's functions of two-dimensional flow share many of the properties of the Green's functions of three-dimensional flow discussed in the preceding sections. The defining equations (3.1.1) and (3.1.2) are applicable, as is the symmetry property (3.1.3) (see problem 3.5.1). In the case of infinite flow, equations (3.2.1) and (3.2.2) express respectively the flow due to a two-dimensional point source and the flow due to a stresslet. The identities (3.2.5) and (3.2.7) stand as well, but the surface integral on the left-hand side of each equation must be replaced by a contour integral over a line enclosing a selected area of flow, and the constants on the right-hand side of (3.2.5) must be replaced by $-4\pi$, $-2\pi$ and 0, and those on the right-hand side of (3.2.7) by $4\pi$, $2\pi$, and 0.

In the rest of this section we present a number of Green's functions for specific domains of flow.

### *A point force above an infinite plane wall*

The Green's function for semi-infinite flow that is bounded by a solid plane wall at $y = w$ is given by

$$\mathbf{G}^{\mathrm{W}}(\mathbf{x}, \mathbf{x}_0) = \mathscr{S}(\hat{\mathbf{x}}) - \mathscr{S}(\hat{\mathbf{X}}) + 2h_0^2\mathbf{G}^{\mathrm{D}}(\hat{\mathbf{X}}) - 2h_0\mathbf{G}^{\mathrm{SD}}(\hat{\mathbf{X}}) \qquad (3.5.1)$$

where $h_0 = y_0 - w$, $\hat{\mathbf{x}} = \mathbf{x} - \mathbf{x}_0$, $\hat{\mathbf{X}} = \mathbf{x} - \mathbf{x}_0^{\mathrm{IM}}$, $\mathbf{x}_0^{\mathrm{IM}} = (x_0, 2w - y_0)$ is the image of the pole with respect to the wall, $\mathscr{S}$ is the two-dimensional Stokeslet, and the superscript W stands for wall (Pozrikidis 1987a). The matrices $\mathbf{G}^{\mathrm{D}}$ and $\mathbf{G}^{\mathrm{SD}}$ contain potential dipoles and Stokeslet doublets respectively, and are given by

$$G_{ij}^{\mathrm{D}}(\mathbf{x}) = \pm\frac{\partial}{\partial x_j}\left(\frac{x_i}{|\mathbf{x}|^2}\right) = \pm\left(\frac{\delta_{ij}}{|\mathbf{x}|^2} - 2\frac{x_i x_j}{|\mathbf{x}|^4}\right) \qquad (3.5.2)$$

$$G_{ij}^{\mathrm{SD}}(\mathbf{x}) = \pm\frac{\partial S_{i2}}{\partial x_j} = x_2 G_{ij}^{\mathrm{D}}(\mathbf{x}) \pm \frac{\delta_{j2}x_i - \delta_{i2}x_j}{|\mathbf{x}|^2} \qquad (3.5.3)$$

with a plus sign for $j = 1$, for the $x$ direction, and a minus sign for $j = 2$, for the $y$ direction. The associated pressure vector is given by

$$\mathbf{p}^{W}(\mathbf{x}, \mathbf{x}_0) = \mathbf{p}^{ST}(\hat{\mathbf{x}}) - \mathbf{p}^{ST}(\hat{\mathbf{X}}) - 2h_0\mathbf{p}^{SD}(\hat{\mathbf{X}}) \tag{3.5.4}$$

where

$$p_i^{ST}(\mathbf{x}) = 2\frac{x_i}{|\mathbf{x}|^2} \qquad \mathbf{p}^{SD}(\mathbf{x}) = -\frac{2}{|\mathbf{x}|^4}(2xy, x^2 - y^2) \tag{3.5.5}$$

whereas the associated stress tensor is given by

$$\mathbf{T}^{W}(\mathbf{x}, \mathbf{x}_0) = \mathbf{T}(\hat{\mathbf{x}}) - \mathbf{T}(\hat{\mathbf{X}}) + 2h_0^2\mathbf{T}^{D}(\hat{\mathbf{X}}) - 2h_0\mathbf{T}^{SD}(\hat{\mathbf{X}}) \tag{3.5.6}$$

where

$$T_{ijk}^{ST}(\mathbf{x}) = -4\frac{x_i x_j x_k}{|\mathbf{x}|^4} \tag{3.5.7}$$

The tensors $\mathbf{T}^{D}$ and $\mathbf{T}^{SD}$ may be computed by straightforward differentiation using the equations

$$T_{ijk}^{D} = \frac{\partial G_{ij}^{D}}{\partial x_k} + \frac{\partial G_{kj}^{D}}{\partial x_i} \qquad T_{ijk}^{SD} = -\delta_{ik}p_j^{SD} + \frac{\partial G_{ij}^{SD}}{\partial x_k} + \frac{\partial G_{kj}^{SD}}{\partial x_i} \tag{3.5.8}$$

The stream functions associated with a point force that is oriented in the $x$ or $y$ direction are given by

$$\psi = -\hat{y}\ln\left(\frac{r}{R}\right) + \frac{2}{R^2}h_0\hat{Y}(\hat{y} + h_0) \qquad \psi = \hat{x}\left[\ln\left(\frac{r}{R}\right) + \frac{2}{R^2}h_0(\hat{y} + h_0)\right] \tag{3.5.9}$$

where $r = |\hat{\mathbf{x}}|$ $\hat{Y} = y - y_0^{IM} = y + y_0 - 2w$, and $R = |\hat{\mathbf{X}}|$ (Liron & Blake 1981).

### *An array of point forces in an infinite fluid*

Next, we consider a Green's function representing an array of point forces with poles at $\mathbf{x}_n = (x_n = x_0 + nl,\ y_n = y_0)$, where $n$ runs between $-N$ and $N$ (Pozrikidis 1987a). At the outset, it will be convenient to reduce all length scales using the wave number $k = 2\pi/l$. Summing the Green's function corresponding to the individual point forces, we obtain

$$G_{ij}^{N}(\mathbf{x}, \mathbf{x}_0) = \sum_{n=-N}^{n=N}\left[-\delta_{ij}\ln r_n + \frac{\hat{x}_{n,i}\hat{x}_{n,j}}{r_n^2}\right] \tag{3.5.10}$$

where $\hat{\mathbf{x}}_n = \mathbf{x} - \mathbf{x}_n, \mathbf{r}_n = |\hat{\mathbf{x}}_n|$.

To derive the Green's function corresponding to an infinite periodic array of point forces, we pass to the limit as $N$ tends to infinity. We observe, however, that in this limit the infinite sum in (3.5.10) diverges. To obtain a regular expression, we use the summation formula

$$A(\hat{\mathbf{x}}_0) \equiv \sum_{n=-\infty}^{\infty} \ln r_n = \tfrac{1}{2}\ln(\cosh\hat{y}_0 - \cos\hat{x}_0) + \tfrac{1}{2}\ln 2 \tag{3.5.11}$$

(Lamb 1932, p. 71). The term $\frac{1}{2}\ln 2$ on the right-hand side of (3.5.11) was added so that close to a pole the Green's function reduces to the Stokeslet. Differentiating (3.5.11), we obtain

$$\frac{\partial A}{\partial \hat{x}_0} \equiv A_x(\hat{\mathbf{x}}_0) = \sum_{n=-\infty}^{\infty} \frac{\hat{x}_n}{r_n^2} = \frac{1}{2} \frac{\sin \hat{x}_0}{\cosh \hat{y}_0 - \cos \hat{x}_0} \tag{3.5.12}$$

$$\frac{\partial A}{\partial \hat{y}_0} \equiv A_y(\hat{\mathbf{x}}_0) = \hat{y}_0 \sum_{n=-\infty}^{\infty} \frac{1}{r_n^2} = \frac{1}{2} \frac{\sinh \hat{y}_0}{\cosh \hat{y}_0 - \cos \hat{x}_0} \tag{3.5.13}$$

Returning to (3.5.10), passing to the limit as $N$ goes to infinity, and using (3.5.11), (3.5.12), and (3.5.13) we obtain the periodic Green's function $\mathbf{G}^{\mathrm{P}}(\hat{\mathbf{x}}_0)$ where

$$\mathbf{G}^{\mathrm{P}}(\mathbf{x}) = \begin{bmatrix} -A - yA_y + 1 & yA_x \\ yA_x & -A + yA_y \end{bmatrix} \tag{3.5.14}$$

and the superscript P stands for periodic. The corresponding pressure and stress fields are given by $\mathbf{p}^{\mathrm{P}}(\hat{\mathbf{x}}_0)$ and $\mathbf{T}^{\mathrm{P}}(\hat{\mathbf{x}}_0)$ where

$$\mathbf{p}^{\mathrm{P}}(\mathbf{x}) = 2[A_x, A_y] \tag{3.5.15}$$

and

$$\begin{bmatrix} T^{\mathrm{P}}_{xxx} & T^{\mathrm{P}}_{xxy} \\ T^{\mathrm{P}}_{yxx} = T^{\mathrm{P}}_{xyx} & T^{\mathrm{P}}_{yxy} = T^{\mathrm{P}}_{xyy} \\ T^{\mathrm{P}}_{yyx} & T^{\mathrm{P}}_{yyy} \end{bmatrix}(\mathbf{x}) = 2\begin{bmatrix} \nabla\cdot(-A, -yA_y) & \nabla\cdot(yA_x, -A) \\ \nabla\cdot(yA_x, -A) & \nabla\cdot(-A, yA_x) \\ \nabla\cdot(-A, yA_x) & \nabla\cdot(-yA_x, -A) \end{bmatrix} \tag{3.5.16}$$

The argument of $A$, and of its derivatives, in (3.5.14), (3.5.15), and (3.5.16) is $\mathbf{x}$.

### *An array of point forces above a plane wall*

Proceeding, we consider the flow due to an infinite array of point forces located above a plane wall. To compute the corresponding Green's function we sum up the individual Green's functions given in (3.5.1), obtaining (Pozrikidis 1987a)

$$\mathbf{G}^{\mathrm{WP}}(\mathbf{x}, \mathbf{x}_0) = \mathbf{G}^{\mathrm{P}}(\hat{\mathbf{x}}) - \mathbf{G}^{\mathrm{P}}(\hat{\mathbf{X}}) + 2h_0^2\mathbf{G}^{\mathrm{DP}}(\hat{\mathbf{X}}) - 2h_0\mathbf{G}^{\mathrm{SDP}}(\hat{\mathbf{X}}) \tag{3.5.17}$$

where $h_0 = y_0 - w$, $\hat{\mathbf{x}} = \mathbf{x} - \mathbf{x}_0$, $\hat{\mathbf{X}} = \mathbf{x} - \mathbf{x}_0^{\mathrm{IM}}$ and $\mathbf{x}_0^{\mathrm{IM}} = (x_0, 2w - y_0)$ is the image of the pole $\mathbf{x}_0$ with respect to the wall, and the superscripts have the following meanings: WP stands for wall periodic, DP stands for potential dipole periodic, and SDP stands for Stokeslet doublet periodic. Using (3.5.2), (3.5.3), and (3.5.11) we find

$$\mathbf{G}^{\mathrm{DP}}(\mathbf{x}) = \begin{bmatrix} A_{yy} & A_{xy} \\ -A_{xy} & A_{yy} \end{bmatrix} \tag{3.5.18}$$

and

$$\mathbf{G}^{\mathrm{SDP}}(\mathbf{x}) = \begin{bmatrix} -yA_{yy} & -A_x - yA_{xy} \\ -A_x + yA_{xy} & -yA_{yy} \end{bmatrix} \tag{3.5.19}$$

The associated pressure fleld is given by

$$\mathbf{p}^{\mathrm{WP}}(\mathbf{x}, \mathbf{x}_0) = \mathbf{p}^{\mathrm{P}}(\hat{\mathbf{x}}) - \mathbf{p}^{\mathrm{P}}(\hat{\mathbf{X}}) - 2h_0\mathbf{p}^{\mathrm{SDP}}(\hat{\mathbf{X}}) \tag{3.5.20}$$

whre $\mathbf{p}^{\mathrm{P}}$ was given in (3.5.15), and

$$\mathbf{p}^{\mathrm{SDP}}(\mathbf{x}) = 2[A_{xy}, - A_{yy}] \tag{3.5.21}$$

It will be noted that (3.5.20) expresses the flow due to an infinite array of two-dimensional point sources in the presence of the wall.

The stress tensor associated with the Green's function is given by

$$\mathbf{T}^{\mathrm{WP}}(\mathbf{x}, \mathbf{x}_0) = \mathbf{T}^{\mathrm{P}}(\hat{\mathbf{x}}) - \mathbf{T}^{\mathrm{P}}(\hat{\mathbf{X}}) + 2h_0^2\mathbf{T}^{\mathrm{DP}}(\hat{\mathbf{X}}) - 2h_0\mathbf{T}^{\mathrm{SDP}}(\hat{\mathbf{X}}) \tag{3.5.22}$$

where $\mathbf{T}^{\mathrm{P}}$ was given in (3.5.16). The tensors $\mathbf{T}^{\mathrm{DP}}$ and $\mathbf{T}^{\mathrm{SDP}}$ may be computed by straightforward differentiation using the equations

$$T_{ijk}^{\mathrm{DP}} = \frac{\partial G_{ij}^{\mathrm{DP}}}{\partial x_k} + \frac{\partial G_{kj}^{\mathrm{DP}}}{\partial x_i} \tag{3.5.23}$$

$$T_{ijk}^{\mathrm{SDP}} = -\delta_{ik}p_j^{\mathrm{SDP}} + \frac{\partial G_{ij}^{\mathrm{SDP}}}{\partial x_k} + \frac{\partial G_{kj}^{\mathrm{SDP}}}{\partial x_i} \tag{3.5.24}$$

The periodic Green's function $\mathbf{T}^{\mathrm{WP}}$ has found applications in a number of numerical computations using boundary element methods (Pozrikidis 1987a, 1988; Yiantsios & Higgins 1989; Newhouse & Pozrikidis 1990).

### *A point force between two parallel plane walls*

Next, we consider the flow due to a point force in a domain that is confined between two parallel plane walls, as illustrated in Figure 3.5.1. It will be convenient to decompose the corresponding Green's function $\mathbf{G}^{2\mathrm{W}}$ into a fundamental component $\mathbf{G}^{\mathrm{F}}$ and a complementary component $\mathbf{G}^{\mathrm{C}}$, setting $\mathbf{G}^{2\mathrm{W}} = \mathbf{G}^{\mathrm{F}} + \mathbf{G}^{\mathrm{C}}$, and to identify the fundamental component with the flow

*Figure 3.5.1.* A two-dimensional point force in a channel confined by two parallel walls.

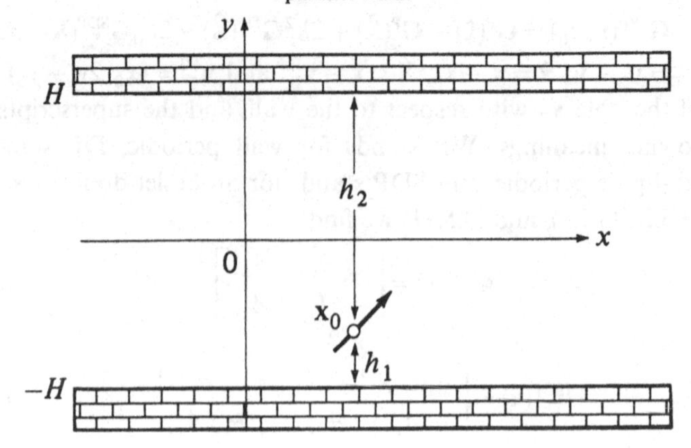

produced by the point force and its image system with respect to the two walls. This image system is composed of two $y$-periodic arrays of point forces separated by a distance equal to $4H$. The first array contains the point force itself, whereas the second array contains the image of the point force with respect to the lower wall. The strength of a point force in the second array is equal and opposite to that of a point force in the first array. Thus, the fundamental component $\mathbf{G}^{\mathrm{F}}$ is given by the limit, as $N$ tends to infinity, of

$$G_{ij}^{\mathrm{F}}(\mathbf{x}, \mathbf{x}_0) = \sum_{n=-N}^{N} \left( -\delta_{ij} \ln r_n + \frac{\hat{x}_{n,i}\hat{x}_{n,j}}{r_n^2} \right) - \sum_{n=-N}^{N} \left( -\delta_{ij} \ln R_n + \frac{\hat{X}_{n,i}\hat{X}_{n,j}}{R_n^2} \right) \tag{3.5.25}$$

where

$$\begin{aligned} \hat{\mathbf{x}}_n &= \mathbf{x} - \mathbf{x}_n & \hat{\mathbf{X}}_n &= \mathbf{x} - \mathbf{x}_n^{\mathrm{IM}} \\ r_n &= |\hat{\mathbf{x}}_n| & R_n &= |\hat{\mathbf{X}}_n| \\ \mathbf{x}_n &= (x_0, y_0 + 4Hn) & \mathbf{x}_n^{\mathrm{IM}} &= (x_0, -y_0 - 2H + 4Hn) \end{aligned} \tag{3.5.26}$$

As $N$ tends to infinity, both sums on the right-hand side of (3.5.25) diverge. To obtain a regular limiting behaviour we use the summation formula (3.5.11), finding

$$\mathbf{G}^{\mathrm{F}}(\mathbf{x}, \mathbf{x}_0) = \mathbf{Q}(\hat{\mathbf{x}}_0) - \mathbf{Q}(\hat{\mathbf{X}}_0) \tag{3.5.27}$$

where

$$\mathbf{Q}(\mathbf{x}) = \begin{bmatrix} -B + xB_x & xB_y \\ xB_y & -B - xB_x \end{bmatrix} \tag{3.5.28}$$

and

$$B(\mathbf{x}) = \tfrac{1}{2} \ln \left[ \cosh\left( \frac{x\pi}{2H} \right) - \cos\left( \frac{y\pi}{2H} \right) \right] \tag{3.5.29}$$

$B_x$ and $B_y$ denote derivatives of $B$ with respect to $x$ and to $y$. Evaluating the fundamental component at the two walls we find

$$G_{xx}^{\mathrm{F}}(y = \pm H) = G_{yy}^{\mathrm{F}}(y = \pm H) = 0 \tag{3.5.30}$$

$$G_{xy}^{\mathrm{F}}(y = H) = G_{yx}^{\mathrm{F}}(y = H) = f(\hat{x}_0, h_2) \tag{3.5.31}$$

$$G_{xy}^{\mathrm{F}}(y = -H) = G_{yx}^{\mathrm{F}}(y = -H) = -f(\hat{x}_0, h_1) \tag{3.5.32}$$

where $h_1 = H + y_0$ and $h_2 = H - y_0$ are the distances of the point force from the lower and the upper wall, and

$$f(x, h) = \frac{\pi}{2} \frac{x}{H} \frac{\sin(\pi h/2H)}{\cosh(\pi x/2H) - \cos(\pi h/2H)} \tag{3.5.33}$$

It will be noted that the diagonal components of the fundamental tensor $\mathbf{G}^{\mathrm{F}}$ vanish over both walls. To ensure that all components of the Green's function $\mathbf{G}^{\mathrm{W}}$ vanish over each wall, we require that the complementary component $\mathbf{G}^{\mathrm{C}}$ satisfies the boundary conditions

$$G_{xx}^{\mathrm{C}}(y=\pm H)=G_{yy}^{\mathrm{C}}(y=\pm H)=0 \tag{3.5.34}$$

$$G_{xy}^{\mathrm{C}}(y=\pm H)=G_{yx}^{\mathrm{C}}(y=\pm H)=-G_{xy}^{\mathrm{F}}(y=\pm H) \tag{3.5.35}$$

To compute $\mathbf{G}^{\mathrm{C}}$ we use the method of Fourier transforms. First, predicting the need for using the $x$-Fourier transform of the boundary values of $\mathbf{G}^{\mathrm{C}}$ over each wall, we compute the $x$-Fourier transform of the function $f$ defined in (3.5.33). Using the method of residues we find

$$\mathscr{L}(\omega,h)\equiv\int_{-\infty}^{\infty} f(x,h)\exp(\mathrm{i}\omega x)\,\mathrm{d}x=2\pi\mathrm{i}\frac{\omega}{|\omega|}g(|\omega|,h) \tag{3.5.36}$$

where

$$g(z,h)=\frac{\mathrm{d}}{\mathrm{d}z}\left\{\frac{\exp[-(4H-h)z]-\exp(-hz)}{1-\exp(-4Hz)}\right\} \tag{3.5.37}$$

Taking the $x$-Fourier transform of the equations of Stokes flow, enforcing the boundary conditions (3.5.34) and (3.5.35), and carrying out the inverse transforms, we obtain

$$\begin{aligned}
G_{xx}^{\mathrm{C}}&=\int_0^{\infty}[y\mathrm{e}^{\omega y},-y\mathrm{e}^{-\omega y},2\mathrm{e}^{\omega y},2\mathrm{e}^{-\omega y}]\cdot\mathbf{a}\cos\omega\hat{x}\,\mathrm{d}\omega\\
G_{yx}^{\mathrm{C}}&=\int_0^{\infty}\left[\left(\frac{\omega y-1}{\omega}\right)\mathrm{e}^{\omega y},\left(\frac{\omega y+1}{\omega}\right)\mathrm{e}^{-\omega y},2\mathrm{e}^{\omega y},-2\mathrm{e}^{-\omega y}\right]\cdot\mathbf{a}\sin\omega\hat{x}\,\mathrm{d}\omega\\
G_{xy}^{\mathrm{C}}&=-\int_0^{\infty}[y\mathrm{e}^{\omega y},-y\mathrm{e}^{-\omega y},2\mathrm{e}^{\omega y},2\mathrm{e}^{-\omega y}]\cdot\mathbf{b}\sin\omega\hat{x}\,\mathrm{d}\omega\\
G_{yy}^{\mathrm{C}}&=\int_0^{\infty}\left[\left(\frac{\omega y-1}{\omega}\right)\mathrm{e}^{\omega y},\left(\frac{\omega y+1}{\omega}\right)\mathrm{e}^{-\omega y},2\mathrm{e}^{\omega y},-2\mathrm{e}^{-\omega y}\right]\cdot\mathbf{b}\cos\omega\hat{x}\,\mathrm{d}\omega
\end{aligned} \tag{3.5.38}$$

where $\hat{x}=x-x_0$. The vectors $\mathbf{a}$ and $\mathbf{b}$ arise as solutions to the linear systems

$$\mathbf{B}\cdot\mathbf{a}=\exp(-\omega H)\mathbf{c}\qquad \mathbf{B}\cdot\mathbf{b}=\exp(-\omega H)\mathbf{d} \tag{3.5.39}$$

where

$$\mathbf{B}=\begin{bmatrix}\frac{1}{2}H & -\frac{1}{2}H\theta & 1 & \theta\\ -\frac{1}{2}H\theta/2 & \frac{1}{2}H & \theta & 1\\ \frac{1}{2}\left(H-\frac{1}{\omega}\right) & \frac{\theta}{2}\left(H+\frac{1}{\omega}\right) & 1 & -\theta\\ -\frac{\theta}{2}\left(H+\frac{1}{\omega}\right) & -\frac{1}{2}\left(H-\frac{1}{\omega}\right) & \theta & -1\end{bmatrix}\qquad \theta=\exp(-2\omega H) \tag{3.5.40}$$

$$\mathbf{c}=[0,0,-g(\omega,h_2),g(\omega,h_1)] \tag{3.5.41}$$

$$\mathbf{d}=[g(\omega,h_2),-g(\omega,h_1),0,0] \tag{3.5.42}$$

### *An array of point forces between two parallel plane walls*

To derive the Green's function $\mathbf{G}^{2WP}$ associated with an infinite $x$-periodic array of point forces in a channel confined by two parallel plane walls, we simply sum the Green's functions corresponding to the individual point forces. By analogy with the case of a single point force, we decompose $\mathbf{G}^{2WP}$ into a fundamental component $\mathbf{G}^{FP}$ and a complementary component $\mathbf{G}^{CP}$ setting $\mathbf{G}^{2WP} = \mathbf{G}^{FP} + \mathbf{G}^{CP}$, where the superscript P stands for periodic. Summing up the contributions of the individual point forces we obtain

$$\mathbf{G}^{FP}(\mathbf{x}, \mathbf{x}_0) = \sum_{n=-\infty}^{\infty} \mathbf{G}^{F}(\mathbf{x}, \mathbf{x}_n) \tag{3.5.43}$$

where $\mathbf{x}_n = (x_0 + nl, y_0)$ and $l$ is the distance between two successive point forces. Furthermore, using (3.5.38) and the identity (3.4.16) we find

$$G^{CP}_{xx} = \frac{2n}{l} \sum_{m=1}^{\infty} [ye^{w_m y}, -ye^{-w_m y}, 2e^{w_m y}, 2e^{-w_m y}]\cdot\mathbf{a} \cos w_m \hat{x}_0 \tag{3.5.44}$$

where $w_m = 2\pi m/l$, as well as three similar expressions for the remaining components of $\mathbf{G}^{C}$. To compute $\mathbf{a}$ and $\mathbf{b}$ we solve the linear systems (3.5.39) with $w_m$ in place of $\omega$.

### *A point force in the exterior of a circular cylinder*

Dorrepaal, O'Neill & Ranger (1984) computed the stream function associated with the Green's function for an infinite domain of flow that is bounded internally by a circular cylinder. Their results are expressed most conveniently in plane polar coordinates $r, \theta$ with origin at the centre of the cylinder as

$$\psi(\mathbf{x}, \mathbf{x}_0) = -r \sin\hat{\theta} \ln\left(\frac{|\hat{\mathbf{x}}|}{|\hat{\mathbf{X}}|}\right) - \frac{a}{2}\frac{(r^2 - a^2)}{|\hat{\mathbf{X}}|^2}\left(1 - \frac{a^2}{r_0^2}\right) \tag{3.5.45}$$

and

$$\psi(\mathbf{x}, \mathbf{x}_0) = (r\cos\hat{\theta} - r_0)\ln\left(\frac{|\hat{\mathbf{x}}|}{|\hat{\mathbf{X}}|}\right) - \left(r_0 - \frac{a^2}{r_0}\right)\ln r - \frac{1}{2}\frac{(r^2 - a^2)}{|\hat{\mathbf{X}}|^2}\left(1 - \frac{a^2}{r_0^2}\right)\left(r\cos\hat{\theta} - \frac{a^2}{r_0}\right) \tag{3.5.46}$$

for point forces that are oriented in the radial and azimuthal directions respectively. In (3.5.45) and (3.5.46) $a$ is the radius of the cylinder, $(r_0, \theta_0)$ are the coordinates of the point force, $\hat{\mathbf{x}} = \mathbf{x} - \mathbf{x}_0$, and $\hat{\mathbf{X}} = \mathbf{x} - \mathbf{x}_0^{IM}$, where $\mathbf{x}_0^{IM}$ is the image of the pole of the point force with respect to the cylinder; in polar coordinates $\mathbf{x}^{IM} = (a^2/r_0, \theta_0)$.

**Problems**

3.5.1 Following a procedure similar to that outlined in section 3.2, prove the symmetry property (3.1.3) for two-dimensional flow. Note that the integrals at infinity will require special consideration, as the flow due to the Green's functions does not necessarily vanish far away from the solid boundary.

3.5.2 Show explicitly that (3.5.1) satisfies the symmetry property (3.1.3).

3.5.3 Show that (3.5.4) expresses the flow due to a two-dimensional point source in the presence of the wall.

3.5.4 Show that in the limit as $\hat{\mathbf{x}}_0$ tends to zero, (3.5.14) yields the two-dimensional Stokeslet.

3.5.5 Prove (3.5.36) using the theory of residues. Note that the result arises in the form of an infinite series which may be summed in closed form.

## 3.6 Unsteady flow

The Green's functions of unsteady Stokes flow share many of the properties of the Green's functions of steady Stokes flow including the defining constraints (3.1.1), (3.1.2) and the symmetry identity (3.1.3). Not surprisingly, the computation of Green's functions for unsteady flow turns out to be considerably more involved than for steady flow.

As an example, we discuss the Green's function $\mathbf{G}^{\mathrm{W}}$ for semi-infinite flow bounded by a plane wall (Pozrikidis 1989a). To construct this Green's function we introduce an oscillating point force and its image with respect to the wall. In this fashion we satisfy part but not all of the required boundary conditions of zero velocity over the wall. To satisfy the remaining boundary conditions we introduce the complementary tensor $\mathbf{G}^{\mathrm{C}}$, defined by the equation

$$\mathbf{G}^{\mathrm{W}}(\mathbf{x}, \mathbf{x}_0) = \mathscr{S}(\hat{\mathbf{x}}) - \mathscr{S}(\hat{\mathbf{X}}) - \mathbf{G}^{\mathrm{C}}(\hat{\mathbf{X}}, h_0) \tag{3.6.1}$$

where $\mathscr{S}$ is the unsteady Stokeslet, $h_0 = x_0 - w$, $\hat{\mathbf{x}} = \mathbf{x} - \mathbf{x}_0$, $\hat{\mathbf{X}} = \mathbf{x} - \mathbf{x}_0^{\mathrm{IM}}$, $\mathbf{x}_0^{\mathrm{IM}} = (2w - x_0, y_0, z_0)$ is the image of the pole with respect to the wall, and the wall is located at $x = w$. The corresponding pressure field is given by

$$\mathbf{p}^{\mathrm{W}}(\mathbf{x}, \mathbf{x}_0) = \mathbf{p}^{\mathrm{S}}(\hat{\mathbf{x}}) - \mathbf{p}^{\mathrm{S}}(\hat{\mathbf{X}}) - \mathbf{p}^{\mathrm{C}}(\hat{\mathbf{X}}, h_0) \tag{3.6.2}$$

where $\mathbf{p}^{\mathrm{S}}$ is the pressure field associated with the unsteady Stokeslet, given in (2.7.12). To compute the complementary tensor $\mathbf{G}^{\mathrm{C}}$ we use the method of Fourier transforms. After a considerable amount of algebra we find that for a point force oriented perpendicular to the wall

$$G_{i1}^{\mathrm{C}}(\mathbf{x}, h) = -\frac{4}{\lambda^4}\left(\delta_{i1}\nabla^2 - \frac{\partial^2}{\partial x_i \partial x_1}\right)F_1(\mathbf{x}, h) \tag{3.6.3}$$

where we have identified the 1-direction with the $x$-axis. The function

$F_1$ is given by

$$F_1(\mathbf{x}, h) = \int_0^\infty \left[\frac{1 - e^{(\omega - \xi)h}}{e^{\omega x_1}} + \frac{1 - e^{(\xi - \omega)h}}{e^{\xi x_1}}\right] J_0[-\xi(x_2^2 + x_3^2)^{1/2}](\omega + \xi)\xi \, d\xi \tag{3.6.4}$$

where $\omega^2 = \xi^2 + \lambda^2$. The corresponding component of the pressure is given by

$$p_1^C(\mathbf{x}, h) = -\frac{4}{\lambda^2}\frac{\partial H}{\partial x_1}(\mathbf{x}, h) \tag{3.6.5}$$

where the function $H$ is given by

$$H(\mathbf{x}, h) = \int_0^\infty \frac{1 - e^{(\xi - \omega)h}}{e^{\xi x_1}} J_0[-\xi(x_2^2 + x_3^2)^{1/2}](\omega + \xi)\xi \, d\xi \tag{3.6.6}$$

In the limit $\lambda \to 0$ we obtain the asymptotic expressions

$$F_1 \approx \lambda^4 \frac{h}{2}(x_1 - h)\int_0^\infty e^{-\xi x_1} J_0[-\xi(x_2^2 + x_3^2)^{1/2}] d\xi = \lambda^4 \frac{h}{2}\frac{x_1 - h}{(x_1^2 + x_2^2 + x_3^2)^{1/2}} \tag{3.6.7}$$

and

$$H \approx \lambda^2 h \int_0^\infty e^{-\xi x_1} J_0[-\xi(x_2^2 + x_3^2)^{1/2}]\xi \, d\xi = \lambda^2 \frac{x_1 h}{(x_1^2 + x_2^2 + x_3^2)^{3/2}} \tag{3.6.8}$$

Substituting (3.6.7) and (3.6.8) into (3.6.2) and (3.6.4) we recover (3.3.7) and (3.3.10) in consistency with the limit of steady Stokes flow.

For a point force oriented parallel to the wall we find

$$G_{ij}^C(\mathbf{x}, h) = -\frac{4}{\lambda^4}\frac{\partial}{\partial x_j}\left[\frac{\partial F_2}{\partial x_i} + \left(\delta_{i1}\nabla^2 - \frac{\partial^2}{\partial x_i \partial x_1}\right)F_3\right](\mathbf{x}, h) \tag{3.6.9}$$

where $j = 2, 3$, and

$$F_2(\mathbf{x}, h) = \int_0^\infty \frac{1 - e^{(\xi - \omega)h}}{e^{\xi x_1}} J_0[-\xi(x_2^2 + x_3^2)^{1/2}]\omega(\omega + \xi) \, d\xi \tag{3.6.10}$$

$$F_3(\mathbf{x}, h) = \int_0^\infty \frac{1 - e^{(\xi - \omega)h}}{e^{\omega x_1}} J_0[-\xi(x_2^2 + x_3^2)^{1/2}](\omega + \xi) \, d\xi \tag{3.6.11}$$

The corresponding components of the complementary pressure are given by

$$p_j^C(\mathbf{x}, h) = -\frac{4}{\lambda^2}\frac{\partial F_2}{\partial x_j}(\mathbf{x}, h) \tag{3.6.12}$$

Passing to the limit $\lambda \to 0$ we obtain

$$F_2 \approx \lambda^2 h \frac{x_1}{|\mathbf{x}|^3} \qquad F_3 \approx -\lambda^2 h \frac{1}{|\mathbf{x}|} \tag{3.6.13}$$

These asymptotic expressions agree with our previous results in section 3.3 for steady flow.

**Problem**

3.6.1 Verify the asymptotic formulae (3.6.7), (3.6.8), and (3.6.13).

# 4

## *Generalized boundary integral methods*

The boundary integral equation discussed in section 2.3, provides us with a representation of a flow in terms of a dual distribution of a Green's function **G** and its associated stress tensor **T**; the two individual distributions are called the single-layer potential and the double-layer potential respectively. Odqvist (1930) noted that each of these potentials expresses an acceptable Stokes flow which, in principle, may be used independently to represent a given flow. This observation provides the basis for a new class of boundary integral methods called *generalized* or *indirect boundary integral methods.* In these methods, the flow is expressed simply in terms of a single-layer or a double-layer potential with an unknown density distribution. Imposing boundary conditions yields integral equations of the first or second kind for the densities of the distributions.

Odqvist (1930) studied the properties of the single-layer and double-layer potentials, and investigated the solutions of the integral equations that arise from generalized boundary integral representations. His work is discussed by Ladyzhenskaya (1969, Chapter 3) and Kim & Karrila (1991, Part 4, Chapter 15). In this chapter we summarize the main properties of the hydrodynamic potentials and then proceed to investigate the extent to which generalized boundary integral methods are capable of representing various types of internal and external flow.

It will be helpful to keep in mind throughout this chapter that boundary integral methods leading to integral equations of the second kind are highly preferable over those leading to integral equations of the first kind. The reason is that the existence and uniqueness of solution of integral equations of the second kind as well as the feasibility of finding solutions using iterative methods may be studied theoretically. In contrast, integral equations of the first kind allow only rudimentary theoretical investigations (Pogorzelski 1966, Chapter VI). Furthermore, as a general rule integral equations of the second kind allow numerical solutions that are more stable than those of integral equations of the first kind (see Chapter 6).

## 4.1 The single-layer potential

The single-layer potential represents the flow due to a surface distribution of point forces, i.e.

$$u_i(\mathbf{x}_0) = \int_D G_{ij}(\mathbf{x}_0, \mathbf{x}) q_j(\mathbf{x}) \, \mathrm{d}S(\mathbf{x}) \tag{4.1.1}$$

where $D$ is the domain of the distribution and $\mathbf{q}$ is the density of the distribution (Figure 4.1.1). It will be noted that because of the defining constraint (3.1.1), $\mathbf{u} = 0$ over the solid boundary $S_B$. To avoid mathematical complications, we shall assume that $D$ is a Lyapunov surface and also that the density $\mathbf{q}$ varies over $D$ in a continuous manner. Under these conditions, the velocity field expressed by (4.1.1) is bounded and continuous throughout the domain of flow as well as across $D$.

Multiplying (4.1.1) by the normal vector, integrating over an arbitrary closed surface $S$, and using (2.1.4), we find

$$\int_S u_i(\mathbf{x}) n_i(\mathbf{x}) \, \mathrm{d}S(\mathbf{x}) = 0 \tag{4.1.2}$$

which ensures that the flow rate through any closed surface is equal to zero.

The pressure and stress fields associated with the flow (4.1.1) are given by the corresponding distributions

$$P(\mathbf{x}_0) = \mu \int_D p_j(\mathbf{x}_0, \mathbf{x}) q_j(\mathbf{x}) \, \mathrm{d}S(\mathbf{x}) \tag{4.1.3}$$

*Figure 4.1.1.* Schematic illustration of the domain of distribution of a hydrodynamic potential with various closed surfaces $S_1, S_2, S_3, S_4$ in the domain of flow; $S_B$ is a solid surface bounding the flow.

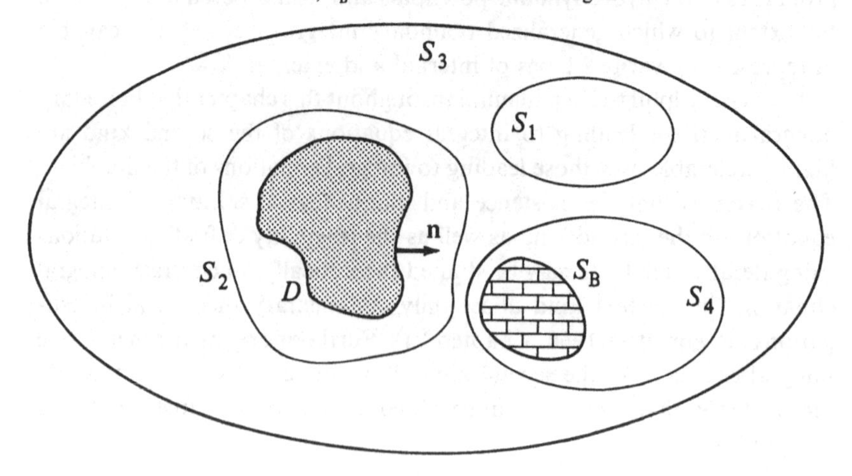

and

$$\sigma_{ik}(\mathbf{x}_0) = \mu \int_D T_{ijk}(\mathbf{x}_0, \mathbf{x}) q_j(\mathbf{x}) \, \mathrm{d}S(\mathbf{x}) \tag{4.1.4}$$

where $\mathbf{p}$ and $\mathbf{T}$ are the pressure vector and stress tensor associated with the Green's function, as discussed in section 2.1

The force and torque exerted on a surface that encloses pure fluid (for instance, the surface $S_1$ shown in Figure 4.1.1) are equal to zero. The force $\mathbf{F}$ and torque $\mathbf{L}$ exerted on a surface that encloses the distribution domain $D$ but not the solid wall $S_B$ (for instance, the surface $S_2$ shown in Figure 4.1.1) are equal to

$$\mathbf{F} = \int_{S_2} \boldsymbol{\sigma} \cdot \mathbf{n} \, \mathrm{d}S = -8\pi\mu \int_D \mathbf{q} \, \mathrm{d}S \tag{4.1.5}$$

and

$$\mathbf{L} = \int_{S_2} \mathbf{x} \times (\boldsymbol{\sigma} \cdot \mathbf{n}) \, \mathrm{d}S = -8\pi\mu \int_D \mathbf{x} \times \mathbf{q} \, \mathrm{d}S \tag{4.1.6}$$

(problem 4.1.1). The force and torque exerted on a surface that encloses $S_B$ and possibly $D$ (for instance the surfaces $S_3$ or $S_4$ shown in Figure 4.1.1) have finite values that depend on the geometry of $S_B$.

The surface force on the two sides of the distribution domain $D$ has two different values. On the external side of $D$, indicated by the direction of the normal vector,

$$f_i^+(\mathbf{x}_0) \equiv \sigma_{ik}^+(\mathbf{x}_0) n_k(\mathbf{x}_0) = -4\pi\mu q_i(\mathbf{x}_0) + \mu n_k(\mathbf{x}_0) \int_D^{\mathscr{PV}} T_{ijk}(\mathbf{x}_0, \mathbf{x}) q_j(\mathbf{x}) \, \mathrm{d}S(\mathbf{x}) \tag{4.1.7}$$

whereas on the internal side of $D$,

$$f_i^-(\mathbf{x}_0) \equiv \sigma_{ik}^-(\mathbf{x}_0) n_k(\mathbf{x}_0) = 4\pi\mu q_i(\mathbf{x}_0) + \mu n_k(\mathbf{x}_0) \int_D^{\mathscr{PV}} T_{ijk}(\mathbf{x}_0, \mathbf{x}) q_j(\mathbf{x}) \, \mathrm{d}S(\mathbf{x}) \tag{4.1.8}$$

(problem 4.1.3). The last term on the right-hand sides of (4.1.7) and (4.1.8) expresses the principal value of the surface force corresponding to the mathematical limit where the point $\mathbf{x}_0$ is located right on the domain of distribution $D$; when $D$ is a Lyapunov surface, the principal value of the surface force is equal to the mean of the two values of the surface force on either side of $D$ (problem 4.1.3). Subtracting (4.1.8) from (4.1.7) we obtain the discontinuity in the surface force across $D$:

$$\Delta\mathbf{f} = \mathbf{f}^+ - \mathbf{f}^- = -8\pi\mu\mathbf{q} \tag{4.1.9}$$

The single-layer potential is a linear integral operator $L^S$, with kernel the Green's function $\mathbf{G}$, operating on the space of density functions $\mathbf{q}$. We note parenthetically that the kernel $K(\eta, \xi, \eta_0, \xi_0)$ of an integral operator defined on a two-dimensional domain $D$ is classified as being weakly singular, if in the limit as $(\eta, \xi)$ tends to $(\eta_0, \xi_0)$,

$$|K(\eta, \xi, \eta_0, \xi_0)| < \frac{A}{(\eta-\eta_0)^2 + (\xi - \xi_0)^2} \tag{4.1.10}$$

where $(\eta, \xi)$ is a set of non-singular surface variables on $D$, and $A$ is a finite positive constant (see, for instance, Pogorzelski 1966, Chapter III; Kress 1989, p. 69). An integral operator with a weakly singular kernel is a completely continuous or compact operator (Stakgold 1967, Vol. I, p. 184). We note that the Green's function has a $1/r$ type of singularity at the pole; this suggests that if the domain of distribution $D$ is a Lyapunov surface, i.e. it has a continuously varying normal vector, the kernel of the single-layer potential is weakly singular and the single-layer operator is compact.

Now, exploiting the symmetry property (3.1.3) we find that for any two density functions $\mathbf{q}_1$ and $\mathbf{q}_2$,

$$(\mathbf{L}^S\mathbf{q}_1, \mathbf{q}_2) = (\mathbf{q}_1, \mathbf{L}^S\mathbf{q}_2) \tag{4.1.11}$$

indicating that the single-layer operator is self-adjoint. The inner product in (4.1.11) is defined as

$$(\mathbf{f}, \mathbf{g}) = \int_D \mathbf{f}(\mathbf{x})\cdot\mathbf{g}^*(\mathbf{x})\, dS(\mathbf{x}) \tag{4.1.12}$$

where an asterisk indicates the complex conjugate. Because the single-layer operator is self-adjoint, it has only real eigenvalues (problem 4.1.4).

### Problems

4.1.1 Use the identities (2.1.12) and (2.1.13) to prove (4.1.5) and (4.1.6).

4.1.2 Show that the velocity field in the interior of a sphere produced by a distribution of Stokeslets of constant density over the surface of the sphere is constant.

4.1.3 Writing

$$f_i(\mathbf{x}_0) \equiv \sigma_{ik}(\mathbf{x}_0) n_k(\mathbf{x}_0) = \mu n_k(\mathbf{x}_0) \int_D T_{ijk}(\mathbf{x}_0, \mathbf{x}) q_j(\mathbf{x})\, dS(\mathbf{x})$$

$$= \mu \int_D [n_k(\mathbf{x}_0) q_j(\mathbf{x}) T_{ijk}(\mathbf{x}_0, \mathbf{x}) - n_k(\mathbf{x}) q_j(\mathbf{x}_0) T^{\mathcal{S}}_{ijk}(\mathbf{x}_0, \mathbf{x})]\, dS(\mathbf{x})$$

$$- \mu q_j(\mathbf{x}_0) \int_D T^{\mathcal{S}}_{ijk}(\mathbf{x}, \mathbf{x}_0) n_k(\mathbf{x})\, dS(\mathbf{x}) \tag{1}$$

where $\mathbf{T}^{\mathcal{S}}$ is the stress tensor associated with the Stokeslet, noting that the integral on the second line of (1) is continuous across $D$, and using (2.1.12), prove (4.1.7) and (4.1.8). Then, show that the principal value of the surface force on $D$ is given by

$$\mathbf{f}^{\mathscr{PV}} = \tfrac{1}{2}(\mathbf{f}^{+} + \mathbf{f}^{-}) = \mathbf{J} + 4\pi\mu\mathbf{q} \tag{2}$$

where $\mathbf{J}$ is the integral on the second line of (1).

4.1.4 Show that the single-layer operator has only real eigenvalues. Note that an eigenvalue $\beta$ and its corresponding eigenfunction $\mathbf{q}$ satisfy the equation $\mathcal{L}^{S}\mathbf{q} = \beta\mathbf{q}$.

## 4.2 Representation of a flow in terms of a single-layer potential

We proceed now to investigate the extent to which the single-layer potential is capable of representing an arbitrary internal or external flow. Specifically, we inquire whether it is possible to find a density $\mathbf{q}$ such that the associated flow, expressed by (4.1.1), will produce a given boundary velocity or surface force over the domain of distribution $D$.

First, we consider the Dirichlet problem, which arises by prescribing the velocity over $D$. In this case, applying (4.1.1) over $D$, we obtain a Fredholm integral equation of the first kind for the density $\mathbf{q}$. To examine the uniqueness of solution of this equation we investigate the existence of eigensolutions of the corresponding homogeneous equation. If $\mathbf{q}''$ is an eigensolution, then

$$\int_D G_{ji}(\mathbf{x}_0, \mathbf{x}) q_i''(\mathbf{x}) \, dS(\mathbf{x}) = 0 \tag{4.2.1}$$

for all points $\mathbf{x}_0$ on $D$. If (4.2.1) does not have an eigensolution, then the problem has at most one solution. If (4.2.1) has an eigensolution then, as we shall see below, for (4.1.1) to have a solution this eigensolution must be orthogonal to the forcing function $\mathbf{u}$ with respect to the inner product defined in (4.1.12).

To investigate the eigensolutions of (4.2.1) we introduce the fictitious velocity field

$$w_j(\mathbf{x}_0) = \int_D G_{ji}(\mathbf{x}_0, \mathbf{x}) q_i''(\mathbf{x}) \, dS(\mathbf{x}) \tag{4.2.2}$$

Clearly, $\mathbf{w}$ vanishes on $S_B$, decays to zero at infinity, and, according to (4.2.1), vanishes over $D$. According to our discussion in section 1.2, $\mathbf{w}$ must vanish at every point in the intervening space as well as within $D$. The pressure must be constant inside and outside $D$, and the surface forces on both sides of $D$ must be proportional to the normal vector. Using (4.1.9) we find $\mathbf{q}'' = (-1/8\pi\mu)(\mathbf{f}^{+} - \mathbf{f}^{-}) = (\Delta P/8\pi\mu)\mathbf{n}$, where $\Delta P$ is the difference between the external and internal pressure.

Now, multiplying (4.1.1) by $\mathbf{q}''$, integrating over $D$, and using (4.2.1) and (3.1.3), we find that one necessary condition for (4.1.1) to have a solution is that $\mathbf{q}''$ is orthogonal to $\mathbf{u}$. Evidently, this requires that the flow rate of the prescribed boundary velocity across the domain of distribution $D$ is equal to zero. When this condition is met, (4.1.1) will have an infinity of solutions; any particular solution may be augmented by the addition of a multiple of the normal vector. In practice, in order to obtain a unique solution, it will be necessary to impose an arbitrary constraint on the density $\mathbf{q}$. We may require, for instance, that $\mathbf{q}$ is orthogonal to the eigensolution $\mathbf{q}''$. Alternatively, when applicable, we may require that $\mathbf{q}$ possesses some sort of spatial symmetry as dictated by the nature of the flow and the geometry of the boundaries.

Computational experience has shown that, in general, numerical solutions of the Fredholm integral equations of the first kind that arise by prescribing the boundary velocity are not notably sensitive to numerical error or to small variations in the prescribed velocity. Oscillatory solutions, however, may occur for certain types of boundary geometries as well as for boundaries with singular or nearly singular points, as will be discussed in Chapter 6.

Next, we consider the Neumann problem that arises by prescribing the surface force on either the external or the internal side of $D$. Considering (4.1.7) or (4.1.8) and enforcing the prescribed boundary condition, we obtain two Fredholm integral equations of the second kind for the density $\mathbf{q}$. The existence and uniqueness of solution of these equations will be investigated in detail in sections 4.5 and 4.6. The results will show that in the case of internal flow a solution will exist only if the force and torque on $D$ produced by the prescribed surface force $\mathbf{f}^-$ are equal to zero. When this constraint is met, the general solution to (4.1.8) will be

$$\mathbf{q} = \mathbf{q}^{\mathrm{P}} + \sum_{j=1}^{6} c_j \mathbf{q}'^{(j)} \tag{4.2.3}$$

where $\mathbf{q}^{\mathrm{P}}$ is a particular solution, the $c_j$ are six arbitrary constants, and the $\mathbf{q}'^{(j)}$ represent the surface forces on the external side of $D$ viewed as the boundary of a body that performs six independent modes of rigid body motion. In the case of external flow, a solution will exist only if the flow rate across $D$ is equal to zero. Fortunately, (4.1.2) ensures that this will always be true. Thus, the general solution to (4.1.7) will be

$$\mathbf{q} = \mathbf{q}^{\mathrm{P}} + c\mathbf{n} \tag{4.2.4}$$

where $\mathbf{q}^{\mathrm{P}}$ is a particular solution. Evidently, any particular solution may be augmented by the addition of a multiple of the normal vector.

**Problems**

4.2.1 Verify by direct substitution that the solution of (4.1.1) with a prescribed boundary condition $\mathbf{u} = \mathbf{U}$ over $D$, where $\mathbf{U}$ is a constant and $D$ is the surface of a sphere, is $\mathbf{q} = 3\mathbf{U}/(16\pi a)$.

4.2.2 Verify that (4.2.3) and (4.2.4) are general solutions of (4.1.8) and (4.1.7) respectively.

## 4.3 The double-layer potential

The double-layer potential represents the flow due to a surface distribution of the stress tensor $\mathbf{T}$ corresponding to a Green's function, namely the stresslet, i.e.

$$u_i(\mathbf{x}_0) = \int_D q_j(\mathbf{x}) T_{jik}(\mathbf{x}, \mathbf{x}_0) n_k(\mathbf{x}) \, \mathrm{d}S(\mathbf{x}) \tag{4.3.1}$$

where $D$ is the domain of the distribution and $\mathbf{q}$ is the density of the distribution. Note that because of (3.1.2), the velocity $\mathbf{u}$ defined in (4.3.1) vanishes over the solid boundary $S_\mathrm{B}$. As in the case of the single-layer potential, it is convenient to assume that $D$ is a Lyapunov surface and also that the density $\mathbf{q}$ varies over $D$ in a continuous manner. Several previous authors have introduced the kernel of the double-layer potential, $K_{ji}(\mathbf{x}, \mathbf{x}_0) = T_{jik}(\mathbf{x}, \mathbf{x}_0) n_k(\mathbf{x})$, writing (4.3.1) in the more compact form

$$u_i(\mathbf{x}_0) = \int_D q_j(\mathbf{x}) K_{ji}(\mathbf{x}, \mathbf{x}_0) \, \mathrm{d}S(\mathbf{x}) \tag{4.3.2}$$

We shall prefer to use an explicit notation in terms of $\mathbf{T}$.

Using the results of section 3.2, we find that when $\mathbf{T}$ corresponds to a Green's function for an infinite domain of flow, the flow (4.3.1) will satisfy the equations of Stokes flow for any arbitrary choice of the density function $\mathbf{q}$. In contrast, when $\mathbf{T}$ corresponds to a Green's function for confined flow, the flow expressed by (4.3.1) will satisfy the equations of Stokes flow only when $\mathbf{q}$ fulfills the integral constraint

$$\int_D q_k(\mathbf{x}) n_k(\mathbf{x}) \, \mathrm{d}S(\mathbf{x}) = 0 \tag{4.3.3}$$

Multiplying (4.3.1) by the normal vector and integrating the product over an arbitrary closed surface $S$, we find that the flow rate $Q$ through $S$ is equal to

$$Q \equiv \int_S u_i(\mathbf{x}_0) n_i(\mathbf{x}_0) \, \mathrm{d}S(\mathbf{x}_0) = \int_D q_j(\mathbf{x}) \varphi_{jk}(\mathbf{x}) n_k(\mathbf{x}) \, \mathrm{d}S(\mathbf{x}) \tag{4.3.4}$$

where

$$\varphi_{jk}(\mathbf{x}) = \int_S T_{jik}(\mathbf{x}, \mathbf{x}_0) n_i(\mathbf{x}_0)\, dS(\mathbf{x}_0)$$

Using (3.2.7) we find that when $S$ is outside but does not enclose $D$ or when $S$ is inside $D$, $\varphi_{jk}(\mathbf{x}) = 0$. In this case, equation (4.3.3) provides us with a statement of conservation of mass, i.e. $Q = 0$. On the other hand, when $S$ encloses $D$ and possibly $S_B$, $\varphi_{jk}(\mathbf{x}) = 8\pi\delta_{jk}$ and therefore

$$Q = 8\pi \int_D q_k(\mathbf{x}) n_k(\mathbf{x})\, dS(\mathbf{x}) \tag{4.3.5}$$

Thus, the double-layer potential is capable of representing flows that contain sources and sinks as well as flows produced by the expansion or contraction of flexible bodies. Using (4.3.3) we find that when $\mathbf{T}$ corresponds to a Green's function for confined flow, $Q = 0$; because the fluid cannot escape to infinity, the flow rate through the domain of distribution $D$ must be equal to zero.

The velocity field (4.3.1) is continuous throughout the domain of flow, but undergoes a discontinuity across the distribution domain $D$. The limiting values of the velocity on either side of $D$ are

$$u_i^+(\mathbf{x}_0) = 4\pi q_i(\mathbf{x}_0) + \int_D^{\mathscr{PV}} q_j(\mathbf{x}) T_{ijk}(\mathbf{x}, \mathbf{x}_0) n_k(\mathbf{x})\, dS(\mathbf{x}) \tag{4.3.6}$$

$$u_i^-(\mathbf{x}_0) = -4\pi q_i(\mathbf{x}_0) + \int_D^{\mathscr{PV}} q_j(\mathbf{x}) T_{jik}(\mathbf{x}, \mathbf{x}_0) n_k(\mathbf{x})\, dS(\mathbf{x}) \tag{4.3.7}$$

where the positive sign is for the external side of $D$, in the direction of the normal vector, and the negative sign is for the internal side (problem 4.3.1). The principal value integral on the right-hand sides of (4.3.6) and (4.3.7) is defined as the value of the improper integral corresponding to the case when the point $\mathbf{x}_0$ is located on $D$, and is equal to the mean of the values of the double-layer potential on either side of $D$ (problem 4.3.1). Substracting (4.3.7) from (4.3.6) we obtain the discontinuity in velocity across $D$:

$$\mathbf{u}^+ - \mathbf{u}^- = 8\pi\mathbf{q} \tag{4.3.8}$$

To understand the nature of the flow produced by the double-layer potential, it is helpful to decompose $\mathbf{T}$ into its constituents using (2.1.8). Exploiting the symmetry identity (3.1.3) we obtain

$$u_j(\mathbf{x}_0) = -\int_D p_j(\mathbf{x}, \mathbf{x}_0) q_i(\mathbf{x}) n_i(\mathbf{x})\, dS(\mathbf{x}) + \int_D \frac{\partial G_{ji}}{\partial x_k}(\mathbf{x}_0, \mathbf{x})[q_i n_k + q_k n_i](\mathbf{x})\, dS(\mathbf{x}) \tag{4.3.9}$$

When **T** corresponds to a Green's function for infinite flow, the first integral on the right-hand side of (4.3.9) represents a distribution of point sources (see section 3.2). The second integral represents a symmetric distribution of point force dipoles.

The pressure field associated with the flow (4.3.1) may be conveniently expressed in terms of the corresponding distribution

$$P(\mathbf{x}_0) = \mu \int_D q_i(\mathbf{x}) \Pi_{ik}(\mathbf{x}_0, \mathbf{x}) n_k(\mathbf{x}) \, \mathrm{d}S(\mathbf{x}) \tag{4.3.10}$$

where $\mathbf{\Pi}$ is the pressure tensor associated with the stresslet, as discussed in section 3.2.

The force and torque due to the flow (4.3.1) exerted on a closed surface that does not enclose the solid boundary $S_B$ are equal to zero. The force and torque exerted on a closed surface that does enclose $S_B$ may have finite values which depend on the detailed geometry of $S_B$.

Next, we consider the behaviour of the surface force on either side of $D$. First, we note that the limit of the surface force on either side of $D$ exists only if the curvature of $D$, as well as the density $\mathbf{q}$ and its first derivative over $D$, vary over $D$ in a continuous manner (Odqvist 1930; Kim & Karrila 1991, Part 4, Chapter 15). Assuming that these restrictions are fulfilled, we apply the boundary integral equation for the flow (4.3.1) in the exterior and interior of $D$, and consider the limit as the field point $\mathbf{x}_0$ approaches $D$ from either side. Neglecting the integrals at infinity we obtain

$$u_i^+(\mathbf{x}_0) = -\frac{1}{4\pi\mu} \int_D f_j^+(\mathbf{x}) G_{ji}(\mathbf{x}, \mathbf{x}_0) \, \mathrm{d}S(\mathbf{x}) + \frac{1}{4\pi} \int_D^{\mathscr{PV}} u_j^+(\mathbf{x}) T_{jik}(\mathbf{x}, \mathbf{x}_0) n_k(\mathbf{x}) \, \mathrm{d}S(\mathbf{x}) \tag{4.3.11}$$

and

$$u_i^-(\mathbf{x}_0) = \frac{1}{4\pi\mu} \int_D f_j^-(\mathbf{x}) G_{ji}(\mathbf{x}, \mathbf{x}_0) \, \mathrm{d}S(\mathbf{x}) - \frac{1}{4\pi} \int_D^{\mathscr{PV}} u_j^-(\mathbf{x}) T_{jik}(\mathbf{x}, \mathbf{x}_0) n_k(\mathbf{x}) \, \mathrm{d}S(\mathbf{x}) \tag{4.3.12}$$

where the normal vector is directed towards the exterior of $D$. Adding (4.3.12) to (4.3.11) we find

$$u_i^+(\mathbf{x}_0) + u_i^-(\mathbf{x}_0) = -\frac{1}{4\pi\mu} \int_D [f_j^+ - f_j^-](\mathbf{x}) G_{ji}(\mathbf{x}, \mathbf{x}_0) \, \mathrm{d}S(\mathbf{x}) + \frac{1}{4\pi} \int_D^{\mathscr{PV}} [u_j^+ - u_j^-](\mathbf{x}) T_{jik}(\mathbf{x}, \mathbf{x}_0) n_k(\mathbf{x}) \, \mathrm{d}S(\mathbf{x}) \tag{4.3.13}$$

Substituting (4.3.8) into the second integral on the right-hand side of (4.3.13) and using (4.3.6) and (4.3.7) we obtain

$$\int_D [f_j^+(\mathbf{x}) - f_j^-(\mathbf{x})] G_{ji}(\mathbf{x}, \mathbf{x}_0)\, dS(\mathbf{x}) = 0 \tag{4.3.14}$$

where we recall that the point $\mathbf{x}_0$ is located right on $D$. Equation (4.3.14) states that the discontinuity in surface force $\Delta\mathbf{f} = \mathbf{f}^+ - \mathbf{f}^-$ is a solution of the homogeneous equation (4.2.1), thereby implying that $\Delta\mathbf{f} = c\mathbf{n}$, where $c$ is a constant. Noting that the magnitude of $\Delta\mathbf{f}$ is independent of the local form of the density distribution $\mathbf{q}$ and setting $\mathbf{q} = 0$ we obtain $\Delta\mathbf{f} = 0$. Thus, we find that the surface force has the same value on either side of the distribution domain $D$. In retrospect, the continuity of the surface force across $D$ is dictated by the absence of point force singularities in the kernel of the double-layer potential.

The double-layer potential is a linear integral operator $\mathrm{L^D}$ with kernel $T_{jik}(\mathbf{x}, \mathbf{x}_0) n_k(\mathbf{x})$, operating on the space of density functions $\mathbf{q}$. If the domain $D$ has a continuous normal vector, i.e. it is a Lyapunov surface, the kernel of the double-layer operator is weakly singular and the double-layer operator is completely continuous or compact (see problem 4.3.4). The adjoint operator $\mathrm{L^{D\text{-}A}}$, defined by the equation

$$(\mathrm{L^D}\mathbf{q}_1, \mathbf{q}_2) = (\mathbf{q}_1, \mathrm{L^{D\text{-}A}}\mathbf{q}_2) \tag{4.3.15}$$

is given by

$$(\mathrm{L^{D\text{-}A}}\mathbf{q})_i = n_k(\mathbf{x}_0) \int_D q_j(\mathbf{x}) T_{ijk}(\mathbf{x}_0, \mathbf{x})\, dS(\mathbf{x}) \tag{4.3.16}$$

where the inner product was defined in (4.1.12). It will be noted that the adjoint operator is obtained by transposing the indices $i$ and $j$ of $\mathbf{T}$, and switching the arguments $\mathbf{x}$ and $\mathbf{x}_0$ of the kernel $T_{jik}(\mathbf{x}, \mathbf{x}_0) n_k(\mathbf{x})$.

### Problems

4.3.1 Writing

$$\begin{aligned} u_i(\mathbf{x}_0) &= \int_D q_j(\mathbf{x}) T_{jik}(\mathbf{x}, \mathbf{x}_0) n_k(\mathbf{x})\, dS(\mathbf{x}) \\ &= \int_D [q_j(\mathbf{x}) - q_j(\mathbf{x}_0)]\, T_{jik}(\mathbf{x}, \mathbf{x}_0) n_k(\mathbf{x})\, dS(\mathbf{x}) \\ &\quad + q_j(\mathbf{x}_0) \int_D T_{jik}(\mathbf{x}, \mathbf{x}_0) n_k(\mathbf{x})\, dS(\mathbf{x}) \end{aligned} \tag{1}$$

noting that the integral in the second line of (1) is continuous across $D$, and then using (2.1.12), prove equations (4.3.6) and (4.3.7). In addition, show that the principal value of the velocity over $D$ is given by

$$\mathbf{u}^{\mathscr{PV}} = \tfrac{1}{2}(\mathbf{u}^{+} + \mathbf{u}^{-}) = \mathbf{Y} - 4\pi\mathbf{q} \tag{2}$$

where $Y_i$ is the first integral on the right-hand side of (1).

4.3.2 Show that the stress field associated with a double-layer distribution of the free-space Green's function is given by

$$\sigma_{jm}(\mathbf{x}_0) = \mu \int_D q_i(\mathbf{x}) \left[ -4\frac{\delta_{jm}\delta_{ik}}{r^3} - 6\frac{\delta_{jk}\hat{x}_i\hat{x}_m + \delta_{mk}\hat{x}_i\hat{x}_j + \delta_{ij}\hat{x}_m\hat{x}_k + \delta_{im}\hat{x}_j\hat{x}_k}{r^5} + 60\frac{\hat{x}_j\hat{x}_i\hat{x}_m\hat{x}_k}{r^7} \right] n_k(\mathbf{x})\,\mathrm{d}S(\mathbf{x})$$

where $\hat{\mathbf{x}} = \mathbf{x} - \mathbf{x}_0$, $r = |\hat{\mathbf{x}}|$, and $\mathbf{q}$ is the density of the distribution.

4.3.3 Verify that (4.3.16) is indeed the adjoint of the double-layer operator.

4.3.4 Show that when $D$ is a Lyapunov surface, the kernel of the double-layer operator is weakly singular.

## 4.4 Representation of a flow in terms of a double-layer potential

In this section we investigate the extent to which the double-layer potential is capable of representing an arbitrary internal or external flow. We focus our attention, in particular, on the Dirichlet problem which arises by prescribing the velocity either on the external side or on the internal side of the domain of distribution $D$. Thus, we consider (4.3.6) or (4.3.7) and treat the boundary velocity as known, obtaining two Fredholm integral equations of the second kind for the density of the distribution $\mathbf{q}$. It will be convenient to unify these equations into the compact form

$$q_i(\mathbf{x}_0) = \frac{\alpha}{4\pi} \int_D^{\mathscr{PV}} q_j(\mathbf{x}) T_{jik}(\mathbf{x}, \mathbf{x}_0) n_k(\mathbf{x})\,\mathrm{d}S(\mathbf{x}) + F_i(\mathbf{x}_0) \tag{4.4.1}$$

where $\mathbf{F} = (-\alpha/4\pi)\mathbf{u}$ is the forcing function and $\alpha$ is a constant; $\alpha = 1$ and $-1$ correspond to internal and external flow, respectively. In Chapter 5 we shall see that intermediate values of $\alpha$ $(-1 < \alpha < 1)$ correspond to integral equations for two-phase flow.

To investigate the properties of (4.4.1) we consider the associated homogeneous equation obtained by setting $\mathbf{F}$ equal to zero, i.e.

$$q''_i(\mathbf{x}_0) = \frac{\beta^*}{4\pi} \int_D^{\mathscr{PV}} q''_j(\mathbf{x}) T_{jik}(\mathbf{x}, \mathbf{x}_0) n_k(\mathbf{x})\,\mathrm{d}S(\mathbf{x}) \tag{4.4.2}$$

as well as its adjoint

$$q'_i(\mathbf{x}_0) = \frac{\beta}{4\pi} n_k(\mathbf{x}_0) \int_D^{\mathscr{PV}} q'_j(\mathbf{x}) T_{ijk}(\mathbf{x}_0, \mathbf{x})\,\mathrm{d}S(\mathbf{x}) \tag{4.4.3}$$

In (4.4.2) and (4.4.3) $\beta$ and $\beta^*$ are complex eigenvalues and $\mathbf{q}''$, $\mathbf{q}'$ are complex eigensolutions; an asterisk indicates the complex conjugate. The general theory of compact operators ensures that (4.4.2) and (4.4.3) have complex conjugate eigenvalues and the same number of eigensolutions and generalized eigensolutions (Stakgold 1979, Chapter 5). According to the Fredholm–Riesz theory of compact operators, valid for weakly singular integral equations, the following theorems are true:

(1) If $\beta^* = \alpha$ is not an eigenvalue of (4.4.2), then (4.4.1) has a unique solution.
(2) If $\beta^* = \alpha$ is an eigenvalue of (4.4.2) then in order for (4.4.1) to have a solution, all eigensolutions of (4.4.3) corresponding to the eigenvalue $\beta$ must be orthogonal to the forcing function $\mathbf{F}$.
(3) Equation (4.4.1) may be solved using the method of successive substitutions as long as the spectral radius of (4.4.2) (defined as $\min\{|\beta^*|\}$) is larger than $|\alpha|$.

It should be noted that in the method of successive substitutions, we guess a solution for $\mathbf{q}$, compute the right-hand side of (4.4.1), and then replace the original $\mathbf{q}$ by the newly computed $\mathbf{q}$. Setting the initial guess equal to zero, we obtain a series of successive approximations that is the Neumann series of (4.4.1). If the Neumann series converges, then the method of successive substitutions converges for any initial choice of $\mathbf{q}$ (problem 4.1.1).

Keeping in mind the above theorems, we shall proceed in the next section to investigate the eigenvalues of the homogeneous equations (4.4.2) and (4.4.3).

**Problem**

4.4.1 Show that if the Neumann series of an integral equation converges, the method of successive substitutions converges for any initial choice of the unknown function $\mathbf{q}$. (Hint: consider the behaviour of the error during the iterations.)

## 4.5 The eigenvalues of the double-layer potential

Motivated by our discussion in the preceding section, we set out to investigate the eigenvalues of the homogeneous integral equations (4.4.2) and (4.4.3). As a preliminary, we multiply (4.4.2) by the normal vector, integrate over $D$, and use (3.2.7) to find

$$(1 - \beta^*) \int_D \mathbf{q}'' \cdot \mathbf{n} \, dS = 0 \tag{4.5.1}$$

Furthermore, we multiply (4.4.3) by $\mathbf{V} + \boldsymbol{\Omega} \times \mathbf{x}$, where $\mathbf{V}$ and $\boldsymbol{\Omega}$ are two arbitrary constant vectors, integrate over $D$, and use (2.1.12) and (2.1.13) to find

$$(1+\beta)\int_D \mathbf{q}'\,\mathrm{d}S = 0 \qquad (1+\beta)\int_D \mathbf{x} \times \mathbf{q}'\,\mathrm{d}S = 0 \tag{4.5.2}$$

Equations (4.5.1) and (4.5.2) impose integral constraints on the eigensolutions of (4.4.2) and (4.4.3).

First, we shall demonstrate that (4.4.3) does not have any real or complex eigenvalues $\beta$ with $|\beta| < 1$ (Pozrikidis 1990a). For this purpose, we introduce the single-layer potential

$$w_i(\mathbf{x}_0) = \frac{1}{4\pi\mu}\int_D G_{ij}(\mathbf{x}_0, \mathbf{x})q'_j(\mathbf{x})\,\mathrm{d}S(\mathbf{x}) \tag{4.5.3}$$

Considering (4.1.7) and (4.1.8), we find that the limiting values of the surface force on either side of $D$ are given by

$$f_i^+(\mathbf{x}_0) = \sigma_{ik}^+(\mathbf{x}_0)n_k(\mathbf{x}_0) = -q'_j(\mathbf{x}_0) + \frac{1}{4\pi}n_k(\mathbf{x}_0)\int_D^{\mathscr{PV}} T_{ijk}(\mathbf{x}_0, \mathbf{x})q'_j(\mathbf{x})\,\mathrm{d}S(\mathbf{x}) \tag{4.5.4}$$

and

$$f_i^-(\mathbf{x}_0) = \sigma_{ik}^-(\mathbf{x}_0)n_k(\mathbf{x}_0) = q'_j(\mathbf{x}_0) + \frac{1}{4\pi}n_k(\mathbf{x}_0)\int_D^{\mathscr{PV}} T_{ijk}(\mathbf{x}_0, \mathbf{x})q'_j(\mathbf{x})\,\mathrm{d}S(\mathbf{x}) \tag{4.5.5}$$

where the superscripts + and − indicate the external and internal sides of $D$, respectively. Recalling that $\mathbf{q}'$ is a solution to (4.4.3), we rewrite (4.5.4) and (4.4.5) in the compact forms

$$\mathbf{f}^+ = \left(\frac{1}{\beta} - 1\right)\mathbf{q}' \tag{4.5.6}$$

and

$$\mathbf{f}^- = \left(\frac{1}{\beta} + 1\right)\mathbf{q}' \tag{4.5.7}$$

Now, we apply the energy balance (1.5.2) individually for the external and internal flows. Noting that $\mathbf{w}$ vanishes at infinity as well as over the solid boundary $S_B$, we obtain

$$2\mu\int_{V^+} e_{ik}e_{ik}^*\,\mathrm{d}V = -\int_{D^+} w_i^* f_i^+\,\mathrm{d}S \tag{4.5.8}$$

and

$$2\mu\int_{V^-} e_{ik}e_{ik}^*\,\mathrm{d}V = \int_{D^-} w_i^* f_i^-\,\mathrm{d}S \tag{4.5.9}$$

where $V^+$ is exterior to $D$, and $V^-$ is the interior of $D$. Substituting (4.5.6) and (4.5.7) into (4.5.8) and (4.5.9) yields

$$\mathscr{D}^+ = \left(-\frac{1}{\beta}+1\right)\int_{D^+} w_i^* q_i' \, \mathrm{d}S \tag{4.5.10}$$

and

$$\mathscr{D}^- = \left(\frac{1}{\beta}+1\right)\int_{D^-} w_i^* q_i' \, \mathrm{d}S \tag{4.5.11}$$

where, for convenience, we have defined

$$\mathscr{D}^\pm \equiv 2\mu \int_{V^\pm} e_{ik} e_{ik}^* \, \mathrm{d}V \tag{4.5.12}$$

Because the integrand on the right-hand side of (4.5.12) is the product of two complex conjugate functions, $\mathscr{D}^+$ and $\mathscr{D}^-$ are real and non-negative.

Now, we note that $\mathbf{w}$ is continuous across $D$ and this suggests that the integrals on the right-hand sides of (4.5.10) and (4.5.11) have the same values. Dividing (4.5.10) by (4.5.11) and rearranging, we obtain $\beta = (1+\delta)/(1-\delta)$, where $\delta = \mathscr{D}^+/\mathscr{D}^-$ is a real positive constant, indicating that $\beta$ is a real number outside the closed interval $[-1, 1]$. Of course, these results are based on the assumption that neither of the integrals appearing in (4.5.10) and (4.5.11) is equal to zero and thus neither the external nor the internal flow $\mathbf{w}$ represents rigid body motion. External rigid body motion is prohibited by the requirement that $\mathbf{w}$ vanishes at infinity. The only remaining possibility is that $\mathbf{w}$ expresses internal rigid body motion, thereby implying that $\mathbf{f}^- = -P\mathbf{n}$ where $P$ is the constant internal pressure. Returning to (4.5.7) we find two possibilities. The first is that $\beta \neq -1$ and $\mathbf{q}' = [-P\beta/(\beta+1)]\mathbf{n}$, in which case using (4.4.3) and (3.2.7) we obtain $\beta = 1$ and $\mathbf{q}' = (-P/2)\mathbf{n}$. It will be noted that this eigensolution certainly satisfies the integral constraint (4.5.2). The second possibility is $\beta = -1$. Due to the association with $\alpha = \pm 1$ in (4.4.1) (corresponding to internal and external flow), these two special cases, $\beta = \pm 1$, deserve further consideration.

### *Internal flow*

Considering first the eigenvalue $\beta = 1$, we examine the associated eigensolution $\mathbf{q}''$. For this purpose we introduce an external flow $\mathbf{v}$ that vanishes at infinity and whose surface force over the external side of $D$ is $\mathbf{f}^+ = c\mathbf{n}$, where $c$ is a constant. Applying the boundary integral equation (2.3.13) for $\mathbf{v}$, using (2.1.4) to eliminate the single-layer integral and comparing the resulting expression with (4.4.2), we may identify the boundary values of $\mathbf{v}$ over $D$ with the eigenfunction $\mathbf{q}''$. Furthermore, invoking the energy balance (1.5.2) we obtain

$$2\mu \int_{V^+} e^*_{ik} e_{ik} \, dV = -c \int_{D^+} q''_i n_i \, dS \tag{4.5.13}$$

If the right-hand side of (4.5.13) were equal to zero, $\mathbf{v}$ would have to express rigid body motion, which contradicts our assumptions. We must conclude that

$$\int_D \mathbf{q}'' \cdot \mathbf{n} \, dS \neq 0 \tag{4.5.14}$$

Unfortunately, we are not able to compute $\mathbf{q}''$ analytically for an arbitrary domain $D$.

An important consequence of (4.5.14) is that the multiplicity of the eigenvalue $\beta = 1$ is exactly equal to one. If the opposite were true, we would be able to find a principal solution (or generalized eigensolution) $\Psi$ that satisfies the equation[1]

$$q''_i(\mathbf{x}_0) = \psi_i(\mathbf{x}_0) - \frac{1}{4\pi} \int_D^{\mathscr{PV}} \psi_j(\mathbf{x}) \, T_{jik}(\mathbf{x}, \mathbf{x}_0) n_k(\mathbf{x}) \, dS(\mathbf{x}) \tag{4.5.15}$$

According to Fredholm's alternative, (4.5.15) will have a solution only if the forcing function $\mathbf{q}''$ is orthogonal to the adjoint eigenfunction $\mathbf{q}' = (-P/2)\mathbf{n}$. This, however, is prohibited by (4.5.14).

Now, returning to (4.4.1), we recall that by setting $\alpha = 1$ and $\mathbf{F} = (-1/4\pi)\mathbf{u}$ we obtain the Dirichlet problem for internal flow. According to Fredholm's alternative, this problem will have a solution only if the forcing function $\mathbf{F} = (-\alpha/4\pi)\mathbf{u}$ is orthogonal to the eigenfunction $\mathbf{q}'$, i.e. if

$$\int_D \mathbf{u} \cdot \mathbf{n} \, dS = 0 \tag{4.5.16}$$

Clearly, (4.5.16) is a natural constraint imposed by conservation of mass.

### *External flow*

Next, we turn our attention to the eigenvalue $\beta = -1$. We recall that in the interior of $D$ the flow $\mathbf{w}$ represents rigid body motion and note that $\mathbf{w}$ is continuous across $D$; this suggests that exterior to $D$ $\mathbf{w}$ represents the flow due to rigid body motion. Substituting $\beta = -1$ into (4.5.6) we find

$$\mathbf{q}' = -\tfrac{1}{2}\mathbf{f}^+ \tag{4.5.17}$$

When the distribution surface $D$ intersects or encloses $S_B$ (Figure 4.5.1), internal rigid body motion is prohibited by the requirement that $\mathbf{w}$

[1] A principal solution of the homogeneous equation $(\beta\mathbf{K} - \mathbf{I}) \cdot \mathbf{v} = 0$ is defined as a solution of the equation $(\beta\mathbf{K} - \mathbf{I})^m \cdot \psi = \mathbf{v}$, where $\mathbf{K}$ is a linear operator, $\mathbf{v}$ is an eigensolution, and $m$ is an integer (Bodewig 1959, p. 91).

must vanish over $S_B$. This implies that $\mathbf{w}$ must vanish inside and outside $D$, and hence that $\mathbf{f}^+ = -P\mathbf{n}$, where $P$ is the constant external pressure. Recalling that the pressure must vanish at infinity we deduce that $P = 0$ and conclude that $\beta = -1$ is not an eigenvalue. On the contrary, when $D$ does not intersect or enclose $S_B$, internal rigid body motion is perfectly acceptable. In this case (4.4.3) has six independent eigensolutions representing the surface force on $D$ when $D$ is viewed as the surface of a body that performs six independent modes of rigid body motion, namely

$$\mathbf{q}'^{(j)} = -\tfrac{1}{2}\mathbf{f}^{+(j)} \tag{4.5.18}$$

for $j = 1,\ldots,6$. One might argue that these surface forces may be enhanced by the addition of an arbitrary multiple of the normal vector (reflecting the level of the ambient pressure), but this ambiguity is dismissed by noting that $\mathbf{q}' = c\mathbf{n}$ is not an eigensolution of (4.4.3). Furthermore, using (2.1.12) and (2.1.13) we find that $\mathbf{q}'' = \mathbf{V} + \mathbf{\Omega} \times \mathbf{x}$ (where $\mathbf{V}$ and $\mathbf{\Omega}$ are constant vectors) satisfies the homogeneous equation (4.4.2) with $\beta = -1$ and thus provides us with six independent eigensolutions $\mathbf{q}''^{(j)}$, $j = 1,\ldots,6$ corresponding to (4.5.18) and representing six independent modes of rigid body motion. It should be noted that each of the $\mathbf{q}''^{(j)}$ satisfies the integral constraint (4.5.1).

Returning to the integral equation (4.4.1), we recall that by setting $\alpha = -1$ and $\mathbf{F} = (1/4\pi)\mathbf{u}$ we obtain the Dirichlet problem for external flow. Combining the above results with Fredholm's alternative, we conclude that when $D$ intersects or encloses $S_B$ this problem will have a unique solution. On the other hand, when $D$ does not intersect or enclose $S_B$, in order for the problem to have a solution the forcing function $\mathbf{F} = (1/4\pi)\mathbf{u}$

*Figure 4.5.1.* Three possible locations of the distribution domain $D$ with respect to the solid boundary $S_B$.

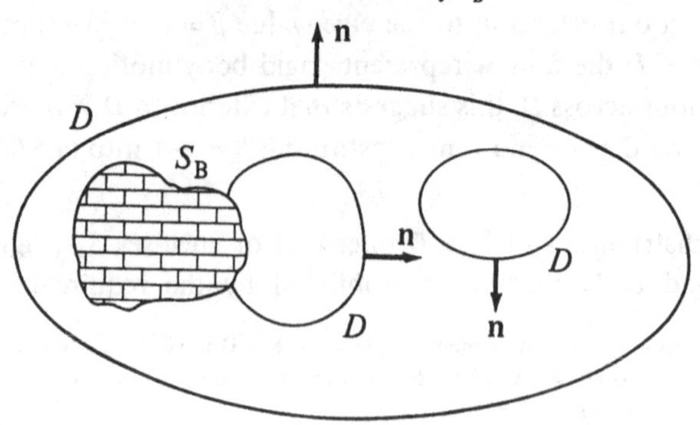

must be orthogonal to all six adjoint eigensolutions $\mathbf{q}'^{(j)}$, i.e.

$$\int_D \mathbf{u}\cdot\mathbf{q}'^{(j)}\,\mathrm{d}S = 0 \qquad (4.5.19)$$

for $j = 1, \ldots, 6$. The constraints expressed by (4.5.19) are not met for any choice of the boundary velocity $\mathbf{u}$. Thus, the double-layer potential is not capable of representing an arbitrary external flow.

One important corollary of (4.5.19) is that the multiplicity of the eigenvalue $\beta = -1$ is exactly equal to six. If this were not true, then we would be able to find a principal solution $\boldsymbol{\psi}$ that satisfies the equation

$$q_i''^{(m)}(\mathbf{x}_0) = \psi_i(\mathbf{x}_0) + \frac{1}{4\pi}\int_D^{\mathscr{PV}} \psi_j(\mathbf{x})\,T_{jik}(\mathbf{x},\mathbf{x}_0)\,n_k(\mathbf{x})\,\mathrm{d}S(\mathbf{x}) \qquad (4.5.20)$$

According to (4.5.19), equation (4.5.20) will have a solution only if the forcing function $\mathbf{q}''^{(m)}$ is orthogonal to each of the adjoint eigenfunctions $\mathbf{q}'^{(j)}$. Since the $\mathbf{q}''^{(m)}$ represent six modes of rigid body motion, this would require that the force and torque on $D$ associated with each of the surface forces $\mathbf{q}'^{(j)}$ be equal to zero. The energy dissipation theorem would then indicate that all $\mathbf{q}'^{(j)}$ are equal to zero, which contradicts our assumptions.

Physically, our inability to represent an arbitrary external flow in terms of a double-layer potential is attributable to the intrinsic constraint that the force and torque exerted on $D$ are equal to zero. In practice, this constraint is fulfilled only when $D$ represents the surface of a force-free and torque-free body such as a neutrally buoyant particle or a freely swimming microorganism. To investigate these special but common cases, let us identify $D$ with the surface of a force-free and torque-free rigid body that translates with velocity $\mathbf{U}$ and rotates with angular velocity $\boldsymbol{\omega}$ under the action of an incident flow $\mathbf{u}^\infty$. Representing the disturbance flow $\mathbf{u}^{\mathrm{D}}$ due to the body in terms of a double-layer potential, enforcing the constraint (4.5.19), and stipulating that on the surface of the body $\mathbf{u}^{\mathrm{D}} = -\mathbf{u}^\infty + \mathbf{U} + \boldsymbol{\omega}\times\mathbf{x}$, we obtain

$$\int_D \mathbf{u}^{\mathrm{D}}\cdot\mathbf{q}'^{(j)}\,\mathrm{d}S = \int_D (-\mathbf{u}^\infty + \mathbf{U} + \boldsymbol{\omega}\times\mathbf{x})\cdot\mathbf{q}'^{(j)}\,\mathrm{d}S = 0 \qquad (4.5.21)$$

After rearranging we find

$$\mathbf{U}\cdot\int_D \mathbf{q}'^{(j)}\,\mathrm{d}S + \boldsymbol{\omega}\cdot\int_D \mathbf{x}\times\mathbf{q}'^{(j)}\,\mathrm{d}S = \int_D \mathbf{u}^\infty\cdot\mathbf{q}'^{(j)}\,\mathrm{d}S \qquad (4.5.22)$$

Choosing the translational and rotational velocities $\mathbf{U}$ and $\boldsymbol{\omega}$ to be the solutions of (4.5.22) guarantees the existence of a six-parameter family of solutions of the integral equation (4.4.1) with $\alpha = -1$ and $\mathbf{F} = (1/4\pi)(-\mathbf{u}^\infty + \mathbf{U} + \boldsymbol{\omega}\times\mathbf{x})$.

To interpret (4.5.22), it is helpful to recall that $\mathbf{q}'^{(j)}$ represents the surface force over $D$ viewed as the boundary of a body that executes rigid body translation or rotation. Applying the reciprocal identity (1.4.5) for the disturbance flow $\mathbf{u}^D$ and the flow produced by the rigid body motion we obtain (1.4.16) and (1.4.17). Setting $\mathbf{F} = \mathbf{L} = 0$ we retrieve the two functional components of (4.5.22) associated with translation and rotation.

***Summary***

In this section we found the following.

(1) The homogeneous integral equation (4.4.2), as well as its adjoint (4.4.3), have only real eigenvalues $\beta$ with $|\beta| \geqslant 1$. Because the double-layer integral operator is compact, an infinite number of eigenvalues cannot accumulate at any point along the real axis.
(2) Representing an internal flow in terms of a double-layer potential leads to an integral equation with a one-parameter family of solutions.
(3) When the boundary $D$ encloses or crosses the solid surface $S_B$, representing an external flow in terms of a double-layer potential leads to an integral equation with a unique solution.
(4) When the boundary $D$ does not enclose or cross the solid boundary $S_B$, representing an external flow with a double-layer potential leads to an integral equation with either no solutions or a six-parameter family of solutions.
(5) The spectral radius of the double-layer operator, defined as the magnitude of the eigenvalue with the minimum norm, is equal to unity. Thus, (4.4.1) may be solved by the method of successive substitutions only when $|\alpha| < 1$.

**Problems**

4.5.1 Verify that when $D$ is the surface of a sphere $\mathbf{q}'' = \mathbf{n}$, where $\mathbf{n}$ is the unit normal vector, is the eigenfunction of (4.4.2) corresponding to the eigenvalue $\beta = 1$.

4.5.2 Verify that when $D$ is the surface of a sphere, $\mathbf{q}' = \mathbf{c}$, where $\mathbf{c}$ is a constant, is an eigensolution of (4.4.3) corresponding to the eigenvalue $\beta = -1$.

## 4.6 Regularizing the double-layer potential: removing the marginal eigenvalues

In section 4.5 we investigated the integral equations that arise by representing a flow in terms of a double-layer potential and prescribing the velocity on the external or internal side of the domain of distribution $D$. We found that in general the corresponding generalized homogeneous equations will have two marginal eigenvalues, namely $\beta = \pm 1$, implying that the integral equations will have either no solutions or an infinity of solutions. This behaviour places severe limitations on the ability of the double-layer potential for representing an internal or external flow.

The difficulties of the double-layer representation, however, are not without a cure. In this section we set out to devise an integral equation that has the same solution as (4.4.1) (which, we recall, arises from the double-layer representation), but whose corresponding generalized homogeneous equation does not possess the marginal eigenvalues $\beta = \pm 1$. Effectively, we seek to redefine the kernel of the integral equation so that the solution remains unchanged but the marginal eigenvalues are removed. It will prove instructive to carry out our procedures in two steps, removing first one and then both of the marginal eigenvalues. Many of the ideas put forward in this as well as in subsequent sections were motivated by the recent work of Kim & Karrila (1991).

### *Removing the $\beta = 1$ eigenvalue*

To set up the grounds for our analysis, it will be helpful to establish a connection with linear algebra (see for instance, Bodewig 1959). Thus, we consider a real square matrix $\mathbf{A}$ along with its transpose $\mathbf{B} = \mathbf{A}^T$. The matrices $\mathbf{A}$ and $\mathbf{B}$ are adjoint to each other, for $(\mathbf{y}, \mathbf{A}\cdot\mathbf{x}) = (\mathbf{B}\cdot\mathbf{y}, \mathbf{x})$ where $\mathbf{y}$ and $\mathbf{x}$ are two arbitrary vectors, and the parenthesis denotes the regular inner vector product. We know that $\mathbf{A}$ and $\mathbf{B}$ have complex conjugate eigenvalues and the same number of eigenvectors and principal vectors. Each of the eigenvectors of $\mathbf{A}$ corresponding to a particular eigenvalue $\lambda_1$ is orthogonal to each of the eigenvectors and principal vectors of $\mathbf{B}$ corresponding to a different eigenvalue $\lambda_2^*$, where an asterisk indicates the complex conjugate (see problem 4.6.1).

Now, let us assume that $\lambda_1$ is a real eigenvalue of $\mathbf{A}$ with multiplicity one, corresponding to the real eigenvector $\mathbf{u}$, i.e. $\mathbf{A}.\mathbf{u} = \mathbf{u}\cdot\mathbf{B} = \lambda_1\mathbf{u}$. Wielandt's theorem states that the modified matrix

$$\mathscr{A}_{ij} \equiv A_{ij} - \lambda_1 u_i w_j \tag{4.6.1}$$

has the same eigenvalues as $\mathbf{A}$, with the exception of $\lambda_1$ which is replaced by zero; here $\mathbf{w}$ is an arbitrary vector that satisfies $\mathbf{w}\cdot\mathbf{u} = 1$ (Bodewig 1959, p. 361; Wilkinson 1965, p. 596; see also problem 4.6.2).

Furthermore, let us assume that $\lambda_2$ is a real eigenvalue of $\mathbf{B}$ with multiplicity one, corresponding to the real eigenvector $\mathbf{v}$, i.e.

$\mathbf{B}\cdot\mathbf{v} = \mathbf{v}\cdot\mathbf{A} = \lambda_2\mathbf{v}$. Wielandt's theorem implies that the modified matrix

$$\mathscr{B}_{ij} \equiv B_{ij} - \lambda_2 v_i z_j \tag{4.6.2}$$

has the same eigenvalues as $\mathbf{B}$, except for $\lambda_2$ which is replaced by zero; here $\mathbf{z}$ is an arbitrary vector that satisfies $\mathbf{z}\cdot\mathbf{v} = 1$. The adjoint (i.e. the transpose) of $\mathscr{B}$, namely

$$\mathscr{A}'_{ij} \equiv A_{ij} - \lambda_2 z_i v_j \tag{4.6.3}$$

has the same eigenvalues as the adjoint to $\mathbf{B}$, which is simply $\mathbf{A}$, with the exception of $\lambda_2$, which is replaced by zero. Thus, $\mathscr{A}'$ has the same eigenvalues as $\mathbf{A}$ except for $\lambda_2$, which is replaced by zero.

To illustrate the relevance of the above results to the solution of the integral equation (4.4.1), we divide the domain $D$ of (4.4.1) into $N$ boundary elements, and assume that $\mathbf{q}$ and $\mathbf{F}$ are constant over an element (see section 6.1). In this manner, we restate (4.4.1) in the discrete form

$$\mathbf{x} = \alpha\mathbf{A}\cdot\mathbf{x} + \mathbf{b} \tag{4.6.4}$$

where the vectors $\mathbf{x}$ and $\mathbf{b}$ contain the constant values of $\mathbf{q}$ and $\mathbf{F}$ over all boundary elements. It will be important to note that the corresponding homogeneous equation (4.4.2) assumes the discrete form $\mathbf{x} = \beta\mathbf{A}\cdot\mathbf{x}$ or, equivalently, $\mathbf{A}\cdot\mathbf{x} = \lambda\mathbf{x}$ where $\lambda = 1/\beta$. Clearly, equation (4.6.4) will have a unique solution only if $1/\alpha$ is not an eigenvalue of $\mathbf{A}$.

Now, instead of solving (4.6.4), we propose to solve the alternative equation

$$\mathbf{x} = \alpha\mathscr{A}'\cdot\mathbf{x} + \mathbf{b}' \tag{4.6.5}$$

where $\mathscr{A}'$ was defined in (4.6.3), and $\mathbf{b}'$ is a constant vector. To ensure that the solution to (4.6.5) also satisfies (4.6.4), we require

$$\mathbf{b}' = \mathbf{b} + \alpha\lambda_2\mathbf{z}(\mathbf{v}\cdot\mathbf{x}) \tag{4.6.6}$$

Multiplying (4.6.4) from the left side by $\mathbf{v}$, we find

$$\mathbf{v}\cdot\mathbf{x} = \alpha(\mathbf{v}\cdot\mathbf{A})\cdot\mathbf{x} + \mathbf{v}\cdot\mathbf{b} \tag{4.6.7}$$

Furthermore, recalling that $\mathbf{v}\cdot\mathbf{A} = \lambda_2\mathbf{v}$ we obtain

$$\mathbf{v}\cdot\mathbf{x} = \frac{1}{1-\alpha\lambda_2}\mathbf{v}\cdot\mathbf{b} \tag{4.6.8}$$

Substituting (4.6.8) into (4.6.6) we find

$$\mathbf{b}' = \mathbf{b} + \frac{\alpha\lambda_2}{1-\alpha\lambda_2}\mathbf{z}(\mathbf{v}\cdot\mathbf{b}) \tag{4.6.9}$$

It will be noted that when $\mathbf{v}\cdot\mathbf{b} = 0$, the last term on the right-hand side of (4.6.9) vanishes and thus $\mathbf{b}' = \mathbf{b}$. Finally, substituting (4.6.9) into (4.6.5) we obtain

$$\mathbf{x} = \alpha\mathscr{A}'\cdot\mathbf{x} + \mathbf{b} + \frac{\alpha\lambda_2}{1-\alpha\lambda_2}\mathbf{z}(\mathbf{v}\cdot\mathbf{b}) \tag{4.6.10}$$

At this point, it is natural to ask what the advantages are of solving (4.6.10) instead of (4.6.4). To respond, we consider the homogeneous equation corresponding to (4.6.10) (without the last two terms on the right-hand side), written in the alternative form $\mathscr{A}' \cdot \mathbf{x} = (1/\alpha)\mathbf{x}$. Equation (4.6.10) will have a unique solution as long as $1/\alpha$ is not an eigenvalue of $\mathscr{A}'$. Recalling that $\lambda_2$ is an eigenvalue of $\mathbf{A}$ but not of $\mathscr{A}$, and comparing (4.6.4) with (4.6.10), we conclude that when $\alpha = 1/\lambda_2$, (4.6.10) will have a unique solution, whereas (4.6.4) will have a multiplicity of solutions. Of course, when $\alpha = 1/\lambda_2$ we must require that $\mathbf{v} \cdot \mathbf{b} = 0$, otherwise the last term on the right-hand side of (4.6.10) becomes infinite.

Now, recalling that the double-layer potential constitutes a compact and, thus, bounded operator, we draw an analogy between the finite-dimensional equations (4.6.4) and (4.6.10) and the infinite-dimensional integral equation (4.4.1). Thus, we set $\lambda_2 = 1/\beta_2 = 1$ and $\mathbf{v} = \mathbf{n}$, and replace (4.4.1) with the modified equation

$$q_i(\mathbf{x}_0) = \frac{\alpha}{4\pi} \int_D^{\mathscr{PV}} q_j(\mathbf{x}) T_{jik}(\mathbf{x}, \mathbf{x}_0) n_k(\mathbf{x}) \, \mathrm{d}S(\mathbf{x}) - \alpha z_i(\mathbf{x}_0) \int_D q_j(\mathbf{x}) n_j(\mathbf{x}) \, \mathrm{d}S(\mathbf{x})$$

$$+ F_i(\mathbf{x}_0) + \frac{\alpha}{1-\alpha} z_i(\mathbf{x}_0) \int_D F_j(\mathbf{x}) n_j(\mathbf{x}) \, \mathrm{d}S(\mathbf{x}) \qquad (4.6.11)$$

where $\mathbf{z}$ is an arbitrary function that satisfies the integral constraint

$$\int_D z_i(\mathbf{x}) n_i(\mathbf{x}) \, \mathrm{d}S(\mathbf{x}) = 1 \qquad (4.6.12)$$

One convenient choice for $\mathbf{z}$ is $\mathbf{n}/S_D$ where $S_D$ is the surface area of $D$ (Varnhorn 1989). To show that a solution to (4.6.11) is also a solution to (4.4.1), we multiply (4.6.11) by the normal vector, integrate over $D$, and use (3.2.7) to find

$$\int_D q_i(\mathbf{x}) n_i(\mathbf{x}) \, \mathrm{d}S(\mathbf{x}) = \frac{1}{1-\alpha} \int_D F_i(\mathbf{x}) n_i(\mathbf{x}) \, \mathrm{d}S(\mathbf{x}) \qquad (4.6.13)$$

Substituting (4.6.13) into the right-hand side of (4.6.11) produces (4.4.1).

To investigate the properties of (4.6.11), we consider the eigenvalues of the associated generalized homogeneous equation

$$q_i''(\mathbf{x}_0) = \frac{\beta}{4\pi} \left[ \int_D^{\mathscr{PV}} q_j''(\mathbf{x}) T_{jik}(\mathbf{x}, \mathbf{x}_0) n_k(\mathbf{x}) \, \mathrm{d}S(\mathbf{x}) - 4\pi z_i(\mathbf{x}_0) \int_D q_j''(\mathbf{x}) n_j(\mathbf{x}) \, \mathrm{d}S(\mathbf{x}) \right] \qquad (4.6.14)$$

Multiplying both sides of (4.6.14) by the normal vector, integrating over $D$, and using (3.2.7), we obtain

$$\int_D q_i''(\mathbf{x}) n_i(\mathbf{x}) \, \mathrm{d}S(\mathbf{x}) = 0 \qquad (4.6.15)$$

indicating that the last term on the right-hand side of (4.6.14) vanishes and that (4.6.14) reduces to (4.4.2). Thus, all eigenvalues and eigensolutions of (4.6.14) are also eigenvalues and eigensolutions of (4.4.2) but not vice versa. In section 4.5 we found that $\beta = 1$ is an eigenvalue of (4.4.2) with multiplicity one. Equation (4.5.14) states that the corresponding eigensolution does not satisfy (4.6.15), thereby implying that $\beta = 1$ is not an eigenvalue of (4.6.14). Furthermore, in the same section we found that when $D$ does not enclose or intersect $S_B$, $\beta = -1$ is an eigenvalue of (4.4.2) with multiplicity six whose corresponding eigensolutions express six independent modes of rigid body motion, i.e. $\mathbf{q}'' = \mathbf{V} + \mathbf{\Omega} \times \mathbf{x}$, where $\mathbf{V}$ and $\mathbf{\Omega}$ are arbitrary constant vectors. These eigensolutions certainly satisfy the integral constraint (4.6.15) and thus are eigensolutions of (4.6.14) as well. Combining these results with Fredholm's theorems (see section 4.4), we conclude that (4.6.11) will have a unique solution for any value of $\alpha$ in the range $(-1, 1]$.

In section 4.4 we saw that equation (4.4.1) with $\alpha = 1$ and $\mathbf{F} = (-1/4\pi)\mathbf{u}$ corresponds to the problem of internal flow. Our results in this section indicate that the corresponding equation (4.6.11) will have a unique solution provided that the flow rate of the prescribed velocity over $D$ is equal to zero, so that the last term on the right-hand side of (4.6.11) may be ignored.

### *Removing both eigenvalues*

We proceed now to remove both marginal eigenvalues $\beta = \pm 1$ from the homogeneous equation (4.4.2). To motivate our procedures, it is again helpful to draw an analogy with linear algebra. Thus, let us assume that $\lambda_1$ is a real eigenvalue of the matrix $\mathbf{A}$ with multiplicity $K$, corresponding to the real eigenvectors $\mathbf{u}^{(k)}$, $k = 1, 2, \dots, K$ and that $\lambda_2$ is another eigenvalue with multiplicity one, corresponding to the real adjoint eigenvector $\mathbf{v}$. We define the modified matrix

$$\mathscr{A}''_{ij} \equiv A_{ij} - \lambda_1 \sum_{k=1}^{K} u_i^{(k)} w_j^{(k)} - \lambda_2 z_i v_j \tag{4.6.16}$$

where the set of vectors $\mathbf{w}^{(k)}$, $k = 1, 2, \dots, K$, is orthonormal to the set of eigenvectors $\mathbf{u}^{(k)}$, i.e. $\mathbf{u}^{(l)} \cdot \mathbf{w}^{(m)} = \delta_{lm}$, and the vector $\mathbf{z}$ satisfies the constraint $\mathbf{z} \cdot \mathbf{v} = 1$. Using Wielandt's theorem, we find that the matrix $\mathscr{A}''$ has the same eigenvalues as $\mathbf{A}$ with the exception of $\lambda_1$ and $\lambda_2$, which are replaced by zero (see problem 4.6.3).

Now, we propose to replace (4.6.4) with the alternative equation

$$\mathbf{x} = \alpha \mathscr{A}'' \cdot \mathbf{x} + \mathbf{b}'' \tag{4.6.17}$$

To ensure that a solution of (4.6.17) is also a solution of (4.6.4) we require

$$\mathbf{b}'' = \mathbf{b}' + \alpha\lambda_1 \sum_{k=1}^{K} \mathbf{u}^{(k)}(\mathbf{w}^{(k)}\cdot\mathbf{x}) \tag{4.6.18}$$

where $\mathbf{b}'$ was defined in (4.6.6) and given explicitly in (4.6.9). Noting that $\mathscr{A}''\cdot\mathbf{u}^{(k)} = 0$, and substituting $\mathbf{u}^{(k)} = (\mathbf{I} - \alpha\mathscr{A})\cdot\mathbf{u}^{(k)}$ into the right-hand side of (4.6.18) and subsequently into (4.6.17), we obtain

$$\mathscr{x} = \alpha\mathscr{A}''\cdot\mathscr{x} + \mathbf{b}' \tag{4.6.19}$$

where

$$\mathscr{x} = \mathbf{x} - \alpha\lambda_1 \sum_{k=1}^{K} \mathbf{u}^{(k)}(\mathbf{w}^{(k)}\cdot\mathbf{x}) \tag{4.6.20}$$

Multiplying (4.6.20) by $\mathbf{w}^{(m)}$ and exploiting the orthogonality condition $\mathbf{u}^{(k)}\cdot\mathbf{w}^{(m)} = \delta_{km}$ we obtain $\mathbf{w}^{(k)}\cdot\mathbf{x} = [1/(1-\alpha\lambda_1)]\mathbf{w}^{(k)}\cdot\mathscr{x}$. Substituting this expression into the sum on the right-hand side of (4.6.20) we find

$$\mathbf{x} = \mathscr{x} + \frac{\alpha\lambda_1}{1-\alpha\lambda_1} \sum_{k=1}^{K} \mathbf{u}^{(k)}(\mathbf{w}^{(k)}\cdot\mathscr{x}) \tag{4.6.21}$$

Finally, we devise the following procedure based on the last three equations. Given $\mathbf{b}$, we use (4.6.9) to compute $\mathbf{b}'$, solve (4.6.19) for $\mathscr{x}$, and then, use (4.6.21) to obtain the desired solution $\mathbf{x}$. To demonstrate the benefits of this seemingly cumbersome procedure, we compare (4.6.19) with the original equation (4.6.4). Considering, in particular, the corresponding homogeneous equations and recalling that $\mathscr{A}''$ has the same eigenvalues as $\mathbf{A}$ with the exception of $\lambda_1$ and $\lambda_2$ each of which is replaced by zero, we conclude:

(1) If one or both of $\lambda_1$ and $\lambda_2$ is an eigenvalue of $\mathbf{A}$ with maximum magnitude, the spectral radius of $\mathbf{A}$, equal to $d = \max\{|\lambda_1|, |\lambda_2|\}$, will be larger than the spectral radius of $\mathscr{A}''$, denoted by $\mathscr{d}$, i.e. $\mathscr{d} < d$. To this end, we recall that (4.6.19) or (4.6.4) may be solved by the method of successive substitutions as long as $|\alpha| < 1/\mathscr{d}$ or $|\alpha| < 1/d$ respectively. This implies that (4.6.19) may be solved with fewer iterations than (4.6.4), and for a wider range of $|\alpha|$.

(2) Unlike (4.6.4), equation (4.6.19) will have a unique solution for both $\alpha = 1/\lambda_1$ and $\alpha = 1/\lambda_2$. When $\alpha = 1/\lambda_2$ we must require $\mathbf{v}\cdot\mathbf{b} = 0$, otherwise $\mathbf{b}'$ on the right-hand side of (4.6.19) will be unbounded. Unfortunately, when $\alpha = 1/\lambda_1$, the right-hand side of (4.6.21) is infinite, yielding pathological behavior.

We turn now to apply the above results to the integral equation (4.4.1). For convenience, we set $\mathbf{z} = \mathbf{v}/|\mathbf{v}|^2$, and identify $\{\mathbf{w}^{(k)}\}$ with $\{\mathbf{u}^{(k)}\}$, thereby requiring that the set of eigenvectors $\{\mathbf{u}^{(k)}\}$ is orthonormal, i.e. $\mathbf{u}^{(l)}\cdot\mathbf{u}^{(m)} = \delta_{lm}$.

In section 4.5 we saw that $\beta_1 = -1$ and $\beta_2 = 1$ are two eigenvalues of (4.4.2) with multiplicity six and one, respectively. The corresponding eigenvalues of the discretized equation are $\lambda_1 = 1/\beta_1 = -1$ and $\lambda_2 = 1/\beta_2 = 1$. The adjoint eigensolution corresponding to $\lambda_2 = 1$ is $\mathbf{q}' = \mathbf{n}$. Identifying $\mathbf{q}'$ with $\mathbf{v}$ and requiring that $\mathbf{z} = \mathbf{v}/|\mathbf{v}|^2$, we find $\mathbf{z} = (1/S_D)\mathbf{n}$, where $S_D$ is the surface area of $D$. The six eigensolutions $\mathbf{q}''^{(k)}$, $k = 1, \ldots, 6$ corresponding to $\lambda_1 = -1$ represent six independent modes of rigid body motion. It will be convenient to decouple translation from rotation, setting

$$\mathbf{q}''^{(n)} = \frac{1}{(S_D)^{1/2}} \mathbf{v}^{(n)} \tag{4.6.22}$$

for $n = 1, 2, 3$, and

$$\mathbf{q}''^{(n)} = \frac{1}{(A_m)^{1/2}} \boldsymbol{\omega}^{(m)} \times \mathbf{X} \tag{4.6.23}$$

for $n = 4, 5, 6$ and $m = n - 3 = 1, 2, 3$. Here $\mathbf{X} = \mathbf{x} - \mathbf{x}_c$, $\mathbf{x}_c$ is the centroid of $D$, defined as

$$\mathbf{x}_c = \frac{1}{S_D} \int_D \mathbf{x}\, \mathrm{d}S \tag{4.6.24}$$

In equation (4.6.22) the $\mathbf{v}^{(n)}$ are three arbitrary but mutually orthogonal unit vectors. In (4.6.23) the $\boldsymbol{\omega}^{(m)}$ are three independent unit vectors and

$$A_m = \int_D [\boldsymbol{\omega}^{(m)} \times \mathbf{X}] \cdot [\boldsymbol{\omega}^{(m)} \times \mathbf{X}]\, \mathrm{d}S \tag{4.6.25}$$

It will be noted that with the above definitions, each $\mathbf{q}''^{(k)}$, $k = 1, \ldots, 6$ is normalized, and furthermore, each $\mathbf{q}''^{(n)}$, $n = 1, 2, 3$ is orthogonal to each vector in the set $\{\mathbf{q}''^{(k)}\}$. To render $\{\mathbf{q}''^{(k)}\}$ orthonormal, we require that the unit normal vectors $\boldsymbol{\omega}^{(m)}$, $m = 1, 2, 3$ satisfy the orthogonality conditions

$$\int_D \mathbf{q}''^{(m)} \cdot \mathbf{q}''^{(n)}\, \mathrm{d}S = \delta_{mn} \tag{4.6.26}$$

for $n, m = 4, 5, 6$. A convenient way of computing $\boldsymbol{\omega}^{(m)}$, $m = 1, 2, 3$, so that (4.6.26) is true is by using the Gram–Schmidt orthogonalization process (Stakgold 1979, Vol. I, p. 123).

Having introduced the necessary set of orthonormal eigensolutions, we identify $\{\mathbf{w}^{(k)}\}$ and $\{\mathbf{u}^{(k)}\}$ with $\{\mathbf{q}''^{(k)}\}$, and write the integral equation counterparts of (4.6.19), (4.6.20), and (4.6.21) in the forms

$$\begin{aligned}
\varphi_i(\mathbf{x}_0) = {} & \frac{\alpha}{4\pi} \mathcal{PV}\!\!\int_D \varphi_j(\mathbf{x}) T_{jik}(\mathbf{x}, \mathbf{x}_0) n_k(\mathbf{x})\, \mathrm{d}S(\mathbf{x}) \\
& + \alpha \sum_{k=1}^{6} q_i''^{(k)}(\mathbf{x}_0) \int_D q_j''^{(k)}(\mathbf{x}) \varphi_j(\mathbf{x})\, \mathrm{d}S(\mathbf{x}) - \frac{\alpha}{S_D} n_i(\mathbf{x}_0) \int_D \varphi_j(\mathbf{x}) n_j(\mathbf{x})\, \mathrm{d}S(\mathbf{x}) \\
& + F_i(\mathbf{x}_0) + \frac{\alpha}{1-\alpha} \frac{1}{S_D} n_i(\mathbf{x}_0) \int_D F_j(\mathbf{x}) n_j(\mathbf{x})\, \mathrm{d}S(\mathbf{x})
\end{aligned} \tag{4.6.27}$$

where

$$\boldsymbol{\varphi} = \mathbf{q} + \alpha \sum_{n=1}^{6} \mathbf{q}''^{(n)} \int_D \mathbf{q}''^{(n)}(\mathbf{x}) \cdot \mathbf{q}(\mathbf{x})\, dS(\mathbf{x}) \tag{4.6.28}$$

and

$$\mathbf{q} = \boldsymbol{\varphi} - \frac{\alpha}{1+\alpha} \sum_{n=1}^{6} \mathbf{q}''^{(n)} \int_D \mathbf{q}''^{(n)}(\mathbf{x}) \cdot \boldsymbol{\varphi}(\mathbf{x})\, dS(\mathbf{x}) \tag{4.6.29}$$

Bearing in mind our discussion of the discretized system, we now devise the following procedure. Given $\mathbf{F}$, we solve (4.6.27) for $\boldsymbol{\varphi}$ and then use (4.6.29) to compute the desired solution $\mathbf{q}$. The advantages of pursuing this procedure instead of solving directly the original equation (4.4.1) become evident when we note that the homogeneous equation corresponding to (4.6.27) does not have any eigenvalues in the closed range $[-1, 1]$. This implies that (4.6.27) will have a unique solution for any value of $\alpha$ within $[-1, 1]$. Unfortunately, when $\alpha = -1$ the right-hand side of (4.6.29) becomes unbounded; despite our best efforts, we are unable to obtain a deflated integral equation for external flow.

It will be noted that to carry out the Wielandt's deflations it was necessary to know the exact multiplicity of both marginal eigenvalues. If the integral equation had generalized solutions as well as eigensolutions, applying Wielandt's deflation with the eigensolutions alone would not be sufficient for removing the marginal eigenvalues.

### Problems

4.6.1 Assume that $\mathbf{u}$ is an eigenvector of the real matrix $\mathbf{A}$ corresponding to the eigenvalue $\lambda_1$, and that $\mathbf{v}$ is an eigenvector (or generalized eigenvector) of the adjoint of $\mathbf{A}$ corresponding to a different eigenvalue $\lambda_2$. Show that $\mathbf{u} \cdot \mathbf{v} = 0$.

4.6.2 Show that the matrix $\mathscr{A} = \mathbf{A} - \lambda_1 \mathbf{u}\mathbf{w}$ defined in (4.6.1) has the same eigenvalues as $\mathbf{A}$ except for $\lambda_1$ which is replaced by zero.

4.6.3 Show that the matrix $\mathscr{A}''$ defined in (4.6.16) has the same eigenvalues as $\mathbf{A}$ with the exception of $\lambda_1$ and $\lambda_2$, which are replaced by zero.

4.6.4 Verify that (4.6.29) is indeed the solution of (4.6.28).

4.6.5 Substituting (4.6.28) into (4.6.27), verify that $\mathbf{q}$ satisfies (4.4.1).

4.6.6 Show that the homogeneous equation corresponding to (4.6.27) does not possess any eigenvalues in the closed interval $[-1, 1]$.

## 4.7 A compound double-layer representation for external flow

In the preceding two sections we found that the double-layer potential alone is not capable of representing an arbitrary external flow. Physically, this behaviour is attributable to the intrinsic restriction that when $D$ does not enclose $S_B$, the force and torque exerted on the distribution domain $D$ are equal to zero. Motivated by this interpretation, and following an analogous strategy for problems of exterior potential flow (Mikhlin 1957, p. 172; Power & Miranda 1987), we introduce the compound representation

$$u_i(\mathbf{x}_0) = \int_D q_j(\mathbf{x}) T_{jik}(\mathbf{x}, \mathbf{x}_0) n_k(\mathbf{x}) \, dS(\mathbf{x}) + \mathscr{V}_i(\mathbf{x}_0) \tag{4.7.1}$$

where $\mathscr{V}$ is a sufficiently general supplementary flow that is required to be regular in the exterior of $D$ and to produce a finite force and torque on $D$. The underlying mathematical motivation for introducing the supplementary flow $\mathscr{V}$ will be discussed shortly.

Power & Miranda (1987) identified $\mathscr{V}$ with the flow produced by a point force and a couplet or rotlet (see (7.2.22)), setting

$$\mathscr{V}_i(\mathbf{x}_0) = G_{il}(\mathbf{x}_0, \mathbf{x}_G)\mathscr{F}_l + G^C_{il}(\mathbf{x}_0, \mathbf{x}_R)\mathscr{L}_l \tag{4.7.2}$$

where $\mathscr{F}$ and $\mathscr{L}$ are constant vectors, and the poles $\mathbf{x}_G$ and $\mathbf{x}_R$ are located within $D$. The force and torque with respect to the point $\mathbf{x}_G$ exerted on $D$ are

$$\mathbf{F} = -8\pi\mu\mathscr{F} \qquad \mathbf{L} = -8\pi\mu\mathscr{L} \tag{4.7.3}$$

Hebeker (1986) identified $\mathscr{V}$ with the flow due to a single-layer potential whose density is proportional to that of the double-layer potential, thus

$$\mathscr{V}_i(\mathbf{x}_0) = \eta \int_D G_{il}(\mathbf{x}_0, \mathbf{x}) q_l(\mathbf{x}) \, dS(\mathbf{x}) \tag{4.7.4}$$

where $\eta$ is an arbitrary positive constant. Generalizing (4.7.4) we write

$$\mathscr{V}_i(\mathbf{x}_0) = \int_D G_{il}(\mathbf{x}_0, \mathbf{x}) \psi_l(\mathbf{x}) \, dS(\mathbf{x}) \tag{4.7.5}$$

where we require that the density of the single-layer potential $\boldsymbol{\psi}$ is a linear function of $\mathbf{q}$. The specific choice (4.7.4) is recovered from (4.7.5) by setting $\boldsymbol{\psi} = \eta\mathbf{q}$. With the definition (4.7.5), the force and torque exerted on $D$ are

$$\mathbf{F} = -8\pi\mu \int_D \boldsymbol{\psi} \, dS, \qquad \mathbf{L} = -8\pi\mu \int_D \mathbf{x} \times \boldsymbol{\psi} \, dS \tag{4.7.6}$$

Regardless of the choice of $\mathscr{V}$, applying (4.7.1) at a point on the external side of $D$ we obtain an integral equation of the second kind for $\mathbf{q}$:

$$u_i(\mathbf{x}_0) = 4\pi q_i(\mathbf{x}_0) + \int_D^{\mathscr{PV}} q_j(\mathbf{x}) T_{jik}(\mathbf{x}, \mathbf{x}_0) n_k(\mathbf{x}) \, dS(\mathbf{x}) + \mathscr{V}_i(\mathbf{x}_0) \tag{4.7.7}$$

We are now in a position to discuss the underlying mathematical motivation for introducing the supplementary flow $\mathscr{V}$. For this purpose, we rewrite (4.7.7) in the equivalent form

$$u_i(\mathbf{x}_0) = \mathscr{L}_{ij}\langle q_j(\mathbf{x})\rangle + \mathscr{V}_i(\mathbf{x}_0) \tag{4.7.8}$$

where $\mathscr{L}$ is a linear integral operator defined by

$$\mathscr{L}_{ij}(\mathbf{x}_0) = 4\pi\delta_{ij} \int_D \delta(\mathbf{x} - \mathbf{x}_0)\,\mathrm{d}S(\mathbf{x}) + \int_D^{\mathscr{P}\mathscr{V}} T_{jik}(\mathbf{x}, \mathbf{x}_0) n_k(\mathbf{x})\,\mathrm{d}S(\mathbf{x}) \quad (4.7.9)$$

and $\delta(\mathbf{x} - \mathbf{x}_0)$ is the two-dimensional delta function. In section 4.5 we saw that the homogeneous equation $\mathscr{L}\langle\mathbf{q}(\mathbf{x})\rangle = 0$ has six independent eigensolutions representing rigid body motion. The general theory of compact linear operators indicates that the range of $\mathscr{L}$ is deficient and, moreover, that the complement of the range of $\mathscr{L}$ is the null space of the adjoint homogeneous equation $\mathscr{L}^{\mathrm{A}}\langle\mathbf{q}(\mathbf{x})\rangle = 0$, where in our case

$$\mathscr{L}^{\mathrm{A}}_{ij}(\mathbf{x}_0) = 4\pi\delta_{ji} \int_D \delta(\mathbf{x} - \mathbf{x}_0)\,\mathrm{d}S(\mathbf{x}) + n_k(\mathbf{x}_0) \int_D^{\mathscr{P}\mathscr{V}} T_{ijk}(\mathbf{x}_0, \mathbf{x})\,\mathrm{d}S(\mathbf{x}) \quad (4.7.10)$$

(Stakgold 1979, p. 320). In section 4.5 we saw that the null space of the adjoint homogeneous equation (4.7.10) contains six independent eigenfunctions expressing the surface force on $D$ viewed as the boundary of a body that executes six independent modes of rigid body motion. It is evident that for (4.7.8) to have a solution for any choice of $\mathbf{u}$, the range of the right-hand side of (4.7.8) must cover the whole space of admissible functions $\mathbf{u}$. The supplementary flow $\mathscr{V}$ is simply meant to compensate for the deficiency of the incomplete range of $\mathscr{L}$.

Proceeding now with the compound representation, we return to (4.7.7) and decompose the velocity $\mathbf{u}$ over $D$ as $\mathbf{V} + \boldsymbol{\Omega} \times \mathbf{X} + \mathbf{u}^{\mathrm{d}}$, where $\mathbf{V} + \boldsymbol{\Omega} \times \mathbf{X}$ and $\mathbf{u}^{\mathrm{d}}$ express rigid body motion and deformation respectively; $\mathbf{X} = \mathbf{x} - \mathbf{x}_{\mathrm{c}}$ where $\mathbf{x}_{\mathrm{c}}$ is the centroid of $D$ defined in (4.6.24). Substituting into (4.7.7) we obtain

$$u^{\mathrm{d}}_i(\mathbf{x}_0) + V_i + \varepsilon_{ijk}\Omega_j X_{0,k} = 4\pi q_i(\mathbf{x}_0) + \int_D^{\mathscr{P}\mathscr{V}} q_j(\mathbf{x}) T_{jik}(\mathbf{x}, \mathbf{x}_0) n_k(\mathbf{x})\,\mathrm{d}S(\mathbf{x}) + \mathscr{V}_i(\mathbf{x}_0) \quad (4.7.11)$$

At this point, we distinguish two classes of problem: the *resistance problem*, where the velocities $\mathbf{V}$ and $\boldsymbol{\Omega}$ are prescribed and the flow $\mathscr{V}$ is to be computed, and the *mobility problem*, where $\mathscr{V}$ is prescribed and $\mathbf{V}$ and $\boldsymbol{\Omega}$ are to be computed. The deformation field $\mathbf{u}^{\mathrm{d}}$ is presumed to be known in both cases. The resistance problem and the mobility problem will be examined separately in the following two sections.

**Problems**

4.7.1 Select $\mathbf{q}$ and $\mathscr{V}$ so that (4.7.1) reduces to the boundary integral equation (2.3.11).

4.7.2 Compute the force and torque exerted on $D$ when $\mathscr{V}$ is identified with the flow due to a finite collection of point forces and couplets with poles in the interior of $D$.

## 4.8 The resistance problem

According to our previous classification, in the resistance problem the left-hand side of (4.7.11) is known, and the supplementary flow $\mathscr{V}$ is to be computed. In the statement of the problem, the functional form of $\mathscr{V}$ may be specified in an arbitrary manner, under the sole restriction that the force and torque acting on $D$ have finite values. Unfortunately, a unified analysis of (4.7.11) valid for an arbitrary selection of $\mathscr{V}$ is not feasible, and we must proceed by examining each case on an individual basis.

First, we consider the supplementary flow $\mathscr{V}$ selected by Power & Miranda (1987) and given in (4.7.2). Substituting (4.7.2) into (4.7.11) we obtain

$$u_i(\mathbf{x}_0) = 4\pi q_i(\mathbf{x}_0) + \int_D^{\mathscr{PV}} q_j(\mathbf{x}) T_{jik}(\mathbf{x}, \mathbf{x}_0) n_k(\mathbf{x}) \, dS(\mathbf{x}) + G_{ij}(\mathbf{x}_0, \mathbf{x}_G)\mathscr{F}_j + G_{ij}^{C}(\mathbf{x}_0, \mathbf{x}_R)\mathscr{L}_j \tag{4.8.1}$$

where $\mathbf{u} = \mathbf{u}^d + \mathbf{V} + \mathbf{\Omega} \times \mathbf{X}$ is the known boundary velocity over $D$. The problem is reduced to computing $\mathbf{q}$ as well as the coefficients $\mathscr{F}$ and $\mathscr{L}$. To remove the ambiguity in the definitions of $\mathscr{F}$ and $\mathscr{L}$, we arbitrarily stipulate that

$$\mathscr{F} = \int_D \mathbf{q} \, dS, \qquad \mathscr{L} = \int_D \mathbf{X} \times \mathbf{q} \, dS \tag{4.8.2}$$

Substituting (4.8.2) into (4.8.1) we obtain a Fredholm integral equation of the second kind for $\mathbf{q}$. To investigate the existence and uniqueness of solution of this equation, we consider the eigenfunctions of the corresponding homogeneous equation

$$q_i''(\mathbf{x}_0) = -\frac{1}{4\pi}\left[\int_D^{\mathscr{PV}} q_j''(\mathbf{x}) T_{jik}(\mathbf{x}, \mathbf{x}_0) n_k(\mathbf{x}) \, dS(\mathbf{x}) + G_{ij}(\mathbf{x}_0, \mathbf{x}_G)\mathscr{F}_j'' + G_{ij}^{C}(\mathbf{x}_0, \mathbf{x}_R)\mathscr{L}_j''\right] \tag{4.8.3}$$

where $\mathscr{F}''$ and $\mathscr{L}''$ are defined as in (4.8.2) but with $\mathbf{q}''$ in place of $\mathbf{q}$. Thus, we introduce the velocity field

$$w_i(\mathbf{x}_0) = \frac{1}{4\pi}\left[\int_D q_j''(\mathbf{x}) T_{jik}(\mathbf{x}, \mathbf{x}_0) n_k(\mathbf{x}) dS(\mathbf{x}) + G_{ij}(\mathbf{x}_0, \mathbf{x}_G)\mathscr{F}_j'' + G_{ij}^{C}(\mathbf{x}_0, \mathbf{x}_R)\mathscr{L}_j''\right] \tag{4.8.4}$$

Clearly, $\mathbf{w}$ vanishes at infinity as well as on the solid boundary $S_B$. Over the external surface of $D$

$$w_i(\mathbf{x}_0) = q_i''(\mathbf{x}_0) + \frac{1}{4\pi}\left[\int_D^{\mathscr{PV}} q_j''(\mathbf{x})T_{jik}(\mathbf{x},\mathbf{x}_0)n_k(\mathbf{x})\,\mathrm{d}S(\mathbf{x}) + G_{ij}(\mathbf{x}_0,\mathbf{x}_G)\mathscr{F}_j'' + G_{ij}^C(\mathbf{x}_0,\mathbf{x}_R)\mathscr{L}_j''\right] \tag{4.8.5}$$

According to (4.8.3), the right-hand side of (4.8.5) is equal to zero. Thus, $\mathbf{w}$ is equal to zero over all boundaries of the flow including $D, S_B$, and a fictitious boundary located at infinity. This requires that $\mathbf{w}$ must vanish throughout the intervening space; as a result, all three terms on the right-hand side of (4.8.4) must be equal to zero. Requiring that the double-layer term vanishes we obtain $\mathbf{q}'' = \mathbf{c} + \mathbf{d} \times \mathbf{X}$, where $\mathbf{c}$ and $\mathbf{d}$ are constant vectors, whereas requiring that the $\mathscr{F}''$ vanishes we find that $\mathbf{c}$ must be equal to zero. Finally, requiring that $\mathscr{L}''$ vanishes we find

$$\left(\delta_{ij}\int_D X_kX_k\,\mathrm{d}S - \int_D X_iX_j\,\mathrm{d}S\right)d_j = 0 \tag{4.8.6}$$

which is a linear homogeneous algebraic system for $\mathbf{d}$. This system will have a non-trivial solution only if the geometry of $D$ is such that the determinant of the matrix that multiplies $\mathbf{d}$ is equal to zero. This matrix, however, is the *moment of inertia tensor* of $D$ when $D$ is viewed as a thin shell composed of a homogeneous material and as such, is guaranteed to have non-zero eigenvalues and a non-vanishing determinant (Goldstein 1980, Chapter 5). Therefore, we conclude that (4.8.3) does not have any non-trivial characteristic solutions, and, hence, that (4.8.1) has a unique solution. Unfortunately, we are not able to assess the magnitude of the eigenvalues of the generalized homogeneous equation corresponding to (4.8.3) (i.e. (4.8.3) with the right-hand side multiplied by the eigenvalue $\beta^*$). Thus, we are not able to assert whether or not (4.8.1) may be solved using the method of successive substitutions.

Considering next a different choice for $\mathscr{V}$, we adopt (4.7.5). To investigate whether (4.7.11) has a unique solution we examine the eigenfunctions of the corresponding homogeneous equation

$$q_i''(\mathbf{x}_0) = -\frac{1}{4\pi}\left[\int_D^{\mathscr{PV}} q_j''(\mathbf{x})T_{jik}(\mathbf{x},\mathbf{x}_0)n_k(\mathbf{x})\,\mathrm{d}S(\mathbf{x}) + \int_D G_{ij}(\mathbf{x}_0,\mathbf{x})\psi_j''(\mathbf{x})\,\mathrm{d}S(\mathbf{x})\right] \tag{4.8.7}$$

Thus, we intoduce the velocity field

$$w_i(\mathbf{x}_0) = \frac{1}{4\pi}\left[\int_D q_j''(\mathbf{x})T_{jik}(\mathbf{x},\mathbf{x}_0)n_k(\mathbf{x})\,\mathrm{d}S(\mathbf{x}) + \int_D G_{ij}(\mathbf{x}_0,\mathbf{x})\psi_j''(\mathbf{x})\,\mathrm{d}S(\mathbf{x})\right] \tag{4.8.8}$$

Working as previously, we find that $\mathbf{w}$ must vanish throughout the exterior of $D$, implying that the pressure $P$ exterior to $D$ must be equal to zero; thus $\mathbf{f}^+ = 0$. Now, we note that due to the double-layer potential on the right-hand side of (4.8.7) $\mathbf{w}$ is discontinuous across $D$, whereas due to the single-layer potential the corresponding surface force $\mathbf{f}$ is discontinuous across $D$. Using (4.1.9) and (4.3.8) we find that the discontinuities in $\mathbf{w}$ and $\mathbf{f}$ are

$$\mathbf{w}^+ - \mathbf{w}^- = 2\mathbf{q}'', \qquad \mathbf{f}^+ - \mathbf{f}^- = -2\mu\boldsymbol{\psi}''. \tag{4.8.9}$$

Recalling that $\mathbf{w}^+ = \mathbf{f}^+ = 0$ we obtain

$$\mathbf{w}^- = -2\mathbf{q}'' \qquad \mathbf{f}^- = 2\mu\boldsymbol{\psi}'' \tag{4.8.10}$$

Finally, we invoke the energy dissipation identity (1.5.2) for the flow in the interior of $D$ and use (4.8.10) to obtain

$$2\mu \int_{V^-} e^*_{ik} e_{ik} \,\mathrm{d}S = \int_{D^-} w^*_i f_i \,\mathrm{d}S = -4\mu \int_{D^-} q''^*_i \psi''_i \,\mathrm{d}S \tag{4.8.11}$$

Observing that the integral on the extreme left-hand side of (4.8.11) is non-negative, we wish to design $\boldsymbol{\psi}''$ so that the integral on the extreme right-hand side is non-positive. Our objective is to generate a discrepancy that would require that $\mathbf{q}'' = 0$. Clearly, one way to ensure that the last integral in (4.8.11) is positive is to set $\boldsymbol{\psi} = \eta\mathbf{q}$, in which case $\boldsymbol{\psi}'' = \eta\mathbf{q}''$, where $\eta$ is a positive constant (Hebeker 1986).

Now, restricting our attention to real eigensolutions, we observe that if $\boldsymbol{\psi}''$ happens to be orthogonal to $\mathbf{q}''$ the last integral in (4.8.11) will be equal to zero. This would imply that the energy dissipation integral vanishes and thus that the internal flow $\mathbf{w}$ expresses rigid body motion, $\mathbf{w}^- = \mathbf{U} + \boldsymbol{\Omega} \times \mathbf{x}$. In this case, the internal pressure $P'$ would be constant, and the surface force on the internal side of $D$ would be $\mathbf{f}^- = -P'\mathbf{n}$. Furthermore, (4.8.10) would yield

$$\mathbf{q}'' = -\tfrac{1}{2}(\mathbf{U} + \boldsymbol{\Omega} \times \mathbf{x}) \qquad \boldsymbol{\psi}'' = \frac{-P'}{2\mu}\mathbf{n} \tag{4.8.12}$$

One way to render $\boldsymbol{\psi}''$ orthogonal to $\mathbf{q}''$ is to set $\boldsymbol{\psi} = \mathbf{b} \times \mathbf{q}$ and thus $\boldsymbol{\psi}'' = \mathbf{b} \times \mathbf{q}''$, where $\mathbf{b}$ is an arbitrary constant. Another way is to set $\boldsymbol{\psi} = \mathbf{x} \times \mathbf{q}$ and thus $\boldsymbol{\psi}'' = \mathbf{x} \times \mathbf{q}''$. With the first choice, (4.8.12) requires $\mathbf{n} = (\mu/P')\mathbf{b} \times (\mathbf{U} + \boldsymbol{\Omega} \times \mathbf{x})$ or, equivalently, $\mathbf{n} = \mathbf{c} + \mathbf{d} \times \mathbf{x}$, where $\mathbf{c}$ and $\mathbf{d}$ are two arbitrary constant vectors. This is a strong geometric constraint on $D$ that appears to be satisfied only by a plane. The second choice also leads to a strong geometric constraint (problem 4.8.2).

In summary, the compound double-layer representation leads to an integral equation with a unique solution for a rather general selection of the supplementary flow $\mathscr{V}$. It is interesting to note that whereas the solution for the density $\mathbf{q}$ depends on $\mathscr{V}$, the computed force and torque exerted on $D$ are independent of the choice of $\mathscr{V}$.

**Problems**

4.8.1 Investigate whether the integral equation that arises by identifying $\mathscr{V}$ with the flow produced by a finite collection of point forces and couplets with poles in the interior of $D$ has a unique solution.

4.8.2 Show that setting $\boldsymbol{\psi} = \mathbf{x} \times \mathbf{q}$ in (4.7.5) yields an integral equation with a unique solution, provided that the domain $D$ is not described by the equation $\mathbf{n} = \mathbf{x} \times (\mathbf{c} + \mathbf{d} \times \mathbf{x})$, where $\mathbf{c}$ and $\mathbf{d}$ are two arbitrary constant vectors.

## 4.9 The mobility problem

Returning to (4.7.11), we now investigate the mobility problem in which $\mathbf{u}^{\mathrm{d}}$ and $\mathscr{V}$ are given and the translational and rotational velocities $\mathbf{V}$ and $\boldsymbol{\Omega}$, as well as the density $\mathbf{q}$, are to be computed. It will prove convenient to write (4.7.11) in the alternative form

$$q_i(\mathbf{x}_0) = -\frac{1}{4\pi}\int_D^{\mathscr{P}\mathscr{V}} q_j(\mathbf{x}) T_{jik}(\mathbf{x}, \mathbf{x}_0) n_k(\mathbf{x})\,\mathrm{d}S(\mathbf{x}) + \mathscr{B}_i(\mathbf{x}_0) + \mathscr{C}_i(\mathbf{x}_0) \quad (4.9.1)$$

where

$$\mathscr{B}_i(\mathbf{x}_0) = \frac{1}{4\pi}[V_i + \varepsilon_{ijk}\Omega_j X_{0,k}] \qquad \mathscr{C}_i(\mathbf{x}_0) = \frac{1}{4\pi}[u_i^{\mathrm{d}}(\mathbf{x}_0) - \mathscr{V}_i(\mathbf{x}_0)] \quad (4.9.2)$$

and $\mathscr{C}$ is presumed to be known.

In section 4.5 we saw that when the domain $D$ does not cross or enclose $S_{\mathrm{B}}$, the homogeneous equation corresponding to (4.9.1) (i.e. (4.9.1) with $\mathscr{B}$ and $\mathscr{C}$ set equal to zero) has six eigensolutions, representing six independent modes of rigid body motion. It will be important to note that $\mathscr{B}$ consists of a linear combination of these eigensolutions. According to Fredholm's alternative, (4.9.1) will have a solution only if the sum $\mathscr{B} + \mathscr{C}$ is orthogonal to the surface force exerted on $D$ where $D$ is viewed as the surface of a body that executes an arbitrary mode of rigid body motion. The key idea is that $\mathbf{V}$ and $\boldsymbol{\Omega}$ should be chosen in such a way as to satisfy this constraint.

It is again helpful to establish a connection with linear algebra. For this purpose, we divide the domain $D$ into a set of $N$ boundary elements and assume that $\mathbf{q}, \mathscr{B}$, and $\mathscr{C}$ are constant over each element. In this fashion, we write (4.9.1) in the discrete form

$$\mathbf{x} = -\mathbf{A}\cdot\mathbf{x} + \mathbf{b} + \mathbf{c} \quad (4.9.3)$$

where the vectors $\mathbf{x}$, $\mathbf{b}$, and $\mathbf{c}$ contain the components of $\mathbf{q}$, $\mathscr{B}$, and $\mathscr{C}$ that are constant over each element. We assume that $\mathbf{u}^{(k)}$, $k = 1, \ldots, 6$, are six independent solutions of the homogeneous equation corresponding to (4.9.3) (i.e. (4.9.3) with $\mathbf{b}$ and $\mathbf{c}$ set equal to zero), and that $\mathbf{b}$ is a linear combination of the $\mathbf{u}^{(k)}$. Clearly, the set $\{\mathbf{u}^{(k)}\}$ contains six eigensolutions of $\mathbf{A}$ corresponding to the eigenvalue $\lambda_1 = -1$. Furthermore, bearing in mind our results in section 4.5, we assume that $\lambda_2 = 1$ is an eigenvalue of $\mathbf{A}$ with multiplicity one, with the real vector $\mathbf{v}$ as corresponding adjoint eigenvector. Since $\mathbf{v} \cdot \mathbf{u}^{(k)} = 0$, and $\mathbf{b}$ is a linear combination of the $\mathbf{u}^{(k)}$, $\mathbf{v} \cdot \mathbf{b} = 0$.

Next, we consider the matrix $\mathscr{A}''$ defined in (4.6.16), with $\lambda_1 = -1$, $\lambda_2 = 1$, and $K = 6$, i.e.

$$\mathscr{A}''_{ij} \equiv A_{ij} + \sum_{k=1}^{6} u_i^{(k)} u_j^{(k)} - \frac{1}{|\mathbf{v}|^2} v_i v_j \tag{4.9.4}$$

where, for simplicity, we have set $\mathbf{z} = \mathbf{v}/|\mathbf{v}|^2$ and in addition, we have identified $\{\mathbf{w}^{(k)}\}$ with $\{\mathbf{u}^{(k)}\}$, thereby requiring that the set of eigenvectors $\{\mathbf{u}^{(k)}\}$ is orthonormal. Wielandt's theorem ensures that $\mathscr{A}''$ has the same eigenvalues as $\mathbf{A}$ except for $\lambda_1 = -1$ and $\lambda_2 = 1$, which are replaced by zero.

Now, instead of solving (4.9.3), we propose to solve the alternative equation

$$\mathbf{x} = -\mathscr{A}'' \cdot \mathbf{x} + \mathbf{b}' \tag{4.9.5}$$

where $\mathbf{b}'$ is a constant vector. Using (4.9.4), we rewrite (4.9.5) in the equivalent form

$$\mathbf{x} = -\mathbf{A} \cdot \mathbf{x} - \sum_{k=1}^{6} \mathbf{u}^{(k)} (\mathbf{u}^{(k)} \cdot \mathbf{x}) + \frac{1}{|\mathbf{v}|^2} \mathbf{v} (\mathbf{v} \cdot \mathbf{x}) + \mathbf{b}' \tag{4.9.6}$$

Multiplying (4.9.3) from the left by $\mathbf{v}$ and using the equations $\mathbf{v} \cdot \mathbf{A} = \mathbf{v}$ and $\mathbf{v} \cdot \mathbf{b} = 0$, we find $\mathbf{v} \cdot \mathbf{x} = \frac{1}{2} \mathbf{v} \cdot \mathbf{c}$. Substituting this expression into the right-hand side of (4.9.6) we obtain

$$\mathbf{x} = -\mathbf{A} \cdot \mathbf{x} - \sum_{k=1}^{6} \mathbf{u}^{(k)} (\mathbf{u}^{(k)} \cdot \mathbf{x}) + \frac{1}{2|\mathbf{v}|^2} \mathbf{v} (\mathbf{v} \cdot \mathbf{c}) + \mathbf{b}' \tag{4.9.7}$$

To render (4.9.7) equivalent to (4.9.3) we require

$$\mathbf{b} = -\sum_{k=1}^{6} \mathbf{u}^{(k)} (\mathbf{u}^{(k)} \cdot \mathbf{x}) \tag{4.9.8}$$

and

$$\mathbf{b}' = \mathbf{c} - \frac{1}{2|\mathbf{v}|^2} \mathbf{v} (\mathbf{v} \cdot \mathbf{c}) \tag{4.9.9}$$

Substituting (4.9.9) into (4.9.6) we obtain

$$\mathbf{x} = -\mathbf{A} \cdot \mathbf{x} - \sum_{k=1}^{6} \mathbf{u}^{(k)} (\mathbf{u}^{(k)} \cdot \mathbf{x}) + \mathbf{c} + \frac{1}{|\mathbf{v}|^2} \mathbf{v} (\mathbf{v} \cdot \mathbf{x}) - \frac{1}{2|\mathbf{v}|^2} \mathbf{v} (\mathbf{v} \cdot \mathbf{c}) \tag{4.9.10}$$

At this point, we compare (4.9.1) with (4.9.3) and, in analogy with (4.9.8), we define

$$\mathscr{B}(\mathbf{x}_0) = -\sum_{k=1}^{6} \mathbf{q}''^{(k)}(\mathbf{x}_0) \int_D \mathbf{q}''^{(k)}(\mathbf{x}) \cdot \mathbf{q}(\mathbf{x})\, dS(\mathbf{x}) \tag{4.9.11}$$

where $\{\mathbf{q}''^{(k)}\}$ is the set of orthonormal eigenvectors given in (4.6.22) and (4.6.23). Substituting (4.9.2), (4.6.22), and (4.6.23) into (4.9.11) we obtain the translational and rotational velocities

$$\mathbf{V} = -\frac{4\pi}{S_D} \int_D \mathbf{q}\, dS \tag{4.9.12}$$

and

$$\mathbf{\Omega} = -4\pi \sum_{m=1}^{3} \frac{1}{A_m} \boldsymbol{\omega}^{(m)} \left( \boldsymbol{\omega}^{(m)} \cdot \int_D \mathbf{X} \times \mathbf{q}\, dS \right) \tag{4.9.13}$$

(problem 4.9.1). Furthermore, motivated by (4.9.10), recalling that $\beta_2 = 1$ is an eigenvalue of the double-layer potential with corresponding adjoint eigenvalue $\mathbf{q}' = \mathbf{n}$, and making an analogy between the role of $\mathbf{v}$ and that of $\mathbf{q}'$, we replace (4.9.1) with the modified equation

$$\begin{aligned} q_i(\mathbf{x}_0) = & -\frac{1}{4\pi} \int_D^{\mathscr{PV}} q_j(\mathbf{x}) T_{jik}(\mathbf{x}, \mathbf{x}_0) n_k(\mathbf{x})\, dS(\mathbf{x}) + \mathscr{B}_i(\mathbf{x}_0) + \mathscr{C}_i(\mathbf{x}_0) \\ & + \frac{1}{S_D} n_i(\mathbf{x}_0) \int_D q_j(\mathbf{x}) n_j(\mathbf{x})\, dS(\mathbf{x}) - \frac{1}{2S_D} n_i(\mathbf{x}_0) \int_D \mathscr{C}_j(\mathbf{x}) n_j(\mathbf{x})\, dS(\mathbf{x}) \end{aligned} \tag{4.9.14}$$

To show that a solution of (4.9.14) also satisfies (4.9.1), we multiply (4.9.14) by the normal vector $\mathbf{n}$, integrate over $D$, and use (3.2.7), finding

$$\int_D q_j(\mathbf{x}) n_j(\mathbf{x})\, dS(\mathbf{x}) = \frac{1}{2} \int_D \mathscr{C}_j(\mathbf{x}) n_j(\mathbf{x})\, dS(\mathbf{x}) \tag{4.9.15}$$

Substituting (4.9.15) into the right-hand side of (4.9.14) produces (4.9.1).

Now, to justify replacing (4.9.1) by the more complicated equation (4.9.14), we wish to demonstrate that (4.9.14) has a unique solution that may be obtained using the method of successive substitutions. For this purpose, we examine the eigenvalues of the associated generalized homogeneous equation that arises by discarding the forcing terms on the right-hand side of (4.9.14) and multiplying the remaining terms by the eigenvalue $\beta^*$. Thus, we consider the equation

$$\begin{aligned} q_i''(\mathbf{x}_0) = \frac{\beta^*}{4\pi} \Bigg[ & \int_D^{\mathscr{PV}} q_j''(\mathbf{x}) T_{jik}(\mathbf{x}, \mathbf{x}_0) n_k(\mathbf{x})\, dS(\mathbf{x}) \\ & - V_i'' - \varepsilon_{ijk} \Omega_j'' X_{0,k} - \frac{4\pi}{S_D} n_i(\mathbf{x}_0) \int_D q_j''(\mathbf{x}) n_j(\mathbf{x})\, dS(\mathbf{x}) \Bigg] \end{aligned} \tag{4.9.16}$$

where $\mathbf{V}''$ and $\mathbf{\Omega}''$ are given by (4.9.12) and (4.9.13) but with $\mathbf{q}''$ in place of $\mathbf{q}$. Our task will be to show that (4.9.16) does not have any eigenvalues $\beta^*$ with $|\beta^*| \leqslant 1$.

First, we multiply (4.9.16) by the normal vector, integrate over $D$, and use (3.2.7) to obtain

$$\int_D \mathbf{q}''(\mathbf{x}) \cdot \mathbf{n}(\mathbf{x}) \, \mathrm{d}S(\mathbf{x}) = 0 \tag{4.9.17}$$

This equation suggests that the last term on the right-hand side of (4.9.16) may be neglected. The eigenfunctions of the resulting simplified equation, however, must be tested against (4.9.17) before they can be counted as acceptable.

Beginning the search for the eigenvalues, we introduce the double-layer potential

$$w_i(\mathbf{x}_0) = \frac{1}{4\pi} \int_D q_j''(\mathbf{x}) T_{jik}(\mathbf{x}, \mathbf{x}_0) n_k(\mathbf{x}) \, \mathrm{d}S(\mathbf{x}) \tag{4.9.18}$$

The limiting values of $\mathbf{w}$ on either side of $D$ are

$$w_i^{\pm}(\mathbf{x}_0) = \pm q_i''(\mathbf{x}_0) + \frac{1}{4\pi} \int_D^{\mathscr{PV}} q_j''(\mathbf{x}) T_{jik}(\mathbf{x}, \mathbf{x}_0) n_k(\mathbf{x}) \, \mathrm{d}S(\mathbf{x}) \tag{4.9.19}$$

where the plus sign applies for the external side of $D$ and the minus sign for the internal side of $D$. Combining (4.9.19) with (4.9.16) and (4.9.17) we find that over $D$,

$$\mathbf{w}^{\pm} = \left( \pm 1 + \frac{1}{\beta^*} \right) \mathbf{q}'' + \frac{1}{4\pi} (\mathbf{V}'' + \mathbf{\Omega}'' \times \mathbf{X}) \tag{4.9.20}$$

Next, we consider the energy balances (4.5.8) and (4.5.9). Using (4.9.20) we obtain

$$\begin{aligned} -8\pi\mu \int_{V^{\pm}} e_{ik} e_{ik}^* \, \mathrm{d}V = 4\pi \left( 1 \pm \frac{1}{\beta^*} \right) \int_{D^{\pm}} q_i''^* f_i \, \mathrm{d}S \\ \pm U_i''^* \int_{D^{\pm}} f_i \mathrm{d}S \pm \Omega_j''^* \int_{D^{\pm}} \varepsilon_{jli} X_l f_i \, \mathrm{d}S \end{aligned} \tag{4.9.21}$$

The last two integrals on the right-hand side of (4.9.21) represent the force and torque exerted on $D$ due to the flow $\mathbf{w}$: we recall that when $D$ does not enclose $S_B$ the force and torque exerted on the domain of distribution of the double-layer potential are equal to zero, which means that both of these integrals vanish. Furthermore, we recall that the surface force of the double-layer potential is continuous across the domain of distribution, which indicates that the value of the right-hand side of (4.9.21) on the external side of $D$ is the same as that on the internal side. Applying (4.9.21) on the internal side and on the external side and finding the ratio of the resulting expressions, we obtain $\beta^* = (1+\delta)/(1-\delta)$, where $\delta = \Delta^+/\Delta^-$, and $\Delta^\pm$ is defined as the left-hand side of (4.9.21). We note that $\Delta^+$ and $\Delta^-$ are real and non-negative, which suggests that the eigenvalues $\beta^*$ are real and outside the range $[-1, 1]$.

Of course, the above conclusions were based on the assumption that neither $\Delta^+$ nor $\Delta^-$ is equal to zero, i.e. $\mathbf{w}$ represents rigid body motion neither in the exterior nor in the exterior of $D$. Rigid body motion in the exterior of $D$ is precluded by the requirement that $\mathbf{w}$ must vanish at finity. The only remaining possibility is that $\mathbf{w}$ represents rigid body motion in the interior of $D$, thereby implying that $\mathbf{f}^- = -P\mathbf{n}$, where $P$ is the constant internal pressure. Since the surface force is continuous across the domain of distribution of the double-layer potential, $\mathbf{f}^+ = -P\mathbf{n}$ as well. Now, applying (4.9.21) for the external flow and using (4.9.17), we find $\Delta^+ = 0$, which indicates that $\mathbf{w}^+ = 0$. Equation (4.9.19) then reveals that $\mathbf{q}''$ is an eigensolution of the double-layer potential with corresponding eigenvalue $\beta^* = -1$. But then, according to our discussion in section 4.3, $\mathbf{q}''$ must represent rigid body motion. Assuming that $\mathbf{q}'' = \mathbf{U} + \boldsymbol{\omega} \times \mathbf{X}$ and substituting into (4.9.16) we find $\mathbf{q}'' = 0$ and conclude that there are no eigenfunctions for which $\mathbf{w}$ represents rigid body motion.

In summary, the homogeneous equation (4.9.16) does not have eigenvalues with magnitude less than or equal to unity, and therefore, (4.9.14) has a unique solution that may be computed using the method of successive substitutions.

The compound double-layer representation discussed in this section may be extended in a straightforward manner to cases of multiply-connected domains $D$. These may represent, for instance, the surfaces of a

*Figure 4.9.1.* Suspended particles moving under the action of an incident flow in the vicinity of a solid wall.

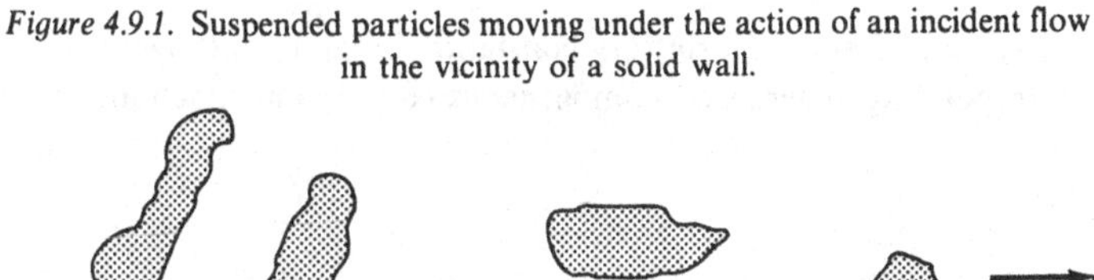

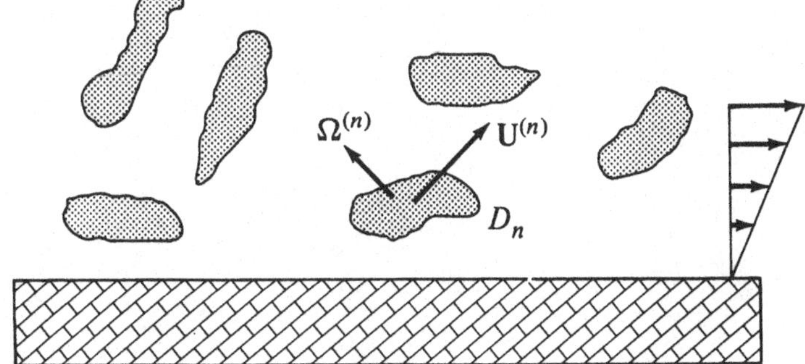

group of suspended particles (Figure 4.9.1), in which case the domain of the integral equation (4.9.14) is the union of the surfaces of the particles. The translational and rotational velocities of the $n$th particle are given by

$$\mathbf{V}^{(n)} = -\frac{4\pi}{S_{D_n}} \int_{D_n} \mathbf{q} \, dS \tag{4.9.22}$$

and

$$\boldsymbol{\Omega}^{(n)} = -4\pi \sum_{m=1}^{3} \frac{1}{A_{mn}} \boldsymbol{\omega}^{(m,n)} \left( \boldsymbol{\omega}^{(m,n)} \cdot \int_{D_n} \mathbf{X}^{(n)} \times \mathbf{q} \, dS \right) \tag{4.9.23}$$

where $D_n$ is the surface of the $n$th particle, $\mathbf{X}^{(n)} = \mathbf{x} - \mathbf{x}_c^{(n)}$, $\mathbf{x}_c^{(n)}$ is the centroid of the surface of the $n$th particle defined as

$$\mathbf{x}_c^{(n)} = \frac{1}{S_{D_n}} \int_{D_n} \mathbf{x} \, dS \tag{4.9.24}$$

and $S_{D_n}$ is the surface area of the $n$th particle. The unit vectors $\boldsymbol{\omega}^{(m,n)}$ and the coefficients $A_{mn}$ are defined by analogy with (4.6.23) and (4.6.25).

Kim & Karrila (1991, Part 4, Chapter 17) present a thorough theoretical investigation of the compound double-layer representation for the general case of flow past a group of suspended particles in a bounded domain of flow.

### Problems

4.9.1 Show that substituting (4.9.2), (4.6.22), and (4.6.23) into (4.9.11) produces (4.9.12) and (4.9.13).

4.9.2 Examine the uniqueness of solution of the integral equation (4.9.14) when $\mathbf{V}$ and $\boldsymbol{\Omega}$ are defined as

$$\mathbf{V} = -\varepsilon \int_D \mathbf{q} \, dS \qquad \boldsymbol{\Omega} = \zeta \int_D \mathbf{X} \times \mathbf{q} \, dS \tag{1}$$

where $\varepsilon$ and $\zeta$ are two arbitrary constants. Compute the eigenvalues of the corresponding generalized homogeneous equation as functions of $\varepsilon$ and $\zeta$.

# 5

# *Flow due to interfaces*

## 5.1 Introduction

Flows involving interfaces between two different fluids occur in a variety of natural, engineering, and biomechanical applications. Two examples are the flow of a suspension of bubbles, drops, or biological cells, and the flow of a liquid film over a solid surface.

The significance of an interface for the behaviour of a flow is two-fold. From a kinematical standpoint, the interface marks the permanent boundary between two adjacent regions of flow with distinct physical constants. Fluid parcels that are located away from the interface are required to reside in the bulk of the flow at all times. From a dynamical standpoint, the interface is a singular surface of concentrated force. To elucidate this interpretation we note that, in general, the surface forces acting on the two sides of an interface have different values. The difference between these values, termed the discontinuity in the surface force, $\Delta\mathbf{f}$, depends upon the physical properties of the fluids and the structure and thermodynamic properties of the interface. This dependence may be expressed in terms of a constitutive relationship that may involve a number of physical constants, including the densities of the fluids, surface tension, surface viscosity, surface elasticity, and surface modules of bending and dilatation. An interface is *active* when $\Delta\mathbf{f}$ is finite, and *inactive* or *passive* when $\Delta\mathbf{f} = 0$. An active interface plays a leading role in determining the dynamics of the flow, whereas a passive interface is simply advected by the ambient flow.

To describe a flow in the presence of an interface we must consider the flow on each side of the interface separately, and then require proper matching conditions for the velocity and surface force. Typically, we require continuity of velocity across the interface, and a constitutive relation for the discontinuity in the interfacial surface force. To construct a boundary integral representation, we write two boundary integral equations (for the flow on either side of the interface) involving the interfacial velocity, the interfacial surface force, or the density of a hydrodynamic potential. Requiring the interfacial boundary conditions, we obtain a system of integral equations for the unknown interfacial functions. While these are the general principles, the details of implementation depend on the chosen bounday integral representations as well as the topology of the domain of flow.

Before proceeding to discuss specific formulations, it is helpful to outline certain basic assumptions adopted in our discussion. For simplicity, we consider a closed interface $S$ in either an unbounded fluid or a fluid that is bounded by a stationary solid surface $S_B$ as illustrated in Figure 5.1.1. In either case we assume that the flow is due exclusively to the interface. Necessary modifications to account for the effect of an ambient flow will be discussed in the problems following each section. Furthermore, in order to account for a body force, we introduce the modified stress tensor $\boldsymbol{\sigma}^{MOD}$ defined with respect to the modified pressure $P^{MOD} = P - \rho \mathbf{b} \cdot \mathbf{x}$. When $\mathbf{b}$ is the acceleration of gravity, $\rho$ is the density of the fluid, whereas when $\mathbf{b}$ is the intensity of an electric field, $\rho$ is the concentration of an electrolyte. The modified pressure and its corresponding stress tensor satisfy the equations of unforced Stokes flow and thus are amenable to the analysis of the previous chapters. For convenience, we shall drop the superscript MOD, but shall tacitly assume that the pressure and stress are enhanced by the addition of a term corresponding to the body force. The distinction between the regular and modified stress will become significant only when we begin developing constitutive relations for the discontinuity in the interfacial surface force (see section 5.5).

To simplify our nomenclature, we shall call the fluid enclosed by the interface a particle, and label the ambient fluid and the particle fluid with

*Figure 5.1.1.* Sketch of an interface $S$ between two fluids labelled 1 and 2, in the presence of a solid boundary $S_B$.

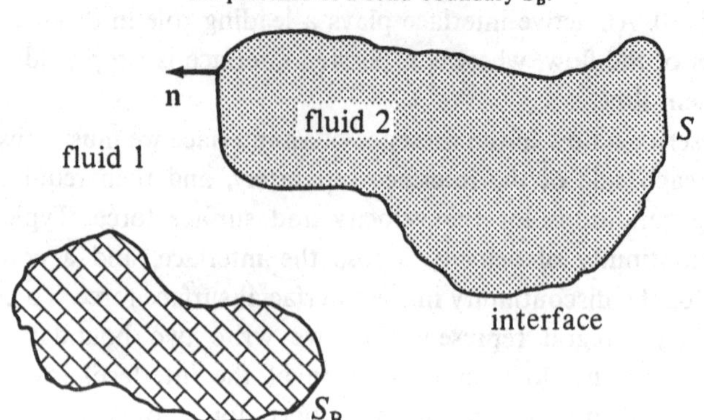

the indices 1 and 2 respectively. Furthermore, we shall denote the viscosity ratio between the internal and the external fluid by $\lambda = \mu_2/\mu_1$. It will be helpful to keep in mind that when $\lambda = 0$ or $\infty$ the particle becomes a frictionless bubble or a rigid body respectively.

## 5.2 The boundary integral formulation

One way to obtain a boundary integral representation of the flow in the exterior of an interface is to use the standard boundary integral equation (2.3.11). Assuming that the flow vanishes at infinity and identifying the boundary $D$ with the interface $S$, we find

$$u_j^{(1)}(\mathbf{x}_0) = -\frac{1}{8\pi\mu_1}\int_S f_i^{(1)}(\mathbf{x})G_{ij}(\mathbf{x},\mathbf{x}_0)\,\mathrm{d}S(\mathbf{x}) + \frac{1}{8\pi}\int_S u_i(\mathbf{x})T_{ijk}(\mathbf{x},\mathbf{x}_0)n_k(\mathbf{x})\,\mathrm{d}S(\mathbf{x}) \tag{5.2.1}$$

It will be noted that the single-layer integral in (5.2.1) involves the surface force over the external side of $S$, indicated by the superscript$^{(1)}$.

Now, to render the boundary integral representation directly accessible to the interfacial boundary conditions, we seek an alternative representation involving the discontinuity in the interfacial surface force $\Delta\mathbf{f}$. Thus, we apply the reciprocal identity (2.3.4) for the internal flow $\mathbf{u}^{(2)}$ at a point $\mathbf{x}_0$ that is located exterior to the particle, obtaining

$$\int_S f_i^{(2)}(\mathbf{x})G_{ij}(\mathbf{x},\mathbf{x}_0)\,\mathrm{d}S(\mathbf{x}) - \mu_2\int_S u_i(\mathbf{x})T_{ijk}(\mathbf{x},\mathbf{x}_0)n_k(\mathbf{x})\,\mathrm{d}S(\mathbf{x}) = 0 \tag{5.2.2}$$

Combining (5.2.1) and (5.2.2) we find

$$u_j^{(1)}(\mathbf{x}_0) = -\frac{1}{8\pi\mu_1}\int_S \Delta f_i(\mathbf{x})G_{ij}(\mathbf{x},\mathbf{x}_0)\,\mathrm{d}S(\mathbf{x}) + \frac{1-\lambda}{8\pi}\int_S u_i(\mathbf{x})T_{ijk}(\mathbf{x},\mathbf{x}_0)n_k(\mathbf{x})\,\mathrm{d}S(\mathbf{x}) \tag{5.2.3}$$

where we recall that $\lambda = \mu_2/\mu_1$. $\Delta\mathbf{f}$ is the discontinuity in the interfacial surface force, defined as

$$\Delta\mathbf{f} = \mathbf{f}^{(1)} - \mathbf{f}^{(2)} = (\boldsymbol{\sigma}^{(1)} - \boldsymbol{\sigma}^{(2)})\cdot\mathbf{n} \tag{5.2.4}$$

where $\boldsymbol{\sigma}$ is the modified stress tensor (defined with respect to the modified pressure). Equation (5.2.3) provides us with a convenient integral representation of the external flow in terms of $\Delta\mathbf{f}$. To obtain a corresponding representation for the pressure, we use (4.1.3) and (4.3.10), finding

$$P^{(1)}(\mathbf{x}_0) = -\frac{1}{8\pi}\int_S \Delta f_i(\mathbf{x})p_i(\mathbf{x}_0,\mathbf{x})\,\mathrm{d}S(\mathbf{x}) + \frac{1-\lambda}{8\pi}\mu_1\int_S u_i(\mathbf{x})\Pi_{ij}(\mathbf{x}_0,\mathbf{x})n_j(\mathbf{x})\,S(\mathbf{x}) \tag{5.2.5}$$

Next, we seek to derive a boundary integral representation for the internal flow. Thus, we apply the boundary integral equation (2.3.11) at a point $\mathbf{x}_0$ that is located in the interior of the particle, obtaining

$$u_j^{(2)}(\mathbf{x}_0) = \frac{1}{8\pi\mu_2}\int_S f_i^{(2)}(\mathbf{x})G_{ij}(\mathbf{x},\mathbf{x}_0)\,dS(\mathbf{x}) - \frac{1}{8\pi}\int_S u_i(\mathbf{x})T_{ijk}(\mathbf{x},\mathbf{x}_0)n_k(\mathbf{x})\,dS(\mathbf{x}) \tag{5.2.6}$$

Furthermore, we apply the reciprocal relation (2.3.4) for the external flow at the point $\mathbf{x}_0$ and note that the external flow vanishes at infinity to obtain

$$\int_S f_i^{(1)}(\mathbf{x})G_{ij}(\mathbf{x},\mathbf{x}_0)\,dS(\mathbf{x}) - \mu_1\int_S u_i(\mathbf{x})T_{ijk}(\mathbf{x},\mathbf{x}_0)n_k(\mathbf{x})\,dS(\mathbf{x}) = 0 \tag{5.2.7}$$

Combining (5.2.6) and (5.2.7) we find

$$u_j^{(2)}(\mathbf{x}_0) = -\frac{1}{8\pi\mu_1\lambda}\int_S \Delta f_i(\mathbf{x})G_{ij}(\mathbf{x},\mathbf{x}_0)\,dS(\mathbf{x}) + \frac{1-\lambda}{8\pi\lambda}\int_S u_i(\mathbf{x})T_{ijk}(\mathbf{x},\mathbf{x}_0)n_k(\mathbf{x})\,dS(\mathbf{x}) \tag{5.2.8}$$

It is instructive to observe that (5.2.8) may be obtained directly from (5.2.3) by dividing the right-hand side by the viscosity ratio $\lambda$. Correspondingly, the internal pressure is obtained from (5.2.5) if we divide the right-hand side by $\lambda$.

Now, letting the point $\mathbf{x}_0$ approach the interface either from the external or from the internal side, and using (4.3.6) and (4.3.7), we find that both (5.2.3) and (5.2.8) reduce to the equation

$$u_j(\mathbf{x}_0) = -\frac{1}{(4\pi\mu_1)(\lambda+1)}\int_S \Delta f_i(\mathbf{x})G_{ij}(\mathbf{x},\mathbf{x}_0)dS(\mathbf{x}) + \frac{\kappa}{4\pi}\int_S^{\mathscr{PV}} u_i(\mathbf{x})T_{ijk}(\mathbf{x},\mathbf{x}_0)n_k(\mathbf{x})\,dS(\mathbf{s}) \tag{5.2.9}$$

where $\kappa = (1-\lambda)/(1+\lambda)$ (Youngren & Acrivos 1976; Rallison & Acrivos 1978). Prescribing $\Delta\mathbf{f}$ renders (5.2.9) a Fredholm integral equation of the second kind for the interfacial velocity $\mathbf{u}$. Once this equation is solved, the internal and external velocity and pressure fields may be computed using the representations (5.2.3), (5.2.5), and (5.2.8) as discussed above.

It will be noted that when the viscosities of the two fluids are equal, i.e. $\lambda = 1$ and $\kappa = 0$, the coefficients of the double-layer integral on the right-hand sides of (5.2.3), (5.2.5), and (5.2.8) vanish, and the flow is expressed merely in terms of a single-layer potential with known density $\Delta\mathbf{f}$.

### Problems

5.2.1 Verify that in the limit as the point $\mathbf{x}_0$ approaches the interface from either side (5.2.3) and (5.2.8) reduce to (5.2.9).

5.2.2 Consider an ambient flow $\mathbf{u}^\infty$ past a particle. Show that the velocity exterior to the particle may be expressed in the form

$$u_j^{(1)}(\mathbf{x}_0) = u_j^\infty(\mathbf{x}_0) - \frac{1}{8\pi\mu_1}\int_S \Delta f_i(\mathbf{x}) G_{ij}(\mathbf{x}, \mathbf{x}_0)\, dS(\mathbf{x})$$
$$+ \frac{1-\lambda}{8\pi}\int_S u_i(\mathbf{x}) T_{ijk}(\mathbf{x}, \mathbf{x}_0) n_k(\mathbf{x})\, dS(\mathbf{x}) \quad (1)$$

and that the velocity in the interior of the particle is given by an identical expression, except that the whole right-hand side, including $u_j^\infty$, is divided by $\lambda$. Taking the limit as the point $\mathbf{x}_0$ approaches $S$ from either side, derive the integral equation

$$u_j(\mathbf{x}_0) = \frac{2}{1+\lambda} u_j^\infty(\mathbf{x}_0) - \frac{1}{(4\pi\mu_1)(\lambda+1)}\int_S \Delta f_i(\mathbf{x}) G_{ij}(\mathbf{x}, \mathbf{x}_0)\, dS(\mathbf{x})$$
$$+ \frac{\kappa}{4\pi}\int_S^{\mathscr{PV}} u_i(\mathbf{x}) T_{ijk}(\mathbf{x}, \mathbf{x}_0) n_k(\mathbf{x})\, dS(\mathbf{x}) \quad (2)$$

(Rallison & Acrivos 1978). Finally, show that far from the particle, the flow is described by the multipole expansion

$$u_j^{(1)}(\mathbf{x}_0) = u_j^\infty(\mathbf{x}_0) - \frac{1}{8\pi\mu_1} G_{ji}(\mathbf{x}_0, \mathbf{x}_c)\int_S \Delta f_i(\mathbf{x})\, dS(\mathbf{x}) - \frac{1}{8\pi\mu_1}\frac{\partial G_{ji}}{\partial x_{c,k}}(\mathbf{x}_0, \mathbf{x}_c)$$
$$\times \int_S [\Delta f_i(\mathbf{x})\hat{x}_k - \mu_1(1-\lambda)(u_i n_k + u_k n_i)(\mathbf{x})]\, dS(\mathbf{x}) + \cdots \quad (3)$$

where $\mathbf{x}_c$ is an arbitrary point in the neighbourhood of the particle and $\hat{\mathbf{x}} = \mathbf{x} - \mathbf{x}_c$.

5.2.3 Show that (2.5.17) may be reduced to the equivalent form

$$\langle \sigma_{ij} \rangle = -\delta_{ij}\langle P \rangle + 2\mu_1 \langle e_{ij} \rangle + \frac{1}{\mathscr{V}}\int_{S_P} [\Delta f_i x_j - \mu_1(1-\lambda)(u_i n_j + u_j n_i)]\, dS$$

where the averages $\langle\, \rangle$ are defined in (2.5.18).

## 5.3 The single-layer formulation

In the previous section we obtained two individual boundary integral representations for the flows outside and inside a particle. In this section we seek to derive a unified representation valid throughout the domain of flow. One way to achieve this is to express the exterior and interior velocities in terms of a single-layer potential as

$$u_i(\mathbf{x}_0) = \int_S G_{ij}(\mathbf{x}_0, \mathbf{x}) q_j(\mathbf{x})\, dS(\mathbf{x}) \quad (5.3.1)$$

where $\mathbf{q}$ is the density of the distribution. It should be noted that due to the properties of the single-layer potential, the velocity field expressed by (5.3.1), is continuous throughout the domain of flow as well as across the interface $S$. The corresponding pressure and stress fields are given by

$$P^{(m)}(\mathbf{x}_0) = \mu_m \int_S p_j(\mathbf{x}_0, \mathbf{x}) q_j(\mathbf{x}) \, dS(\mathbf{x}) \qquad (5.3.2)$$

and

$$\sigma_{ik}^{(m)}(\mathbf{x}_0) = \mu_m \int_S T_{ijk}(\mathbf{x}_0, \mathbf{x}) q_j(\mathbf{x}) \, dS(\mathbf{x}) \qquad (5.3.3)$$

respectively, where $m = 1, 2$ for the external and for the internal flow. Using (4.1.4) we find that the surface forces exerted on either side of $S$ are given by

$$f_i^{(1)}(\mathbf{x}_0) \equiv \sigma_{ik}^{(1)}(\mathbf{x}_0) n_k(\mathbf{x}_0) = \mu_1 n_k(\mathbf{x}_0) \int_{S^+} T_{ijk}(\mathbf{x}_0, \mathbf{x}) q_j(\mathbf{x}) \, dS(\mathbf{x}) \quad (5.3.4)$$

and

$$f_i^{(2)}(\mathbf{x}_0) \equiv \sigma_{ik}^{(2)}(\mathbf{x}_0) n_k(\mathbf{x}_0) = \mu_2 n_k(\mathbf{x}_0) \int_{S^-} T_{ijk}(\mathbf{x}_0, \mathbf{x}) q_j(\mathbf{x}) \, dS(\mathbf{x}) \quad (5.3.5)$$

Taking the limit as the point $\mathbf{x}_0$ approaches the interface either from the external or from the internal side, and expressing the limits of the surface forces given by (5.3.4) and (5.3.5) in terms of their principal values using (4.1.7) and (4.1.8), we find

$$f_i^{(1)}(\mathbf{x}_0) = -4\pi\mu_1 q_i(\mathbf{x}_0) + \mu_1 n_k(\mathbf{x}_0) \int_S^{\mathscr{PV}} T_{ijk}(\mathbf{x}_0, \mathbf{x}) q_j(\mathbf{x}) \, dS(\mathbf{x}) \qquad (5.3.6)$$

and

$$f_i^{(2)}(\mathbf{x}_0) = 4\pi\mu_2 q_i(\mathbf{x}_0) + \mu_2 n_k(\mathbf{x}_0) \int_S^{\mathscr{PV}} T_{ijk}(\mathbf{x}_0, \mathbf{x}) q_j(\mathbf{x}) \, dS(\mathbf{x}) \qquad (5.3.7)$$

Subtracting (5.3.7) from (5.3.6) yields

$$\Delta f_i(\mathbf{x}_0) \equiv [f_i^{(1)} - f_i^{(2)}](\mathbf{x}_0)$$

$$= -4\pi(\mu_1 + \mu_2) q_i(\mathbf{x}_0) + (\mu_1 - \mu_2) n_k(\mathbf{x}_0) \int_S^{\mathscr{PV}} T_{ijk}(\mathbf{x}_0, \mathbf{x}) q_j(\mathbf{x}) \, dS(\mathbf{x}) \qquad (5.3.8)$$

Rearranging this equation we obtain

$$q_i(\mathbf{x}_0) = -\frac{1}{(4\pi\mu_1)(1+\lambda)} \Delta f_i(\mathbf{x}_0) + \frac{\kappa}{4\pi} n_k(\mathbf{x}_0) \int_S^{\mathscr{PV}} T_{ijk}(\mathbf{x}_0, \mathbf{x}) q_j(\mathbf{x}) dS(\mathbf{x}) \quad (5.3.9)$$

where $\kappa = (1 - \lambda)/(1 + \lambda)$. Prescribing $\Delta\mathbf{f}$ renders (5.3.9) a Fredholm integral equation of the second kind for the density of the single-layer potential $\mathbf{q}$. Once this equation is solved, the velocity, pressure, and stress field may be computed using the representations (5.3.1), (5.3.2), and (5.3.3).

It will be noted that when $\lambda = 1$ or $k = 0$, (5.3.9) yields $\mathbf{q} = (-1/8\pi\mu)\Delta\mathbf{f}$. Substituting this expression into (5.3.1) we recover the boundary integral equation (5.2.9).

**Problems**

5.3.1 Integrating (5.3.9) over the interface, and using (2.1.12) and (2.1.13) show that

$$\int_S \mathbf{q}\, dS = -\frac{1}{8\pi\mu_1}\int_S \Delta\mathbf{f}\, dS \qquad \int_S \mathbf{x}\times\mathbf{q}\, dS = -\frac{1}{8\pi\mu_1}\int_S \mathbf{x}\times\Delta\mathbf{f}\, dS$$

Discuss the implications of these equations for the behaviour of the far flow for a force-free, torque-free particle.

5.3.2 Consider an ambient flow $\mathbf{u}^\infty$ past a particle, and introduce the boundary integral representation

$$u_i(\mathbf{x}_0) = u_i^\infty(\mathbf{x}_0) + \int_S G_{ij}(\mathbf{x}_0, \mathbf{x}) q_j(\mathbf{x})\, dS(\mathbf{x})$$

where the external and internal stress fields are given by

$$\sigma_{ij}^{(1)}(\mathbf{x}_0) = \sigma_{ij}^\infty(\mathbf{x}_0) + \mu_1 \int_S T_{ijk}(\mathbf{x}_0, \mathbf{x}) q_j(\mathbf{x})\, dS(\mathbf{x})$$

and

$$\sigma_{ij}^{(2)}(\mathbf{x}_0) = \lambda\sigma_{ij}^\infty(\mathbf{x}_0) + \mu_2 \int_S T_{ijk}(\mathbf{x}_0, \mathbf{x}) q_j(\mathbf{x})\, dS(\mathbf{x})$$

respectively. Using the procedures outlined in this section, derive the integral equation for a point $\mathbf{x}_0$ on $S$:

$$q_i(\mathbf{x}_0) = -\frac{1}{4\pi\mu_1}\left[\frac{1}{1+\lambda}\Delta f_i - \kappa f_i^\infty\right](\mathbf{x}_0) + \frac{\kappa}{4\pi} n_k(\mathbf{x}_0) \int_S^{\mathscr{PV}} T_{ijk}(\mathbf{x}_0, \mathbf{x}) q_j(\mathbf{x})\, dS(\mathbf{x})$$

## 5.4 Investigation of the integral equations

In the preceding two sections we discussed two alternative integral formulations, one based on the boundary integral equation, and the second based on a single-layer representation. Each of these formulations produced a Fredholm integral equation of the second kind, the first for the interfacial velocity $\mathbf{u}$, and the second for the density of the single-layer potential $\mathbf{q}$, namely equations (5.2.9) and (5.3.9) (and the more general ones accounting for the presence of incident flow discussed in problems 5.2.2 and 5.3.2). To assess the validity of these formulations, we now proceed to examine the properties of the corresponding integral equations. Thus, we consider the associated homogeneous equations

$$u_i(\mathbf{x}_0) = \frac{\beta}{4\pi}\int_S^{\mathscr{PV}} u_j(\mathbf{x}) T_{jik}(\mathbf{x}_0, \mathbf{x}) n_k(\mathbf{x})\, dS(\mathbf{x}) \tag{5.4.1}$$

and

$$q_i(\mathbf{x}_0) = \frac{\beta^*}{4\pi} n_k(\mathbf{x}_0) \int_S^{\mathscr{PV}} T_{ijk}(\mathbf{x}_0, \mathbf{x}) q_j(\mathbf{x})\, dS(\mathbf{x}) \tag{5.4.2}$$

and inquire whether they have any non-trivial cigensolutions. It will be noted that (5.4.1) and (5.4.2) are adjoint to each other and, hence, have complex conjugate eigenvalues and the same number of eigensolutions. In section 4.5 we showed that:

(1) Equations (5.4.1) and (5.4.2) have only real eigenvalues with magnitude larger than or equal to unity, i.e. $|\beta| \geqq 1$.
(2) $\beta = 1$ is an eigenvalue with multiplicity one.
(3) Provided that the interface does not enclose or cross the solid boundary $S_B$, $\beta = -1$ is another eigenvalue, with multiplicity six.

Using these results and invoking Fredholm's theorems we find that equations (5.2.9) and (5.3.9) will have a unique solution as long as $\lambda$ is not equal to zero or infinity, and furthermore, the solution may be computed using the method of successive substitutions.

When $\lambda = \infty$ or $\lambda = 0$ (in which cases the particle has become respectively a solid particle or an inviscid bubble), $\kappa$ becomes respectively equal to $-1$ or 1, and (5.2.9) and (5.3.9) have respectively either none or an infinity of solutions. The limit corresponding to a solid particle was discussed in detail in section 4.7. Considering the limit corresponding to the inviscid bubble, we find that a necessary condition for (5.2.9) to have a solution is that the forcing term, expressed by the single-layer integral, should be orthogonal to the adjoint eigenfunction $\mathbf{n}$ (see section 4.5). This, however, is guaranteed by (2.1.4) for any choice of $\Delta\mathbf{f}$. On the other hand, for (5.3.9) to have a solution, $\Delta\mathbf{f}$ must be orthogonal to the eigenfunction of the corresponding homogeneous equation. This condition will not be met for an arbitrary choice of $\Delta\mathbf{f}$, and is certainly not met when $\Delta\mathbf{f}$ is proportional to $\mathbf{n}$ (see (4.5.14)). Physically, our inability to represent a flow with $\lambda = 0$ in terms of a single-layer potential is attributable to the intrinsic restriction that the rate of expansion of the inviscid fluid composing the particle is equal to zero.

Considering (5.2.9), we note that when the interface is inactive, $\Delta\mathbf{f} = 0$, the single-layer integral vanishes, and we obtain a homogeneous integral equation for the interfacial velocity $\mathbf{u}$ that is identical to (4.4.2). As discussed above, eigensolutions to this equation will exist only in the special cases $\kappa = -1, 1$ corresponding to $\lambda = \infty, 0$ respectively. We recall, in particular, that in the former case there are six independent eigensolutions representing six independent modes of rigid body motion. This behaviour is consistent with our previous remark that the mathematical limit $\lambda \to \infty$ leads us to the physical limit of a solid particle. Overall, we conclude that in the absence of an incident flow, an inactive interface between two fluids with finite viscosities is not capable of producing a flow.

Considering next (5.3.9), we note that when $\Delta\mathbf{f}=0$, we obtain a homogeneous integral equation for $\mathbf{q}$ that is identical to (4.4.3). Clearly, this equation will have eigensolutions only in the special cases $\kappa=-1, 1$ corresponding to $\lambda=\infty, 0$, respectively. In the case $\lambda=0$ there is a single eigensolution $\mathbf{q}=c\mathbf{n}$ where $c$ is a constant; substituting into (5.3.1) and using (2.1.4) as well as the symmetry property (3.1.3) we find that the corresponding velocity field vanishes. In the case $\lambda=\infty$, the eigensolutions represent the surface force over the external side of the interface arising when the particle translates as a rigid body. This interpretation is certainly consistent with the simplified boundary integral equation (2.3.27). Physically, these results indicate that in the absence of an incident flow, the single-layer representation of a flow due to an inactive interface will be meaningful only in the limiting case where the particle reduces to a rigid body.

### Problems

5.4.1 Derive a boundary integral representation of a two-dimensional flow due to the instability of a liquid layer resting on a plane wall below another liquid of higher density (Yiantsios & Higgins 1989; Newhouse & Pozrikidis 1990).

5.4.2 Derive a boundary integral representation of the flow due to the motion of a rigid particle in the vicinity of an interface.

5.4.3 Derive a boundary integral representation of the two-dimensional gravity-driven flow of a liquid film along an inclined wall containing a solitary projection. In your formulation, use the Green's function for flow bounded by a plane wall. Next, consider the case of flow over a wall containing a depression. Discuss whether it will be possible to use the same Green's function.

## 5.5 The discontinuity in the interfacial surface force

To complete the boundary integral representation, we must introduce a constitutive relation for the discontinuity of the interfacial surface force $\Delta\mathbf{f}$ defined in (5.2.4). As mentioned in the introduction, this relation will depend upon the density of the fluids on either side of the interface as well as on the composition and structural characteristics of the interface. To decouple the effect of the fluids from that of the interface, we recall that $\boldsymbol{\sigma}$ is the modified stress tensor defined with respect to the modified pressure, and write

$$\Delta\mathbf{f}=\Delta\rho\,\mathbf{g}\cdot\mathbf{x}\,\mathbf{n}+\Delta\{-P\mathbf{I}+\mu[\nabla\mathbf{u}+(\nabla\mathbf{u})^{T}]\}\cdot\mathbf{n} \tag{5.5.1}$$

where $\Delta[\ ]=[\ ]_1-[\ ]_2$ and $\mathbf{g}$ is the acceleration of gravity. For convenience, we shall denote the second term on the right-hand side of (5.5.1) by $\Delta\mathscr{f}$, setting

$$\Delta \mathbf{f} = \Delta\rho\, \mathbf{g}\cdot\mathbf{x}\,\mathbf{n} + \Delta\mathscr{f} \qquad (5.5.2)$$

The reader should note that the discontinuity in the surface force $\Delta\mathscr{f}$ depends exclusively on the properties of the interface and is independent of the densities of the fluids.

Constitutive relations for $\Delta\mathscr{f}$ have been discussed by several previous authors including Scriven (1960), Secomb & Skalak (1982), Waxman (1984), Sapir & Nir (1985), Zinemanas & Nir (1988), and Aris (1989). In the rest of this section we shall discuss certain simple cases reflecting common types of interfacial behavior.

### *Interfaces with isotropic tension*

Clean interfaces that contain no impurities or surfactants are characterized by an isotropic surface tension $\gamma$ acting in the plane of the interface. To obtain a relation between $\gamma$ and $\Delta\mathscr{f}$, we consider an elemental surface area of the interface $\Delta A$ that is enclosed by the contour $C$, and write the force balance

$$\int_{\Delta A} \Delta\mathscr{f}\, \mathrm{d}S + \int_C \gamma \mathbf{b}\, \mathrm{d}l = 0 \qquad (5.5.3)$$

where the unit vector $\mathbf{b}$ lies in the plane of the interface and is normal to $C$ (Figure 5.5.1). Using the identity $\mathbf{b} = \mathbf{t}\times\mathbf{n}$, we rewrite (5.5.3) in the

*Figure 5.5.1.* Schematic illustration of a section of a three-dimensional interface.

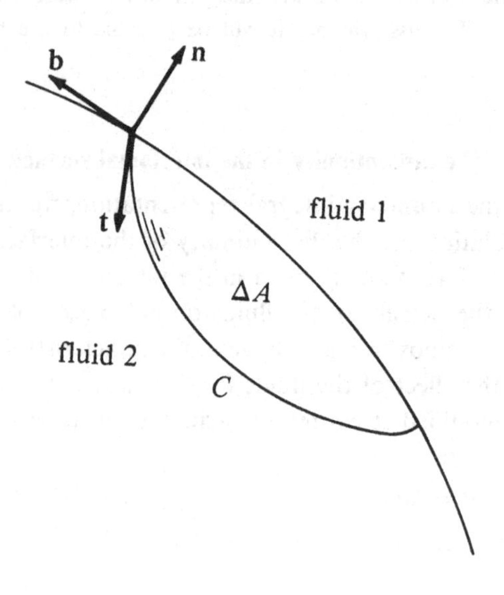

equivalent form

$$\int_{\Delta A} \Delta f \, dS = \int_C \gamma \mathbf{n} \times \mathbf{t} \, dl \tag{5.5.4}$$

where $\mathbf{n}$ and $\mathbf{t}$ are unit vectors normal to the interface and tangential to $C$ respectively. To convert the contour integral on the right-hand side of (5.5.4) into an area integral, we use a variation of Stokes' theorem stating

$$\int_C \mathbf{F} \times \mathbf{t} \, dl = \int_{\Delta A} (\mathbf{n}\nabla \cdot \mathbf{F} - \nabla \mathbf{F} \cdot \mathbf{n}) \, dS \tag{5.5.5}$$

where $\mathbf{F}$ is a differentiable function defined over the whole space (see problem 5.5.1). Extending the domain of definition of the normal vector $\mathbf{n}$ and of the surface tension $\gamma$ away from the interface into the whole space, setting $\mathbf{F} = \gamma \mathbf{n}$, and using (5.5.5) we obtain

$$\int_C \gamma \mathbf{n} \times \mathbf{t} \, dl = \int_{\Delta A} [\mathbf{n}\nabla \cdot (\gamma \mathbf{n}) - \nabla(\gamma \mathbf{n}) \cdot \mathbf{n}] \, dS \tag{5.5.6}$$

Substituting (5.5.6) into (5.5.4), noting that $\nabla \mathbf{n} \cdot \mathbf{n} = \frac{1}{2}\nabla(\mathbf{n} \cdot \mathbf{n}) = 0$, and letting $\Delta A$ tend to zero we obtain

$$\Delta f = \gamma \mathbf{n} \nabla \cdot \mathbf{n} - (\mathbf{I} - \mathbf{nn}) \cdot \nabla \gamma \tag{5.5.7}$$

It is instructive to note that the projection matrix $\mathbf{I} - \mathbf{nn}$ extracts the tangential component of the gradient of the surface tension.

To simplify (5.5.7), we introduce a Cartesian coordinate system $(x', y', z')$ that has two axes tangential to the interface at its origin, as illustrated in Figure 5.5.2. Clearly, at the origin $n_{x'} = n_{z'} = 0$, and $n_{y'} = 1$. Using the identity $\nabla(\mathbf{n} \cdot \mathbf{n}) = 0$ we find $\partial n_{y'}/\partial y' = 0$. Substituting these values into (5.5.7)

*Figure* 5.5.2. A local Cartesian coordinate system tangential to a three-dimensional interface.

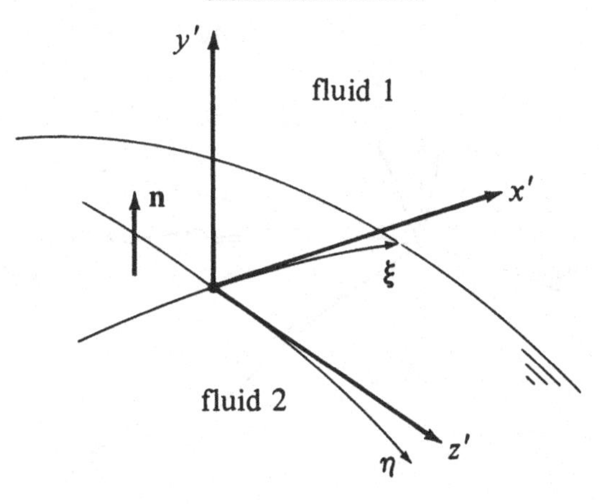

we obtain the simplified expression

$$\Delta \mathbf{f} = \gamma \mathbf{n} \nabla' \cdot \mathbf{n} - \nabla' \gamma \tag{5.5.8}$$

where $\nabla'$ is the *surface gradient* $(\partial/\partial x', \partial/\partial z')$. The divergence

$$\nabla' \cdot \mathbf{n} = \frac{\partial n_{x'}}{\partial x'} + \frac{\partial n_{z'}}{\partial z'} = 2\, k_m \tag{5.5.9}$$

is equal to twice the mean curvature $k_m$ of the interface. More specifically,

$$k_x = \frac{1}{R_x} = \frac{\partial n_{x'}}{\partial x'} \quad \text{and} \quad k_z = \frac{1}{R_z} = \frac{\partial n_{z'}}{\partial z'} \tag{5.5.10}$$

represent the curvatures of the trace of the interface in the $x'y'$-plane and in the $y'z'$-plane, as illustrated in Figure 5.5.2; $R_x$ and $R_z$ are the corresponding radii of curvature.

To compute the surface derivatives in (5.5.8), we draw two curvilinear axes $\xi$ and $\eta$ that lie in the interface and are tangential to the $x'$-axis and to the $z'$-axis at the origin respectively, as illustrated in Figure 5.5.3. Clearly, at the origin

$$\partial\gamma/\partial z' = \partial\gamma/\partial\eta \qquad \partial\gamma/\partial x' = \partial\gamma/\partial\xi \tag{5.5.11}$$

To calculate $k_z$ and $k_x$ we use Meunier's theorem:

$$k_z = \frac{\partial n_{z'}}{\partial z'} = \frac{\partial n_\eta}{\partial \eta} = k_\eta \mathbf{n} \cdot \mathbf{n}_\eta \qquad k_x = \frac{\partial n_{x'}}{\partial x'} = \frac{\partial n_\xi}{\partial \xi} = k_\xi \mathbf{n} \cdot \mathbf{n}_\xi \tag{5.5.12}$$

where $k_\eta$ and $k_\xi$ are the curvatures of the $\eta$-axis and of the $\xi$-axis, and $\mathbf{n}_\eta$ and $\mathbf{n}_\xi$ are the unit vectors normal to the $\eta$-axis and to the $\xi$-axis. as

*Figure 5.5.3.* A system of two curvilinear axes $(\eta, \xi)$ on a three-dimensional interface, tangential to the $z'$-axis and $x'$-axis at the origin; $\mathbf{n}, \mathbf{n}_\eta$, and $\mathbf{n}_\xi$ are the unit vectors normal to the interface, to the $\eta$-axis, and to the $\xi$-axis, respectively.

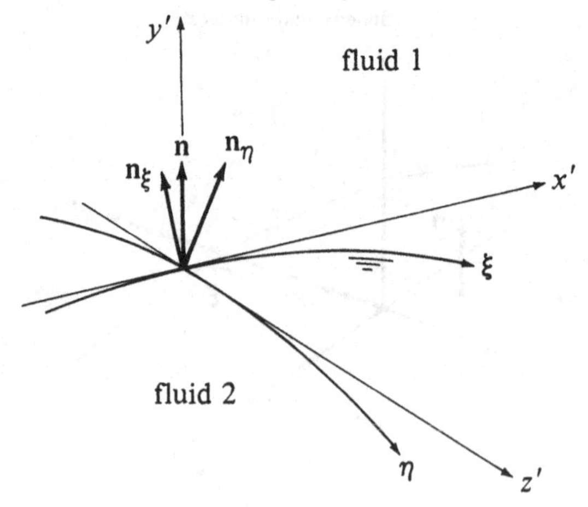

indicated in Figure 5.5.3 (Weatherburn 1961, Vol. I, p. 62). Equations (5.5.11) and (5.5.12) provide us with convenient formulae for computing the surface derivatives merely by differentiating along two curvilinear axes.

As an application of the above formulae, we consider the simple case of a cylindrical interface whose generators are parallel to the $z$-axis, as illustrated in Figure 5.5.4. Identifying $\eta$ with $z$, and $\xi$ with the arc length $s$ along the trace of the interface in the $xy$-plane, and neglecting surface derivatives with respect to $\eta$, we reduce (5.5.8) to the simplified form

$$\Delta \mathbf{f} = \gamma k \mathbf{n} - \frac{\partial \gamma}{\partial s} \mathbf{t} \tag{5.5.13}$$

where $k$ is the curvature of the trace of the interface in the $xy$-plane and is given in (6.2.29).

As a second application, we consider the axisymmetric interface illustrated in Figure 5.5.5. Identifying $\eta$ with the arc length $s$ of the trace of the interface in the $\phi = 0$ azimuthal plane, and $\xi$ with the azimuthal angle $\phi$, and neglecting surface derivatives with respect to $\xi$, we reduce (5.5.8) to the simplified form

$$\Delta \mathbf{f} = 2\gamma k_{\mathrm{m}} \mathbf{n} - \frac{\partial \gamma}{\partial s} \mathbf{t}_s \tag{5.5.14}$$

The mean curvature $k_{\mathrm{m}}$ is given in (6.2.35).

### *Interfaces with elastic behaviour*

Proceeding to more complex behaviour, we consider interfaces that are composed of an isotropic material with elastic properties. For simplicity, we restrict our attention to cylindrical and axisymmetric shapes.

*Figure 5.5.4.* The trace of a two-dimensional interface in a plane normal to the generators; $s$ is the length along the trace of the interface, and $\beta$ is a parametric variable that increases monotonically along the interface.

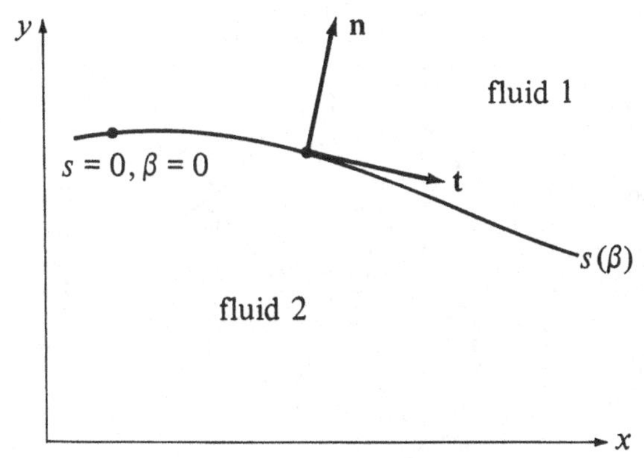

Considering first a cylindrical interface, we introduce a Lagrangian variable $\beta$ marking Lagrangian point particles along the trace of the interface in the $xy$-plane, as illustrated in Figure 5.5.4, and define the extension ratio $\lambda_s$

$$\lambda_s = (\partial s/\partial \beta)_t/(\partial s/\partial \beta)_{t=0}, \tag{5.5.15}$$

where the interface is assumed to be unstressed at $t = 0$. Requiring a linearly elastic response, we set

$$\gamma = E(\lambda_s - 1), \tag{5.5.16}$$

where $\gamma$ is the surface tension, and $E$ is the modulus of elasticity. Note that $\gamma$ is positive when the interface is stretched ($\lambda_s > 0$) and negative when the interface is compressed ($\lambda_s < 0$). Substituting (5.5.16) into (5.5.13) we obtain the desired constitutive relation

$$\Delta \mathbf{f} = E\left[(\lambda_s - 1)k\mathbf{n} - \frac{\partial \lambda_s}{\partial s}\mathbf{t}\right] \tag{5.5.17}$$

Considering next an axisymmetric interface, we note that the surface tension is a rank two tensor whose principal axes are oriented in the meridional and azimuthal directions, as indicated in Figure 5.5.6 (Green & Adkins 1960, p. 148). Writing a force balance over a section of the interface we obtain

$$\Delta \mathbf{f}(\sigma\, \mathrm{d}\phi\, \mathrm{d}s) + \frac{\partial}{\partial s}(\mathbf{t}_s \tau_{ss} \sigma\, \mathrm{d}\phi)\, \mathrm{d}s + \frac{\partial}{\partial \phi}(\mathbf{t}_\phi \tau_{\phi\phi} \mathrm{d}s)\, \mathrm{d}\phi = 0 \tag{5.5.18}$$

*Figure* 5.5.5. Schematic illustration of an axisymmetric interface; $\beta$ is a Lagrangian variable marking point particles along the trace of the interface in a meridional plane. The arc length is measured in a counterclockwise direction from $\beta = 0$.

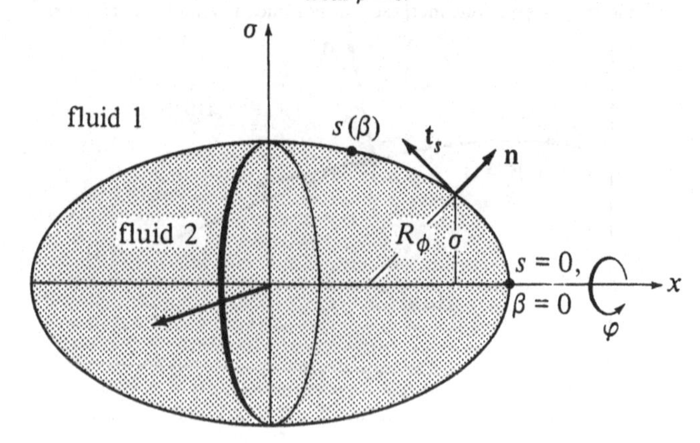

or

$$\Delta f = -\left[\mathbf{t}_s \frac{1}{\sigma}\frac{\partial(\tau_{ss}\sigma)}{\partial s} + \tau_{ss}\frac{\partial \mathbf{t}_s}{\partial s} + \frac{1}{\sigma}\tau_{\phi\phi}\frac{\partial \mathbf{t}_\phi}{\partial \phi}\right] \tag{5.5.19}$$

where $\tau_{ss}$ and $\tau_{\phi\phi}$ are the principal surface tensions in the meridional and azimuthal direction. Using the Frenet formulae

$$\partial \mathbf{t}_s/\partial s = -k_s \mathbf{n} \qquad \partial \mathbf{t}_\phi/\partial \phi = -\mathbf{e}_\sigma \tag{5.5.20}$$

where $k_s$ is the curvature of the interface in a meridional plane and $\mathbf{e}_\sigma$ is the unit vector in the $\sigma$ direction (Stoker 1989, p. 58), we obtain

$$\Delta f = -\frac{1}{\sigma}\frac{\partial(\tau_{ss}\sigma)}{\partial s}\mathbf{t}_s + \tau_{ss}k_s\mathbf{n} + \frac{1}{\sigma}\tau_{\phi\phi}\mathbf{e}_\sigma \tag{5.5.21}$$

Finally, using the relations $\mathbf{t}_s \cdot \mathbf{e}_\sigma = \partial\sigma/\partial s$ and $\mathbf{n}\cdot\mathbf{e}_\sigma = \sigma k_\phi$ (where the second relation is a result of Meunier's theorem), we express $\Delta f$ in terms of its normal and tangential components, namely

$$\Delta f = (k_s\tau_{ss} + k_\phi\tau_{\phi\phi})\mathbf{n} - \left[\frac{\partial\tau_{ss}}{\partial s} + \frac{1}{\sigma}\frac{\partial\sigma}{\partial s}(\tau_{ss} - \tau_{\phi\phi})\right]\mathbf{t}_s \tag{5.5.22}$$

It will be noted that in the limit of isotropic tension, i.e. $\tau_{ss} = \tau_{\phi\phi} = \gamma$, (5.5.22) reduces to (5.5.14).

So far we have made no stipulations of elastic behaviour. To proceed further, it is necessary to introduce a Lagrangian variable $\beta$ marking Lagrangian point particles along the trace of the interface in a meridional plane, as illustrated in Figure 5.5.5, and to define the principal extension ratios $\lambda_s$ and $\lambda_\phi$

$$\lambda_s = (\partial s/\partial\beta)_t/(\partial s/\partial\beta)_{t=0}, \qquad \lambda_\phi = \sigma_t/\sigma_{t=0} \tag{5.5.23}$$

where $s(\beta, t)$ is the arc length in the meridional plane measured in the counter-clockwise sense. The interface is assumed to be unstressed at $t = 0$.

*Figure 5.5.6.* A force balance on a small portion of an axisymmetric interface.

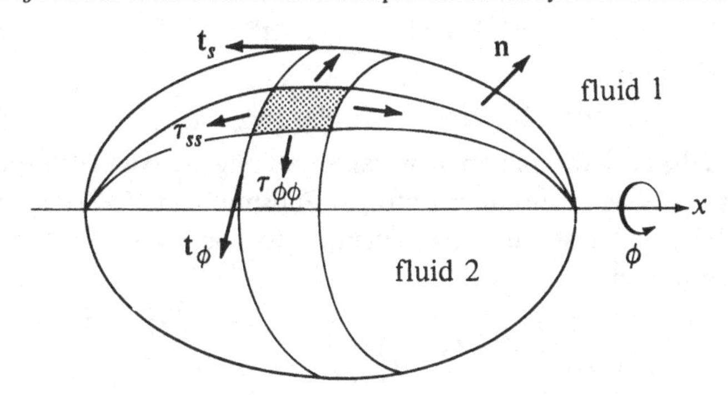

To obtain elastic behaviour, we set

$$\tau_{ss} = 4h_0 \frac{1}{\lambda_s \lambda_\phi} \left( \lambda_s^2 - \frac{1}{\lambda_s^2 \lambda_\phi^2} \right) \left( \frac{\partial W}{\partial I_1} + \lambda_\phi^2 \frac{\partial W}{\partial I_2} \right) \tag{5.5.24}$$

$$\tau_{\phi\phi} = 4h_0 \frac{1}{\lambda_s \lambda_\phi} \left( \lambda_\phi^2 - \frac{1}{\lambda_s^2 \lambda_\phi^2} \right) \left( \frac{\partial W}{\partial I_1} + \lambda_s^2 \frac{\partial W}{\partial I_2} \right) \tag{5.5.25}$$

where $W(I_1, I_2)$ is a strain-energy function, $2h_0$ is the effective thickness of the interface, and $I_1$ and $I_2$ are two invariants of the deformation defined as

$$I_1 = \lambda_s^2 + \lambda_\phi^2 + \frac{1}{\lambda_s^2 \lambda_\phi^2} \qquad I_2 = \frac{1}{\lambda_s^2} + \frac{1}{\lambda_\phi^2} + \lambda_s^2 \lambda_\phi^2 \tag{5.5.26}$$

The strain-energy function $W$ may be expressed by Mooney's constitutive law as

$$W = \frac{E}{12h_0} [(1 - \alpha)(I_1 - 3) + \alpha(I_2 - 3)] \tag{5.5.27}$$

where $E$ is the surface modulus of elasticity, and $\alpha$ is a physical parameter that varies between zero and unity according to whether the interface exhibits linear or non-linear behaviour (Li, Barthes-Biesel & Helmy 1988). Substituting (5.5.27) into (5.5.24) and (5.5.25) we obtain the specific expressions

$$\tau_{ss} = \frac{E}{3} \frac{1}{\lambda_s \lambda_\varphi} \left( \lambda_s^2 - \frac{1}{\lambda_s^2 \lambda_\phi^2} \right) [1 + \alpha(\lambda_\phi^2 - 1)] \tag{5.5.28}$$

$$\tau_{\phi\phi} = \frac{E}{3} \frac{1}{\lambda_s \lambda_\phi} \left( \lambda_\phi^2 - \frac{1}{\lambda_s^2 \lambda_\phi^2} \right) [1 + \alpha(\lambda_s^2 - 1)] \tag{5.5.29}$$

Finally, substituting (5.5.28) and (5.5.29) into (5.5.22) we obtain a constitutive relation for $\Delta\mathbf{f}$.

Now, for future use, we express the principal tensions in terms of their mean and deviatoric components as

$$\tau_{ss} = \tau^{\mathrm{m}} + \tau^{\mathrm{d}}, \qquad \tau_{\phi\phi} = \tau^{\mathrm{m}} - \tau^{\mathrm{d}} \tag{5.5.30}$$

where, evidently,

$$\tau^{\mathrm{m}} = \tfrac{1}{2}(\tau_{ss} + \tau_{\phi\phi}) \qquad \tau^{\mathrm{d}} = \tfrac{1}{2}(\tau_{ss} - \tau_{\phi\phi}) \tag{5.5.31}$$

Note that the deviatoric component vanishes along the axis of deformation, where the interfacial tension is isotropic. Substituting (5.5.28) and (5.5.29) into (5.5.31) and restricting our attention to linearly elastic behaviour ($\alpha = 0$) we obtain

$$\tau^{\mathrm{m}} = \frac{E}{6} \frac{1}{\lambda_s \lambda_\varphi} \left( \lambda_s^2 + \lambda_\phi^2 - \frac{2}{\lambda_s^2 \lambda_\phi^2} \right) \qquad \tau^{\mathrm{d}} = \frac{E}{6} \left( \frac{\lambda_s^2 - \lambda_\phi^2}{\lambda_s \lambda_\varphi} \right) \tag{5.5.32}$$

Substituting (5.5.30) into (5.5.22) we obtain $\Delta\mathbf{f}$ in terms of the mean and deviatoric tensions, namely

$$\Delta\mathbf{f} = [(k_s + k_\phi)\tau^{\mathrm{m}} + (k_s - k_\phi)\tau^{\mathrm{d}}]\mathbf{n} - \left(\frac{\partial \tau^{\mathrm{m}}}{\partial s} + \frac{\partial \tau^{\mathrm{d}}}{\partial s} + \frac{2}{\sigma}\frac{\partial \sigma}{\partial s}\tau^{\mathrm{d}}\right)\mathbf{t}_s \tag{5.5.33}$$

### *Incompressible interfaces*

As a last case, we consider an incompressible interface composed of an inextensible material such as the membrane enclosing a red blood cell (Evans & Skalak 1980).

Considering first a cylindrical interface we state the incompressibility condition in terms of the extension ratio $\lambda_s$ defined in (5.5.15), as

$$\left(\frac{\partial \lambda_s}{\partial t}\right)_\beta = 0 \tag{5.5.34}$$

Carrying out the differentiation we obtain the equivalent expression

$$\mathbf{t}\cdot\frac{\partial \mathbf{u}}{\partial s} = \frac{\partial}{\partial s}(\mathbf{u}\cdot\mathbf{t}) + k\mathbf{u}\cdot\mathbf{n} = 0 \tag{5.5.35}$$

(see (6.2.27)). Representing the velocity in terms of a boundary integral involving $\Delta\mathbf{f}$, using (5.5.13) to express $\Delta\mathbf{f}$ in terms of $\gamma$, and substituting the resulting equation into (5.5.35), we obtain an integral constitutive relation for $\gamma$.

Next, we turn our attention to an axisymmetric interface. Thus, we consider (5.5.33) and require two scalar constitutive relations for $\tau^{\mathrm{m}}$ and $\tau^{\mathrm{d}}$. One relation arises by stipulating that the surface area of the membrane remains locally and globally constant in time. Noting that the area of an axisymmetric section of the interface contained between two planes that are perpendicular to the $x$-axis at $\beta$ and at $\beta = 0$ is equal to

$$A(\beta, t) = 2\pi \int_0^\beta \sigma \frac{\partial s}{\partial \beta'} \mathrm{d}\beta' = 2\pi \int_0^\beta \left(\sigma \frac{\partial s}{\partial \beta'}\right)_{t=0} \lambda_s \lambda_\phi \mathrm{d}\beta' \tag{5.5.36}$$

where $\lambda_s$ and $\lambda_\phi$ are the principal extension ratios defined in (5.5.23), we find that in order for the area of the membrane to remain locally constant in time, $\lambda_s\lambda_\phi = 1$ for any value of $\beta$ or $(\partial\lambda_s\lambda_\phi/\partial t)_\beta = 0$. Carrying out the differentiation we obtain

$$\mathbf{u}\cdot e_\sigma + \sigma \mathbf{t}_s \cdot \frac{\partial \mathbf{u}}{\partial s} = 0 \tag{5.5.37}$$

where $e_\sigma$ is the unit vector in the $\sigma$ direction (Pozrikidis 1990c). Proceeding, we represent the velocity in terms of a boundary integral involving $\Delta\mathbf{f}$, use (5.5.33) to express $\Delta\mathbf{f}$ in terms of $\tau^{\mathrm{m}}$ and $\tau^{\mathrm{d}}$, and substitute the resulting equation into (5.5.37) to obtain an integral relation between $\tau^{\mathrm{m}}$ and $\tau^{\mathrm{d}}$.

This relation plays the role of a constitutive equation for $\Delta f$. To derive a second constitutive equation, we might assume that the deviatoric component of the principal stresses is identical to that for a linearly elastic interface. Substituting the condition for incompressibility, $\lambda_s \lambda_\phi = 1$, into the second equation in (5.5.32) we obtain

$$\tau^{d} = \frac{E}{6}\left(\lambda_s^2 - \frac{1}{\lambda_s^2}\right) \tag{5.5.38}$$

concluding the definition of $\Delta f$.

### Problem

5.5.1 Using Stokes' theorem

$$\int_C \mathbf{u}\cdot\mathbf{t}\,dl = \int_S (\nabla\times\mathbf{u})\cdot\mathbf{n}\,dS$$

where $\mathbf{u}$ is a differentiable function, prove the identity (5.5.5).

## 5.6 Computing the shape of interfaces

### *Evolving interfaces*

One way to follow the motion of an evolving interface is to mark the location of the interface using a set of marker points, and to describe the evolution of the interface by computing the trajectories of the marker points. To facilitate the logistics, it is convenient to identify the marker points using two Lagrangian marker variables $\alpha$ and $\beta$, and to consider the velocity $\mathbf{v}$ of a marker point as a function of $\alpha$, $\beta$, and time:

$$\frac{\partial \mathbf{x}}{\partial t}(\alpha, \beta, t) = \mathbf{v}(\alpha, \beta, t) \tag{5.6.1}$$

Clearly, the velocity of the interface will be dependent upon the velocity of the fluid. Interpreting the interface simply as the boundary between two fluids, we allow the marker points to slip along the interface, but require that they move normally to the interface at a velocity equal to that of the fluid. Thus, we set

$$\mathbf{v} = [\mathbf{u}(\mathbf{x}, t)\cdot\mathbf{n}]\mathbf{n} + \delta\mathbf{t} \tag{5.6.2}$$

where $\mathbf{u}$ is the velocity of the fluid over the interface, $\mathbf{t}$ is an arbitrary tangent vector, and $\delta$ is an arbitrary constant; $\mathbf{t}$ and $\delta$ may be chosen arbitrarily and individually for each marker point. It will be noted that if we set $\delta\mathbf{t} = \mathbf{u}\cdot(\mathbf{I} - \mathbf{nn})$ we shall identify the velocity of the marker points with the velocity of the fluid. Regardless of the choice of $\delta$ and $\mathbf{t}$, if we combine (5.6.1) and (5.6.2) and note that the velocity $\mathbf{u}$ depends on the current configuration of the interface, i.e. it is a function of $\mathbf{x}(\alpha, \beta)$, we shall obtain a system of non-linear ordinary differential equations describing the trajectories of the marker points:

$$\frac{\partial \mathbf{x}}{\partial t}(\alpha, \beta, t) = [\mathbf{u}(\mathbf{x}, t) \cdot \mathbf{n}]\mathbf{n} + \delta \mathbf{t} \tag{5.6.3}$$

### *Stationary interfaces*

To compute the shape of a stationary interface, it is convenient to describe the location of the interface in terms of an unknown function $g$ by setting, for instance, $\mathbf{x} = (g, \alpha, \beta)$. Decomposing the velocity at the interface into its normal and tangential components, we obtain

$$\mathbf{u} = u^{NR}\mathbf{n} + u_1^{TN}\mathbf{t}_1 + u_2^{TN}\mathbf{t}_2 \tag{5.6.4}$$

where

$$u^{NR} = \mathbf{u} \cdot \mathbf{n} \qquad u_1^{TN}\mathbf{t}_1 + u_2^{TN}\mathbf{t}_2 = \mathbf{u} \cdot (\mathbf{I} - \mathbf{nn})$$

where $\mathbf{t}_1$ and $\mathbf{t}_2$ are two unit vectors tangential to the interface. Since the interface is stationary, the component of the velocity normal to the interface must be equal to zero, i.e. $u^{NR} = 0$. (If $\mathbf{u}$ represents a disturbance flow, $\mathbf{u}^{NR}$ is not equal to zero; instead, its value depends on the basic flow and the shape of the interface, i.e. it is a function of $g$.) Noting that $\Delta \mathbf{f}$ depends on the shape of the interface and is thus a function of $g$, substituting (5.6.4) into the integral equation (5.2.9), and requiring that $u^{NR} = 0$, we obtain a system of three scalar integral equations for $u_1^{TN}$, $u_2^{TN}$, and $g$. It should be noted that this system is linear with respect to $u_1^{TN}$ and $u_2^{TN}$, but non-linear with respect to the unknown shape function $g$.

One standard way of solving the above system is by using the Newton–Raphson method. Another way is to guess the shape function $g$, solve two different sets of the scalar integral equations for $u_1^{TN}$ and $u_2^{TN}$, and then adjust $g$ so that partial sets of the two solutions have the same values. Experience has shown that the success of this method depends on the selection of the two sets of integral equations and of the subset of partial solutions used for adjusting $g$ (Youngren & Acrivos 1976, Pozrikidis 1988).

In practice, instead of incorporating all the boundary conditions into the integral equation, it may be more efficient to incorporate only a partial set and to use the remaining conditions as constraints. To implement this strategy, we decompose $\Delta \mathbf{f}$ into a normal and a tangential component given by

$$\Delta \mathbf{f}^{NR} = (\Delta \mathbf{f} \cdot \mathbf{n})\mathbf{n} \tag{5.6.5}$$

$$\Delta \mathbf{f}^{TN} = \Delta \mathbf{f} \cdot (\mathbf{I} - \mathbf{n}\mathbf{n}) \tag{5.6.6}$$

In the *normal stress method* we guess $g$, require $u^{NR} = 0$, enforce (5.6.6), solve the integral equations for $|\Delta \mathbf{f}^{NR}|$, compare the computed values of $|\Delta \mathbf{f}^{NR}|$ with those required by (5.6.5), and adjust $g$ to obtain agreement. In the *shear stress method* we assume $g$, require $u^{NR} = 0$, enforce (5.6.5), assume that $\Delta \mathbf{f}^{TN}$ is in the direction of the tangent vector on the right-hand side of (5.6.6), solve the integral equation for $|\Delta \mathbf{f}^{TN}|$, compare the values of $|\Delta \mathbf{f}^{TN}|$ with those required by (5.6.6), and adjust $g$ to obtain agreement. Finally, in the *normal velocity method*, we assume $g$, enforce (5.6.5) and (5.6.6), compute the velocity at the interface, and adjust $g$ so that $u^{NR} = 0$. Experience has shown that the normal stress method is suitable for interfaces with high tension, whereas the normal velocity method is suitable for interfaces with low tension (Silliman & Scriven 1980; Reinelt 1987; Pozrikidis 1988).

# 6

# *The boundary element method*

## 6.1 General procedures

Analytical solutions to the Fredholm integral equations that arise from boundary integral representations are feasible only for a limited number of boundary geometries and types of flow. To compute the solution under general conditions, we must resort to approximate methods. Reviewing the available strategies for solving the Fredholm integral equations of mathematical physics, we find a variety of approximate functional and numerical methods having varying degrees of accuracy and sophistication (Kantorovich & Krylov 1958, Chapter 2; Atkinson 1976, 1980, 1990; Baker, 1977, Delves & Mohamed 1985). In a popular class of methods, originated by Nyström (1930), we produce a numerical solution by carrying out the following steps.

In the first step, we trace the domain of the integral equation $D$ with a network of marker points or nodes, and approximate $D$ using a set of $N$ boundary elements that are defined with respect to the nodes (see section 6.2).

In the second step, we approximate the unknown boundary velocity $\mathbf{u}$, surface force $\mathbf{f}$, or density of the hydrodynamic potential $\mathbf{q}$ over each boundary element using a truncated polynomial expansion in terms of properly defined surface variables. The set of all local expansions contains $M$ unknown coefficients. When the unknown boundary function is assumed to be constant over each element, $M$ is equal to the number of elements $N$; otherwise, $M$ is greater than $N$. To facilitate the procedure, it is convenient to identify the coefficients of the expansion over an element with the values of the unknown function at the corresponding nodes. In this case, the functions multiplying the terms in the polynomial expansion are called the *local basis functions* of the element. When the expansion of the unknown function is identical to that of the Cartesian coordinates describing the element (see for instance (6.2.39) or (6.2.44)), we obtain an *isoparametric* representation. Adding the local expansions and collecting the coefficients that multiply the values of the unknown function at a particular node, we obtain an expansion in terms of *global basis functions*.

At this point, it should be noted that certain types of local expansions, although superficially innocuous, may provide an entry for numerical instabilities. One example is linear expansion of the density of the single-layer potential over two-dimensional (planar) elements, using as coefficients in the local expansions the values of the unknown density function at the nodes. This approximation leads to saw-tooth type oscillations in the numerical solution.

In the third step, we substitute the collection of the local expansions of the unknown function into the integral equation, and extract the $M$ constant coefficients from the single-layer and double-layer potentials. In this manner we convert the integral equation into an algebraic equation involving $M$ coefficients. The factors multiplying these coefficients are integrals of the single-layer or double-layer potential over selected boundary elements. In practice, it might be more expedient to compute the factors that multiply the coefficients using numerical differentiation. Thus, assuming that the discretized integral equation may be cast in the form $F(c_1, c_2, \dots c_M) = 0$, where $c_m, m = 1, \dots, M$ are the unknown coefficients, and recalling that $F$ is a linear function of $c_m$, we set $F(c_1, c_2, \dots, c_M) = A_1 c_1 + A_2 c_2 + \cdots + A_M c_M + B$ where $B$ is a constant, and compute $A_m = F(c_1 = 0, c_2 = 0, \dots, c_m = 1, \dots, c_M = 0) - B$ (Pozrikidis & Thoroddsen 1991).

In the fourth step, we compute the coefficients of the local expansions using either the *collocation* method, or the method of *weighted residuals*. In the collocation method, we enforce the discretized integral equation at $M$ collocation points over the boundary; in the method of weighted residuals, we multiply the discretized integral equation by a set of $M$ weighting functions, and integrate the product over the domain $D$ of the integral equation. The goal in both cases is to produce a system of $M$ linear algebraic equations for the coefficients of the local expansions. Considering the method of weighted residuals, we note that different choices for the weighting functions produce different schemes with varying degrees of complexity. Identifying the weighting functions with the global basis functions we obtain *Galerkin's method*, while identifying the weighting functions with delta functions having poles at selected points over the boundary $D$ we recover the collocation method. Due to increased computational requirements, Galerkin's method is recommended only for problems with notable geometrical simplicity (Atkinson 1980; Zick & Homsy 1982).

Finally, we solve the derived system of linear algebraic equations for the $M$ coefficients of the local expansions. Since, in general, this system will be dense and non-symmetric, we must use a general-purpose numerical method. Two possible choices are the Gauss elimination method or an iterative method such as Jacobi's method or the method of conjugate gradients (Golub & Van Loan 1989). For systems of large size, iterative methods are preferable over direct methods, as they require a reduced computational effort. Indeed, whereas a direct method requires a number of multiplications of order $\frac{1}{3}K^3$, an iterative method requires a reduced number of order $K^2 I$, where $K$ is the size of linear system, and $I$ is the number of necessary iterations. Several efficient iterative methods for solving integral equations of the second kind are discussed by Atkinson (1973) and Rokhlin (1990).

Now, in solving the final system of linear equations for the coefficients of the local expansions, the issues of well-posedness, existence, and uniqueness of solution of the integral equation become important. Ill-posed integral equations or integral equations with no solution or many solutions lead to ill-conditioned linear systems with faulty and oscillatory solutions. To this end, we must note that, in general, the solution of integral equations of the first kind is susceptible to oscillations due to numerical error (Delves & Mohamed 1985, Chapter 12). Fortunately, experience has shown that integral equations of the first kind arising from boundary integral representations exhibit fairly regular behaviour. While this is the general rule, when the boundaries of the flow contain singular points or sharp corners, or when the solution exhibits strong variations, the numerical results may exhibit oscillatory behaviour and slow convergence (see section 6.5).

An important variation of the above procedure, applicable only for integral equations of the second kind, exploits the feasibility of computing the solution using the method of successive substitutions. The advantages are reduced algebraic effort associated with grouping of the unknown coefficients in the local expansions, and reduced computer memory requirements and programming effort. To illustrate the method, let us consider the prototypical equation $\mathbf{q} = \alpha \mathrm{L}\mathbf{q} + \mathbf{f}$ where L is the double-layer operator, $\mathbf{f}$ is a forcing function, and $\mathbf{q}$ is the unknown density function. To implement the method, we approximate the double-layer integral $\mathrm{L}\mathbf{q}$ using a quadrature over each boundary element, obtaining

$$\mathbf{q}(\mathbf{x}) = \alpha \sum_{\substack{\text{Elements,} \\ n=1,\dots,N}} \sum_{\substack{\text{Quadrature points,} \\ k=1,\dots,K}} \mathbf{A}_{nk}(\mathbf{x}, \mathbf{x}_{nk}) \cdot \mathbf{q}(\mathbf{x}_{nk}) + \mathbf{f}(\mathbf{x}) \qquad (6.1.1)$$

where $\{\mathbf{x}_{nk}\}$ is the set of the quadrature base points over all elements. Applying (6.1.1) at each base point we obtain a system of $3NK$ linear algebraic equations for the three components of each unknown $\mathbf{q}(\mathbf{x}_{nk})$

which we subsequently solve using an iterative procedure. Adopting Jacobi's method, for instance, we guess the values of $\mathbf{q}(\mathbf{x}_{nk})$, compute the right-hand side of (6.1.1), and then replace the original $\mathbf{q}(\mathbf{x}_{nk})$ by the newly computed $\mathbf{q}(\mathbf{x}_{nk})$. To reduce the cost of the computations, it is advisable to compute the coefficients $\mathbf{A}_{nk}$ once and then store and recall them during subsequent iterations. It should be noted that the iterations will converge only if $|\alpha|$ is less than the spectral radius of the double-layer potential (see section 4.5).

In a slight variation of the above iterative procedure, we express the unknown density $\mathbf{q}$ as a polynomial over each element in terms of properly defined surface variables, and identify the coefficients of the polynomial with the values of $\mathbf{q}$ at the corresponding nodes. We assign initial values to $\mathbf{q}$ at the nodes, evaluate the polynomials over all elements, and apply the integral equation at the nodes. Finally, we compute the double-layer integral and add it to the forcing function $\mathbf{f}$, thereby producing new values for $\mathbf{q}$ at the nodes. We then repeat this process until the values of $\mathbf{q}$ at all nodes change by less than a preset minimum after one iteration. The main advantage of this method is that it produces the value of $\mathbf{q}$ at the nodes in a direct manner, circumventing the need for further interpolations. This feature is particularly convenient when $\mathbf{q}$ represents a primary variable such as the boundary velocity (Pozrikidis 1990a, b; Newhouse & Pozrikidis 1990; Kennedy 1991).

The creative numerical analyst will devise a number of alternative iterative procedures with the object of reducing the cost of the computations. One modification stems from the observation that the method of successive substitutions parallels Jacobi's method for solving systems of linear equations, implying that a variation in the spirit of Gauss–Seidel might produce the solution in a faster and more economical fashion. The idea is to perform iterations in groups of nodes using parallel computation in which each processor iterates on a different set of nodes. The frequency of communication among the processors for updating the values of $\mathbf{q}$ then becomes an issue of practical concern (Karrila et al. 1989). The modified method of successive substitutions is particularly effective when the surface area of the domain of the integral equation $D$ is multiply connected or excessively large.

## 6.2 Boundary element representations

One important stage in the numerical implementation of a boundary integral method is the representation of the boundary of the flow by a set of boundary elements. It will be noted that this representation not only allows the automated computation of the required boundary integrals but also provides a framework for expanding the unknown boundary function in a series of local basis functions.

In this section we shall discuss the general principles underlying boundary discretizations, and provide specific examples of boundary element representations. Since we want to apply the boundary integral method to problems involving unsteady interfacial motions, we shall consider, in some detail, adaptive representations.

During our discussion, we shall find it necessary to use certain fundamental concepts from the differential geometry of lines and surfaces. The reader may obtain further information from standard texts on differential geometry such as those by Stoker (1989; see, in particular, Chapter 2 on the differential geometry of planar lines), Struik (1961), and Weatherburn (1961). Comprehensive compilations of and formulae for various boundary elements may be found in standard monographs of boundary element and finite element methods, such as those by Zienkiewicz (1971), Banerjee & Butterfield (1981), Bathe (1982), and Schwarz (1988).

### *Planar boundaries*

The boundary of an axisymmetric or two-dimensional flow may be represented simply by a planar line that resides in a meridional plane or in the plane of the flow. To set up a boundary element representation, we trace the planar line with a set of marker points $\{\mathbf{x}_i\}$, $i = 1, \ldots, N$, numbered in the counter-clockwise direction as shown in Figure 6.2.1. In the simplest

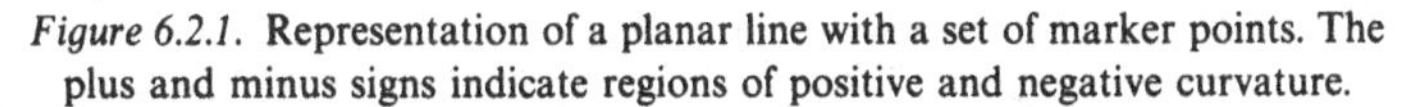
*Figure 6.2.1.* Representation of a planar line with a set of marker points. The plus and minus signs indicate regions of positive and negative curvature.

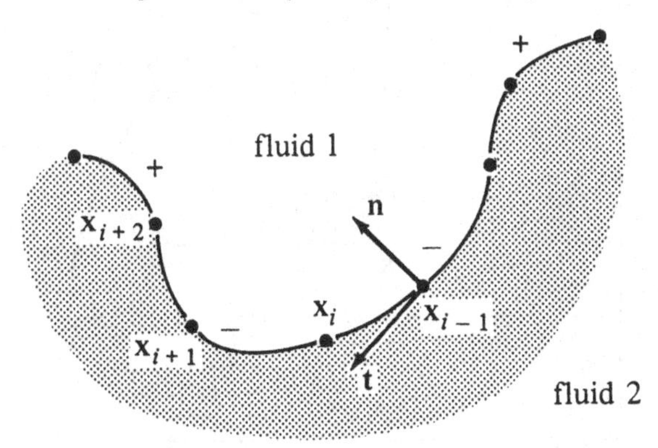

approximation, we represent the line with a set of straight segments that pass through successive pairs of marker points. An elementary computation shows that the straight segment passing through the points $\mathbf{x}_i$ and $\mathbf{x}_{i+1}$ is described by the equation $y = a_i x + b_i$ where $a_i = (y_{i+1} - y_i)/(x_{i+1} - x_i)$, and $b_i = y_i - a_i x_i$. The differential arc length over the $i$th segment is given by

$$dl = (a_i^2 + 1)^{1/2} |dx| = \alpha (a_i^2 + 1)^{1/2} \, dx \tag{6.2.1}$$

where $\alpha = (x_{i+1} - x_i)/|x_{i+1} - x_i|$. The line integral of a function $f$ over the $i$th straight segment is given by

$$\int_{S_i} f(x, y) \, dl = \alpha (a_i^2 + 1)^{1/2} \int_{x_i}^{x_{i+1}} f(x, a_i x + b_i) \, dx \tag{6.2.2}$$

and the unit normal and tangent vectors are given by

$$\mathbf{t} = \frac{\alpha}{(1 + a_i^2)^{1/2}} (1, a_i) \qquad \mathbf{n} = \frac{\alpha}{(1 + a_i^2)^{1/2}} (a_i, -1) \tag{6.2.3}$$

respectively. Unfortunately, due to its inability to estimate the curvature of the boundary, the straight-segment representation is inadequate for problems involving two-phase flow (see Chapter 5).

Proceeding to a second-order representation, we consider quadratic elements passing through trios of successive marker points. One obvious choice is parabolic elements, defined by the equation $y = ax^2 + bx + c$. This representation, however, is meaningful only when $y$ is a one-to-one

*Figure 6.2.2.* Local approximation of a planar line with a circular arc passing through three successive marker points.

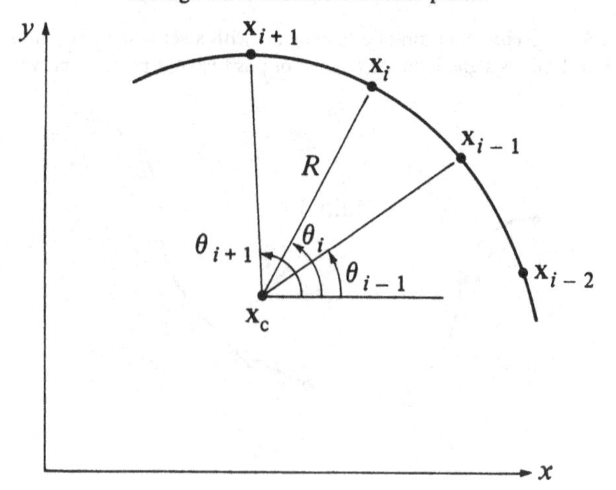

function of $x$, i.e. when the boundary does not turn back upon itself towards the negative $x$-axis. To ensure that this restriction is met, we must introduce individual coordinate systems for each element in which the $x$-axis passes through two of the three marker points that define the element.

To avoid the complications of the parabolic approximation, we introduce an alternative representation in polar coordinates. Thus, we approximate the boundary by a set of circular arcs passing through successive trios of marker points, as illustrated in Figure 6.2.2. It will be interesting to note that the curvature of a circular arc is less than the maximum and greater than the minimum curvature of any other quadratic element that passes through three marker points, thereby suggesting that circular arcs provide us with the smoothest second-order approximation (Higdon & Pozrikidis 1985). Due to its conceptual and practical simplicity, the circular arc has become a popular boundary element in a variety of problems involving interfacial flow (Meiburg & Homsy 1988; Dritschel 1989; Pozrikidis 1990a, b).

Turning to the details of the arc representation, we refer to the polar system shown in Figure 6.2.2, and define an arc by its center $\mathbf{x}_c$, its radius $R$, and the polar angles corresponding to each of the three nodes $\mathbf{x}_{i-1}$, $\mathbf{x}_i, \mathbf{x}_{i+1}$. In practice, it is convenient to compute $\mathbf{x}_c$ as the point of intersection of two lines, one of which is perpendicular to the segment $\mathbf{x}_i - \mathbf{x}_{i-1}$ at the midpoint, and the second of which is perpendicular to the segment $\mathbf{x}_{i+1} - \mathbf{x}_i$ at the midpoint. The radius $R$ follows immediately as the distance between the center $\mathbf{x}_c$ and one of the nodes. In order to keep track of the orientation of the arc, we introduce an index $\alpha$ so that $\alpha = 1$ and $\alpha = -1$ indicate counterclockwise and clockwise orientation, respectively. For this purpose, we compute the mixed vector product $[(\mathbf{x}_i - \mathbf{x}_{i-1}) \times (\mathbf{x}_{i+1} - \mathbf{x}_{i-1})] \cdot \mathbf{k}$, where $\mathbf{k}$ is the unit vector normal to the plane of flow. If this is a positive number we set $\alpha = 1$; otherwise, we set $\alpha = -1$. To compute the polar angles subtended by the marker points we use the equations

$$\begin{aligned} \theta_{i-1} &= \beta \arccos[(x_{i-1} - x_c)/|\mathbf{x}_{i-1} - \mathbf{x}_c|] \\ \theta_i &= \theta_{i-1} + \alpha \arccos[(\mathbf{x}_i - \mathbf{x}_c)\cdot(\mathbf{x}_{i-1} - \mathbf{x}_c)/R^2] \\ \theta_{i+1} &= \theta_i + \alpha \arccos[(\mathbf{x}_{i+1} - \mathbf{x}_c)\cdot(\mathbf{x}_i - \mathbf{x}_c)/R^2] \end{aligned} \tag{6.2.4}$$

where $\beta = (y_{i-1} - y_c)/|y_{i-1} - y_c|$ (Figure 6.2.2). To avoid numerical complications associated with the inverse cosines, we must ensure that the total angle subtended by the arc $\Delta\theta = |\theta_{i+1} - \theta_{i-1}|$ is less than $\pi$.

To describe a circular arc, we use a parametric representation in terms of the polar angle $\theta$, setting

$$x = x_c + R\cos\theta, \qquad y = y_c + R\sin\theta \tag{6.2.5}$$

where $\theta_{i-1} < \theta < \theta_{i+1}$. The differential arc length is simply $dl = R|d\theta| = \alpha R\, d\theta$, whereas the integral of a function $f$ over arc is given by

$$\int_{S_i} f(x, y)\, dl = \alpha R \int_{\theta_{i-1}}^{\theta_{i+1}} f(x_c + R\cos\theta, y_c + R\sin\theta)\, d\theta \tag{6.2.6}$$

The unit tangent and unit normal vectors, and the curvature of the arc, are given by

$$\mathbf{t} = \alpha(-\sin\theta, \cos\theta) \qquad \mathbf{n} = \alpha(\cos\theta, \sin\theta) \qquad k = \frac{\alpha}{R} \tag{6.2.7}$$

It will be noted that the curvature is positive for an arc with counter-clockwise orientation, and negative for an arc with clockwise orientation. The mean curvature of an axisymmetric surface whose trace in a meridional plane is approximated by an arc will be discussed in problem 6.2.2.

To increase the accuracy of the boundary element representation beyond the second order, we must proceed to cubic or higher-order elements. To define these elements, we introduce a parameter $\beta$ that increases monotonically along the boundary, and describe the element in the parametric form as $\mathbf{x} = \mathbf{f}(\beta)$ where $0 < \beta < B$. For this representation to be meaningful, the first few derivatives of the function $\mathbf{f}$ must be positive. This will be true if the boundary is sufficiently smooth and the function $\beta$ has some sort of physical significance. In practice, it is convenient to define $\beta$ as the polygonal arc length, i.e. the arc length of a polygonal line that connects successive marker points, as illustrated in Figure 6.2.3(*a*).

*Figure 6.2.3.* (*a*) Parametrization of a planar line using the polygonal arc length; (*b*) a pathological case where $\phi > \pi/2$.

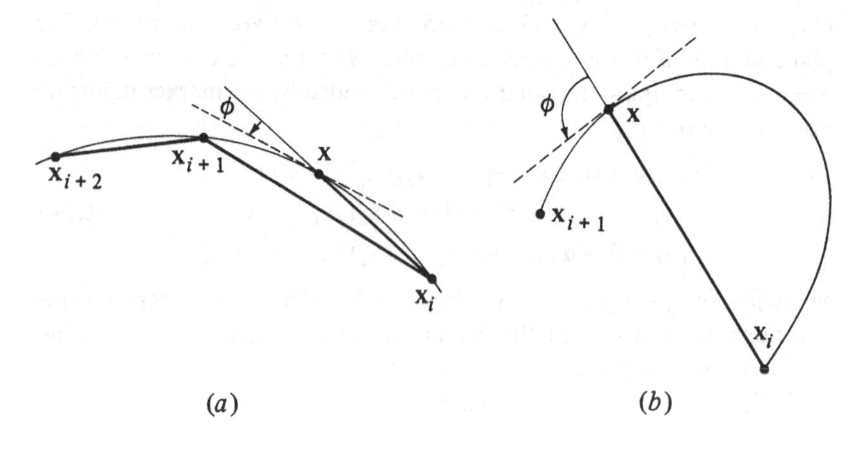

At the first marker point $\beta(\mathbf{x}_1)=0$, whereas at subsequent points

$$\beta(\mathbf{x}_i)=\sum_{j=2}^{i}|\mathbf{x}_j-\mathbf{x}_{j-1}| \tag{6.2.8}$$

for $i=2,3,\ldots$ For a point $\mathbf{x}$ which is located between $\mathbf{x}_i$ and $\mathbf{x}_{i+1}$,

$$\beta(\mathbf{x})=|\mathbf{x}-\mathbf{x}_i|+\sum_{j=2}^{i}|\mathbf{x}_j-\mathbf{x}_{j-1}| \tag{6.2.9}$$

A simple computation reveals that $d\beta/ds=\cos\phi$, where $s$ is the arc length along the boundary, and $\phi$ is the angle subtended by the vector $\mathbf{x}-\mathbf{x}_i$ and the tangent to the boundary at the point $\mathbf{x}$, as illustrated in Figure 6.2.3(*a*). For $\beta$ to be an acceptable parameter $d\beta/ds$ must be positive, i.e. $\phi$ must be less than $\pi/2$. An example of an unacceptable situation where $\phi$ is larger than $\pi/2$ is illustrated in Figure 6.2.3(*b*).

Given the pairs $(\mathbf{x}_i,\beta_i)$ we use standard interpolation to describe the shape of and to compute the normal vector, curvature, and other geometrical features of the element (Carnahan *et al.* 1969). The accuracy of the resulting representation will depend upon the type of interpolation and the number of marker points involved in the interpolation. Spline interpolation, in particular, will provide us with a smooth boundary with continuous first-, second-, and higher-order derivatives (De Boor 1978).

### *Three-dimensional boundaries*

We proceed now to discuss discrete representations of three-dimensional boundaries. Before considering specific discretizations, it is necessary to review certain fundamental concepts from the differential geometry of surfaces.

One way to describe a three-dimensional surface is to introduce a right-handed curvilinear, but not necessarily orthogonal, coordinate system $(\eta,\xi)$ described over the surface, and to view the position of a point $\mathbf{x}$ on the surface as a function of $(\eta,\xi)$, namely $\mathbf{x}(\eta,\xi)$ (Figure 6.2.4). A unit vector tangential to the surface is given by

$$\mathbf{t}=\frac{1}{h_t}\left(\frac{\partial\mathbf{x}}{\partial\eta}+\omega\frac{\partial\mathbf{x}}{\partial\xi}\right)\qquad h_t=\left|\frac{\partial\mathbf{x}}{\partial\eta}+\omega\frac{\partial\mathbf{x}}{\partial\xi}\right| \tag{6.2.10}$$

where $\omega$ is an arbitrary constant. The tangent vectors pointing toward the positive direction of an $\eta$ or a $\xi$ line are denoted $\mathbf{t}_\eta$ and $\mathbf{t}_\xi$ and correspond to $\omega=0$ or $\omega=\infty$ respectively. Two directions corresponding to $\omega_1$ and $\omega_2$ are orthogonal if

$$\left(\frac{\partial\mathbf{x}}{\partial\eta}+\omega_1\frac{\partial\mathbf{x}}{\partial\xi}\right)\cdot\left(\frac{\partial\mathbf{x}}{\partial\eta}+\omega_2\frac{\partial\mathbf{x}}{\partial\xi}\right)=0 \tag{6.2.11}$$

The unit vector normal to the surface is given by

$$\mathbf{n} = \frac{1}{h_n}\frac{\partial \mathbf{x}}{\partial \eta} \times \frac{\partial \mathbf{x}}{\partial \xi} \qquad h_n = \left|\frac{\partial \mathbf{x}}{\partial \eta} \times \frac{\partial \mathbf{x}}{\partial \xi}\right| \tag{6.2.12}$$

The area of a differential surface element is equal to

$$\mathrm{d}S = h_n \,\mathrm{d}\eta\,\mathrm{d}\xi \tag{6.2.13}$$

The rate of change of the area (dilatation) of a differential surface element is

$$\Delta \equiv \frac{\partial h_n}{\partial t} = \mathbf{n}\cdot\frac{\partial}{\partial t}\left(\frac{\partial \mathbf{x}}{\partial \eta} \times \frac{\partial \mathbf{x}}{\partial \xi}\right) = \mathbf{n}\cdot\left(\frac{\partial \mathbf{v}}{\partial \eta} \times \frac{\partial \mathbf{x}}{\partial \xi} + \frac{\partial \mathbf{x}}{\partial \eta} \times \frac{\partial \mathbf{v}}{\partial \xi}\right) \tag{6.2.14}$$

where $\mathbf{v} = (\partial \mathbf{x}/\partial t)_{\eta,\xi}$. For an incompressible surface whose area remains locally and globally constant in time, $\Delta = 0$.

The integral of a function $f$ over the surface is given by

$$\int_D f(\mathbf{x})\,\mathrm{d}S(\mathbf{x}) = \int_D f(\mathbf{x}(\eta,\xi))h_n\,\mathrm{d}\eta\,\mathrm{d}\xi \tag{6.2.15}$$

The differential length of an infinitesimal line element on the surface is given by

$$\begin{aligned} \mathrm{d}l^2 = \mathrm{d}\mathbf{x}\cdot\mathrm{d}\mathbf{x} &= \left(\frac{\partial \mathbf{x}}{\partial \eta}\mathrm{d}\eta + \frac{\partial \mathbf{x}}{\partial \xi}\mathrm{d}\xi\right)\cdot\left(\frac{\partial \mathbf{x}}{\partial \eta}\mathrm{d}\eta + \frac{\partial \mathbf{x}}{\partial \xi}\mathrm{d}\xi\right) \\ &= \frac{\partial \mathbf{x}}{\partial \eta}\cdot\frac{\partial \mathbf{x}}{\partial \eta}\mathrm{d}\eta^2 + 2\frac{\partial \mathbf{x}}{\partial \eta}\cdot\frac{\partial \mathbf{x}}{\partial \xi}\mathrm{d}\eta\,\mathrm{d}\xi + \frac{\partial \mathbf{x}}{\partial \xi}\cdot\frac{\partial \mathbf{x}}{\partial \xi}\mathrm{d}\xi^2 \end{aligned} \tag{6.2.16}$$

To simplify our notation, it is useful to introduce the metric tensor $\mathbf{a}$ defined by the equation

$$\mathrm{d}l^2 \equiv a_{\eta\eta}\,\mathrm{d}\eta^2 + 2a_{\eta\xi}\,\mathrm{d}\eta\,\mathrm{d}\xi + a_{\xi\xi}\,\mathrm{d}\xi^2 \tag{6.2.17}$$

*Figure 6.2.4.* Parametric representation of a three-dimensional surface using two surface variables $(\eta, \xi)$.

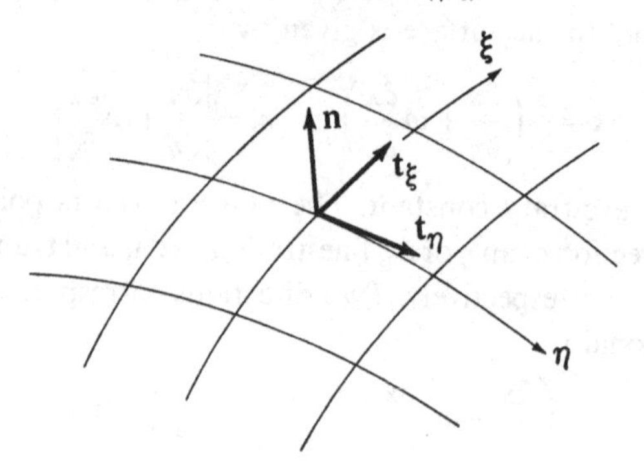

Comparing (6.2.16) with (6.2.17) we find

$$a_{\eta\eta} = \frac{\partial \mathbf{x}}{\partial \eta}\cdot\frac{\partial \mathbf{x}}{\partial \eta} \qquad a_{\eta\xi} = a_{\xi\eta} = \frac{\partial \mathbf{x}}{\partial \eta}\cdot\frac{\partial \mathbf{x}}{\partial \xi} \qquad a_{\xi\xi} = \frac{\partial \mathbf{x}}{\partial \xi}\cdot\frac{\partial \mathbf{x}}{\partial \xi} \tag{6.2.18}$$

Clearly, if $a_{\eta\xi} = 0$, the axes $\eta$ and $\xi$ are orthogonal. Using (6.2.17) we may write the differential arc length of a line in the alternative form

$$\mathrm{d}l \equiv h_{\mathrm{t}}|\mathrm{d}\eta| \tag{6.2.19}$$

where

$$h_{\mathrm{t}} = (a_{\eta\eta} + 2a_{\eta\xi}\omega + a_{\xi\xi}\omega^2)^{1/2} \tag{6.2.20}$$

and $\omega$ was introduced previously in (6.2.10). To ensure that the quantity within the radical in (6.2.20) is positive for any choice of $\omega$, we require that the determinant of the metric tensor is positive, i.e. $a_{\eta\eta}a_{\xi\xi} - a_{\eta\xi}^2 > 0$.

The curvature of the trace of the surface in a plane that is perpendicular to the surface and tangential to a line of constant $\omega$ is given by

$$k(\omega) = \left(\frac{\partial \mathbf{n}}{\partial l}\right)_{\omega}\cdot\mathbf{t}(\omega) = \frac{1}{h_{\mathrm{t}}}\left(\frac{\partial \mathbf{n}}{\partial \eta} + \omega\frac{\partial \mathbf{n}}{\partial \xi}\right)\cdot\mathbf{t}(\omega) \tag{6.2.21}$$

Substituting (6.2.12) into (6.2.21) we obtain

$$k(\omega) = \frac{1}{h_{\mathrm{t}}h_{\mathrm{n}}}\left[\left(\frac{\partial}{\partial \eta} + \omega\frac{\partial}{\partial \xi}\right)\left(\frac{\partial \mathbf{x}}{\partial \eta} \times \frac{\partial \mathbf{x}}{\partial \xi}\right)\right]\cdot\mathbf{t}(\omega) \tag{6.2.22}$$

For the special cases $\omega = 0$ or $\infty$, corresponding to the directions of the $\eta$- or $\xi$-axis, we obtain

$$k(0) = \frac{1}{a_{\eta\eta}}\frac{\partial \mathbf{n}}{\partial \eta}\cdot\frac{\partial \mathbf{x}}{\partial \eta} \qquad k(\infty) = \frac{1}{a_{\xi\xi}}\frac{\partial \mathbf{n}}{\partial \xi}\cdot\frac{\partial \mathbf{x}}{\partial \xi} \tag{6.2.23}$$

The mean curvature of the surface $k_{\mathrm{m}}$ is equal to the mean of the values of the curvature in two arbitrary but mutually perpendicular directions. Using (6.2.11) and (6.2.18) we find that

$$\omega = \omega_{\eta} = -\frac{a_{\eta\eta}}{a_{\eta\xi}} \qquad \omega = \omega_{\xi} = -\frac{a_{\eta\xi}}{a_{\xi\xi}} \tag{6.2.24}$$

define two directions that are perpendicular to the $\eta$ and $\xi$ axes respectively. Thus,

$$k_{\mathrm{m}} = \tfrac{1}{2}[k(0) + k(\omega_{\eta})] = \tfrac{1}{2}[k(\infty) + k(\omega_{\xi})] \tag{6.2.25}$$

When the axes $(\eta, \xi)$ are orthogonal we obtain

$$k_{\mathrm{m}} = \tfrac{1}{2}[k(0) + k(\infty)] \tag{6.2.26}$$

As an application of the above formulae, we consider a cylindrical surface whose generators are perpendicular to the $xy$-plane, as illustrated in Figure 5.5.4. We identify $\eta$ with $z$, and $\xi$ with the arc length $s$ of the trace of the surface in the $xy$-plane. Noting that $\partial\mathbf{x}/\partial\eta = (0, 0, 1)$ and

$\partial \mathbf{t}/\partial s = -k\mathbf{n}$, assuming that $\partial \mathbf{v}/\partial \eta = 0$, and using (6.2.14), we compute the dilatation of the surface as

$$\Delta = \mathbf{t}\cdot\frac{\partial \mathbf{v}}{\partial s} = \frac{\partial}{\partial s}(\mathbf{v}\cdot\mathbf{t}) + k\mathbf{v}\cdot\mathbf{n} \tag{6.2.27}$$

where $k$ is the curvature of the trace of the surface in the $xy$-plane. Furthermore, we identify $\eta$ with $z$, and $\xi$ with $x$, and assume that the surface may be described in terms of a function $f$ in the form $\mathbf{x} = (x, f(x), z)$. Using (6.2.12) and (6.2.23), and noting that $\partial \mathbf{x}/\partial \eta = (0, 0, 1)$, $\partial \mathbf{x}/\partial \xi = (1, f', 0)$, and $a_{\eta\xi} = 0$, $a_{\eta\eta} = 1$, $a_{\xi\xi} = 1 + f'^2$, we compute the normal vector as

$$\mathbf{n} = \frac{1}{(1 + f'^2)^{1/2}}(-f', 1, 0) \tag{6.2.28}$$

and the curvature of the trace of the surface in the $xy$-plane as

$$k = \frac{1}{1 + f'^2}\frac{\mathrm{d}}{\mathrm{d}x}\left[\frac{-f'}{(1 + f'^2)^{1/2}}\right] + f'\frac{\mathrm{d}}{\mathrm{d}x}\left[\frac{1}{(1 + f'^2)^{1/2}}\right] = -\frac{f''}{(1 + f'^2)^{3/2}} \tag{6.2.29}$$

As a second application, we consider the axisymmetric surface illustrated in Figure 6.2.5, and identify the $\eta$ variable with the arc length $s$ of the trace of the surface in the azimuthal plane $\phi = 0$, and the $\xi$ variable with the azimuthal angle $\phi$. Assuming that the cylindrical polar components of the surface velocity $\mathbf{v}$ are not functions of $\phi$, we obtain for a point in

*Figure 6.2.5.* Parametric description of an axisymmetric surface using as surface variables the arc length $s$ and the azimuthal angle $\phi$.

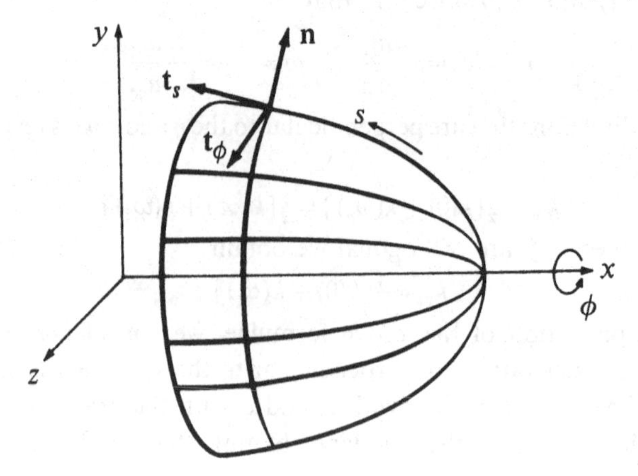

the $\phi = 0$ plane

$$\frac{\partial z}{\partial s} = \frac{\partial v_z}{\partial s} = 0 \qquad \frac{\partial x}{\partial \phi} = \frac{\partial y}{\partial \phi} = \frac{\partial v_x}{\partial \phi} = \frac{\partial v_y}{\partial \phi} = 0 \qquad \frac{\partial z}{\partial \phi} = \sigma \qquad \frac{\partial v_z}{\partial \sigma} = v_\sigma \tag{6.2.30}$$

Substituting (6.2.30) into (6.2.14) we compute the dilatation of the surface as

$$\Delta = \sigma\left(n_x \frac{\partial v_\sigma}{\partial s} - n_\sigma \frac{\partial v_x}{\partial s}\right) + v_\sigma\left(n_x \frac{\partial \sigma}{\partial s} - n_\sigma \frac{\partial x}{\partial s}\right) = \sigma\, \mathbf{t}_s \cdot \frac{\partial \mathbf{v}}{\partial s} + v_\sigma \tag{6.2.31}$$

Now, switching our parametric representation, we identify the $\eta$ variable with $\sigma$ and the $\xi$ variable with $\phi$, and describe the surface in terms of a function $f$ in the form $\mathbf{x} = (f(\sigma), \sigma \cos \phi, \sigma \sin \phi)$. Then we compute

$$\partial \mathbf{x}/\partial \sigma = (f', \cos \phi, \sin \phi) \qquad \partial \mathbf{x}/\partial \phi = \sigma(0, -\sin \phi, \cos \phi),$$

$$(a_{\sigma\sigma}, a_{\sigma\phi}, a_{\phi\phi}) = (1 + f'^2, 0, \sigma^2) \tag{6.2.32}$$

Using (6.2.12) and (6.2.23), we calculate the normal vector

$$\mathbf{n} = \frac{1}{(1 + f'^2)^{1/2}}(1, -f' \cos \phi, -f' \sin \phi) \tag{6.2.33}$$

and the curvatures of the surface in the meridian and azimuthal directions

$$k(0) = -\frac{f''}{(1 + f'^2)^{3/2}} \qquad k(\infty) = -\frac{f'}{\sigma(1 + f'^2)^{1/2}} \tag{6.2.34}$$

The mean curvature of the surface is simply

$$k_m = \tfrac{1}{2}[k(0) + k(\infty)] = -\frac{1}{2\sigma}\frac{\partial}{\partial \sigma}\left[\frac{\sigma f'}{(1 + f'^2)^{1/2}}\right] \tag{6.2.35}$$

### *Three-dimensional boundary elements*

Having discussed the parametric description of three-dimensional surfaces, we proceed now to consider discrete representations of three-dimensional boundaries. Our general strategy is to trace the boundaries with a network of marker points or nodes, and to represent the boundaries with a collection of elements that are defined in terms of a set of neighbouring nodes. The elements may be chosen to have a variety of shapes including flat triangles, curved triangles, and planar or curved polygons.

Now, the success of covering a three-dimensional surface with elements of a particular type will not always be guaranteed. Fortunately, a classical result of differential geometry ensures that a surface may always be triangulated, i.e. divided into a set of flat or curved triangles (Stoker 1989, p. 211). The division may be effected in many different ways, but in all cases, the number of triangles (faces) $F$, the number of sides (edges) $E$, and the number of vertices of the triangles $V$, must satisfy the equation $\chi = F - E + V$, where $\chi$ is the Euler characteristic. Two surfaces with identical Euler characteristic have similar topological characteristics.

Beginning with the simplest approximation, we consider boundary elements in the shape of flat triangles. We identify each triangle by its three vertices $\mathbf{x}_i$, $i = 1, 2, 3$ (numbered in the counterclockwise sense, as shown on the right of Figure 6.2.6), and describe the surface of a triangle using the parametric representation

$$\mathbf{x}(\eta, \xi) = \mathbf{a} + \mathbf{b}\eta + \mathbf{c}\xi \tag{6.2.36}$$

where $\eta$ and $\xi$ are two surface variables that take values over the right isosceles triangle shown in Figure 6.2.6. To ensure that the triangle passes through the three marker points, we require

$$\mathbf{x}(0,0) = \mathbf{x}_1 \quad \mathbf{x}(1,0) = \mathbf{x}_2 \quad \mathbf{x}(0,1) = \mathbf{x}_3 \tag{6.2.37}$$

obtaining a system of linear equations for the unknowns $\mathbf{a}, \mathbf{b}, \mathbf{c}$. Solving this system and substituting the results into (6.2.36) we obtain a local representation in terms of three local basis functions

$$\mathbf{x}(\eta, \xi) = \sum_{i=1}^{3} \mathbf{x}_i \varphi_i(\eta, \xi) \tag{6.2.38}$$

where

$$\phi_1 = 1 - \eta - \xi, \qquad \phi_2 = \eta, \qquad \phi_3 = \xi \tag{6.2.39}$$

Equations (6.2.36) and (6.2.38) define a mapping from physical space to the parametric $(\eta, \xi)$ space. In general, this mapping will not be conformal, for the angles of the triangle in physical space will not be equal to those in the parametric space, namely $\pi/2$, $\pi/4$, and $\pi/4$. We note, however, that all partial derivatives of the basis functions in (6.2.38) with

*Figure 6.2.6.* Mapping of a planar triangle in physical space onto a right triangle in the $\eta\xi$-plane.

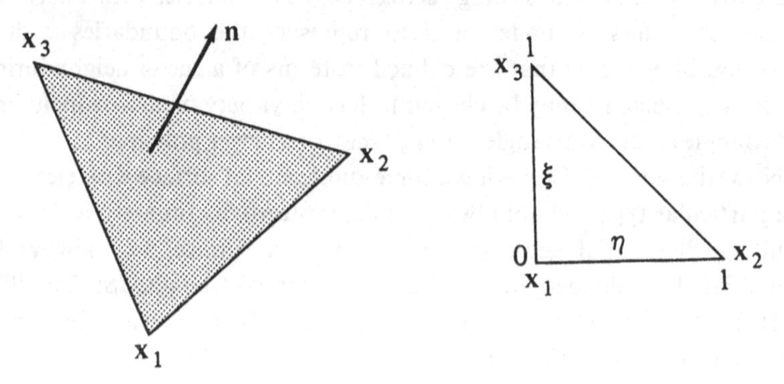

respect to $\eta$ and $\xi$ are constant and this suggests that the mapping will be non-singular even when the triangle is notably skewed.

Proceeding next to a second-order approximation, we discretize the boundary into a set of curved triangles. Each triangle is defined by six marker points $\mathbf{x}_i$, $i = 1, \ldots, 6$, as illustrated in Figure 6.2.7. To describe the surface of a triangle we use a parametric representation in terms of two variables $\eta$ and $\xi$ as

$$\mathbf{x}(\eta, \xi) = \mathbf{a} + \mathbf{b}\eta + \mathbf{c}\xi + \mathbf{d}\eta^2 + \mathbf{e}\xi^2 + \mathbf{f}\eta\xi \tag{6.2.40}$$

where $\eta$ and $\xi$ take values over the right isosceles triangle depicted on the right of Figure 6.2.7. Equation (6.2.40) is a consistent second-order Taylor expansion with respect to $\eta$ and $\xi$. To ensure that the curved triangle passes through the six marker points, we require

$$\begin{aligned} &\mathbf{x}(1,0) = \mathbf{x}_1 \qquad &\mathbf{x}(0,1) = \mathbf{x}_2 \qquad &\mathbf{x}(0,0) = \mathbf{x}_3 \\ &\mathbf{x}(\gamma, 1-\gamma) = \mathbf{x}_4 \qquad &\mathbf{x}(0,\beta) = \mathbf{x}_5 \qquad &\mathbf{x}(\alpha,0) = \mathbf{x}_6 \end{aligned} \tag{6.2.41}$$

where the three scalar coefficients $\alpha$, $\beta$, $\gamma$ are defined as

$$\alpha = \frac{1}{1 + \dfrac{|\mathbf{x}_6 - \mathbf{x}_1|}{|\mathbf{x}_6 - \mathbf{x}_3|}} \qquad \beta = \frac{1}{1 + \dfrac{|\mathbf{x}_5 - \mathbf{x}_2|}{|\mathbf{x}_5 - \mathbf{x}_3|}} \qquad \gamma = \frac{1}{1 + \dfrac{|\mathbf{x}_4 - \mathbf{x}_1|}{|\mathbf{x}_4 - \mathbf{x}_2|}} \tag{6.2.42}$$

It will be noted that the 3-1 side of the triangle is mapped onto the $\eta$-axis ($\xi = 0$ or $\omega = 0$), the 2-3 side is mapped onto the $\xi$-axis ($\eta = 0$ or $\omega = \infty$), and the 1-2 side is mapped onto the diagonal line ($\omega = -1$). Substituting (6.2.40) into (6.2.41), solving for the unknown coefficients **a**–**f**, and substituting the results into (6.2.40) we obtain a representation in terms of six local basis functions, namely

$$\mathbf{x}(\eta, \xi) = \sum_{i=1}^{6} \mathbf{x}_i \varphi_i(\eta, \xi) \tag{6.2.43}$$

*Figure 6.2.7.* Mapping of a curved triangle in physical space onto a right triangle in the $\eta\xi$-plane.

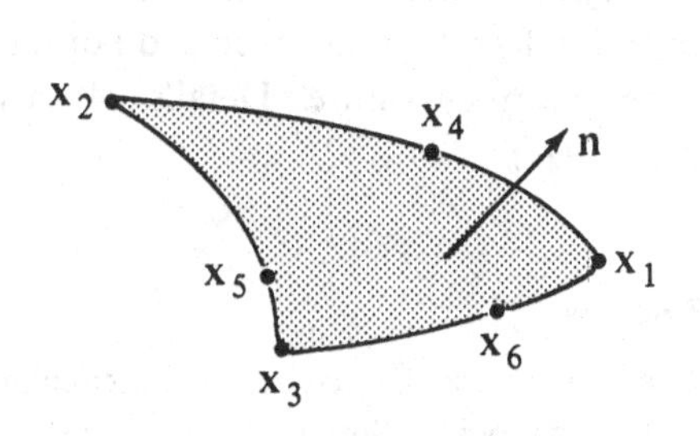


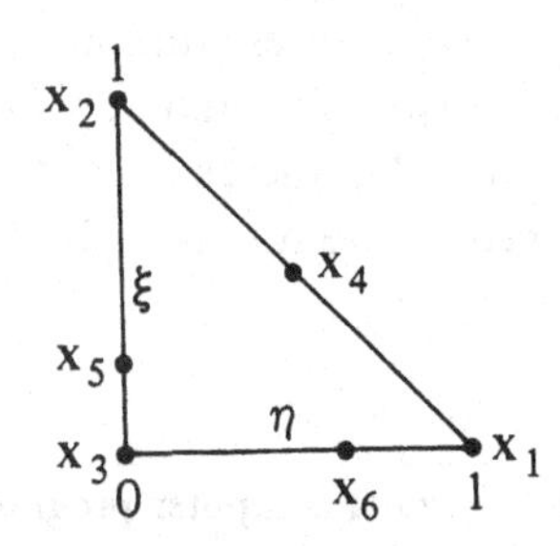

where

$$\phi_1 = \frac{\alpha}{\alpha - 1}\eta\left[\frac{1}{\alpha}\eta - 1 + \frac{(\alpha - \gamma)}{\alpha(1 - \gamma)}\xi\right] \quad \phi_2 = \frac{\beta}{\beta - 1}\xi\left[\frac{1}{\beta}\xi - 1 + \frac{(\beta + \gamma - 1)}{\beta\gamma}\eta\right]$$

$$\phi_4 = \frac{1}{\gamma(1 - \gamma)}\eta\xi \qquad \phi_5 = \frac{1}{\beta(1 - \beta)}\xi(1 - \eta - \xi)$$

$$\phi_6 = \frac{1}{\alpha(1 - \alpha)}\eta(1 - \eta - \xi) \qquad \phi_3 = 1 - \phi_1 - \phi_2 - \phi_4 - \phi_5 - \phi_6$$

(6.2.44)

It is important to note that the mapping defined by (6.2.40) may contain singular points due to the touching or crossing of two lines with different values of $\eta$ or $\xi$. This will happen, in particular, when the triangle in the physical space is markedly skewed, or when one middle marker point along a side of the triangle does not lie sufficiently close to the straight line that connects the corresponding vertices. In practice, in order to guarantee that the mapping of an element is non-singular, it will be sufficient to ensure that the determinant of the metric tensor at several points over the element is a positive number not very close to zero.

Using (6.2.43) and the formulae presented in the preceding subsection, we may compute various geometrical characteristics of a curved triangle in a straightforward manner. Exploratory computations have shown that while the calculations of the normal vector and of the curvature of a line that passes through a vertex and crosses the triangle are accurate and reliable, the calculation of the curvature of a line that passes through a vertex but does not cross the triangle is inaccurate and non-converging (Kennedy 1991). This pathological behaviour is due to the turning, twisting, and folding of the triangle outside the region confined by the marker points. Our inability to compute the curvature of two perpendicular lines passing through a vertex frustrates the computation of the mean curvature. One successful remedy is to compute the average values of the normal vector at the six nodal points of a triangle (by averaging the limits of the normal vectors from neighboring triangles), introduce a parametric representation of the averaged normal vectors in terms of the surface variables, and approximate and differentiate the averaged normal vector field, thereby producing the mean curvature. Details of this procedure are discussed by Kennedy (1991).

## Problems

6.2.1 Write a computer program that computes the characteristics of a circular arc that passes through three given nodal points. Specifically, the program should compute the centre, the radius, and the orientation index of the arc, as well as the three polar angles corresponding to the nodal points.

6.2.2 Show that the mean curvature of an axisymmetric surface whose trace in a meridional plane is approximated with a circular arc is given by

$$k_m = \frac{\alpha}{2}\left(\frac{1}{R} + \frac{\sin\theta}{y_c + R\sin\theta}\right)$$

where the $y$-axis represents the radial ($\sigma$) direction in the $xy$ meridional plane.

6.2.3 Write a computer program that, given a collection of marker points on an ellipse, computes the shape of the ellipse using the parametric representation $\mathbf{x} = \mathbf{f}(\beta)$ where $\beta$ is the polygonal arc length. It will be convenient to normalize $\beta$ so that it varies between 0 and 1. Specifically, the program should be able to compute the $x$ and $y$ coordinates of a point on the ellipse for a given value of $\beta$ in the range $[0, 1]$. The necessary interpolations should be performed using cubic splines.

6.2.4 Compute the normal vector and the mean curvature of an axisymmetric surface that is described by the equation $\sigma = g(x)$.

6.2.5 Compute the local dilatation of a surface that is expanding while preserving its shape so that $\mathbf{x} = a(t)\mathbf{f}(\eta, \xi)$ where $a(t)$ is a function of time.

6.2.6 Compute the normal vector and the mean curvature of an ellipsoid. Verify that in the limit as the three axes of the ellipsoid become identical, the mean curvature becomes a constant.

*Figure 6.3.1.* Adaptive description of a two-dimensional boundary; (*a*) when the distance between two marker points becomes excessively large, an additional point (indicated by an open circle) is introduced midway between the two old points; when the distance becomes excessively small, the points $i$ and $i-1$ are eliminated and a new point (indicated by an open circle) is introduced midway between the two points; (*b*) when the total angle $\Delta\theta$ subtended by the $i$th arc becomes larger than a present maximum, the point $i$ is removed, and two new points (indicated by open circles) are introduced at evenly spaced intervals along the arc.

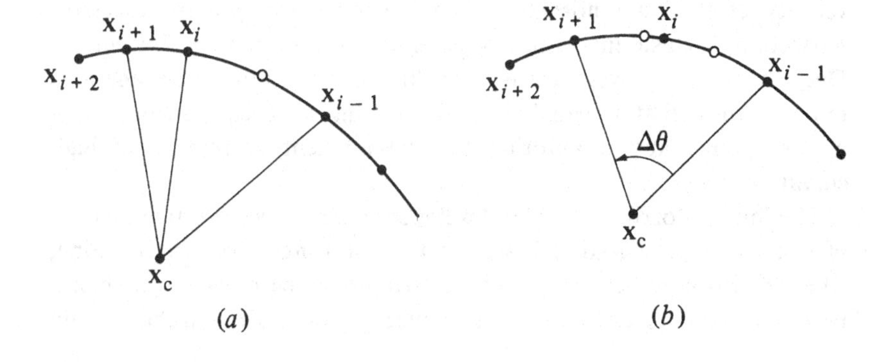

6.2.7 Verify that the Euler characteristic is $\chi = 2$ for a sphere, and $\chi = 0$ for a torus.

6.2.8 Derive expressions for the normal vector, the area, and the perimeter of a planar triangle in terms of the coordinates of the three vertices.

6.2.9 Write a computer program that computes the curvature of an arbitrary line that lies over a curved triangular element. Test the reliability of your program by performing computations for a triangle whose nodes lie on a sphere.

6.2.10 Euler's theorem (Struik 1961, p. 81) states that the curvature $k$ of the trace $C$ of a surface in a plane that is perpendicular to the surface at a point is given by

$$k = k_{max} \cos^2 \alpha + k_{min} \sin^2 \alpha$$

where $k_{max}$ and $k_{min}$ are the maximum and minimum values of curvature, and $\alpha$ is the angle between the tangent vector to $C$ and the tangent vector corresponding to the direction of maximum curvature. Show that Euler's theorem implies that the mean curvature of the surface is equal to the average of the value of the curvatures of the trace of the surface on any two perpendicular planes that are normal to the surface.

## 6.3 Adaptive representation of evolving planar boundaries

Evolving boundaries are the necessary ingredient of unsteady flows involving fluid interfaces or growing and dissolving walls. To study the dynamics of these flows we must pursue the evolution of the boundaries, or from a numerical perspective, we must compute the trajectories of the nodes that mark the location of the boundaries (see section 5.6).

Now, in the course of the evolution some marker points may move close to or farther apart from each other, thereby decreasing the accuracy of the boundary element representation. In addition, the total angle of a circular arc that is subtended by three successive marker points may become excessively large, thereby indicating the development of regions of high curvature, and placing limits on the reliability of the boundary element representation. Forcing the marker points to move only with the velocity of the fluid normal to the boundary will prevent excessive separation or clustering at the expense, however, of numerical accuracy. One way to circumvent these difficulties is to introduce an adaptive representation that is capable of maintaining an even distribution of marker points and monitoring the development of regions of high curvature.

Higdon & Pozrikidis (1985) developed a simple yet effective method of adaptive representation based on the following three criteria. First, when the distance between two successive points becomes larger than a preset maximum, a new point is introduced midway between the two old points. This new point is taken to be on either the backward or the forward arc that passes through the two points (Figure 6.3.1(*a*)). Second, when the angle subtended by an arc becomes larger than a preset maximum, the middle point is eliminated, and two new points are introduced at even intervals along the arc (Figure 6.3.1(*b*)). Third, when two points move very close to each other, they are forced to merge, and are replaced by a single point that is located midway between the two old points. This last operation is permitted only if the resulting point distribution does not violate the first two criteria. Extensions and refinements of this procedure are discussed by Dritschel (1989).

## 6.4 Numerical computation of the boundary integrals

Inspecting the various stages of the boundary element method outlined in section 6.1, we identify several necessary intermediate tasks. One important task is the computation of the single-layer and double-layer integrals over the boundary elements.

To compute the boundary integrals at a point located away from an element, we may use a standard numerical method, preferably a Gaussian quadrature (Stroud & Secrest 1966, Abramowitz & Stegun 1972, Stroud 1971, Zienkiewicz 1971, Banerjee & Butterfield 1981). One example of a Gaussian quadrature suitable for computing a boundary integral over a planar or curved triangle is illustrated in Figure 6.4.1. The triangle is

*Figure 6.4.1.* A cubic quadrature suitable for computing a two-dimensional integral over a planar or curved triangle using seven base points. The weights corresponding to the seven base points are enclosed by ellipses.

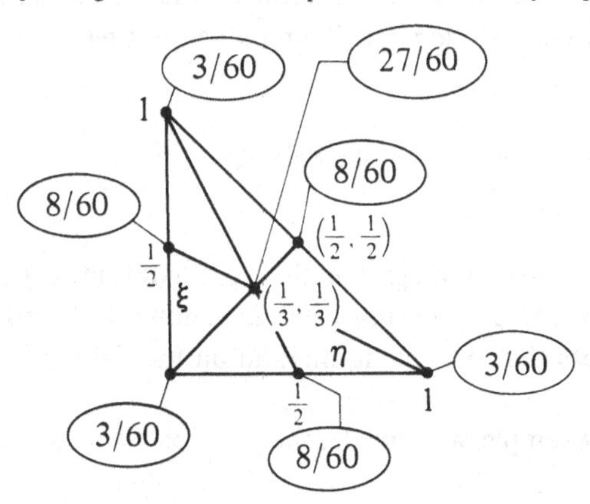

mapped from the physical to the $(\eta, \xi)$ space, and the boundary integral is approximated with a finite weighted sum as

$$\int_{\text{triangle}} f(\mathbf{x})\,\mathrm{d}S(\mathbf{x}) = \int_{\text{triangle}} g(\eta, \xi)\,\mathrm{d}\eta\,\mathrm{d}\xi \approx \frac{1}{2}\sum_{n=1}^{7} g(\eta_n, \xi_n) w_n \qquad (6.4.1)$$

where $g = f h_n$. The coordinates of the base points and the corresponding weights are indicated in Figure 6.4.1.

To compute accurately a boundary integral at a point located on an element, we must account for the singularity of the corresponding kernel at the pole. Three methods for computing singular integrals are:

(1) the use of a Gaussian quadrature that is specifically designed for the type of singularity under consideration. Stroud & Secrest (1966) provide several quadratures for one-dimensional singular integrals including integrals with a logarithmic singularity at the pole. Lean & Wexler (1985) discuss the implementation of these quadratures in the context of boundary integral methods.
(2) the removal of the singularity using a suitable identity such as (2.1.4) or (2.1.12).
(3) the isolation, simplification, and analytical or numerical integration of the singularity over the host boundary element.

To illustrate the second method, we consider the single-layer integral

$$I_j^{\mathrm{S}}(\mathbf{x}_0) = \int_D q_i(\mathbf{x}) G_{ij}(\mathbf{x}, \mathbf{x}_0)\,\mathrm{d}S(\mathbf{x}) \qquad (6.4.2)$$

over a closed surface $D$, and stipulate that the density $\mathbf{q}$ is proportional to the normal vector $\mathbf{n}, \mathbf{q} = g(\mathbf{x})\mathbf{n}$. This type of behaviour occurs when $\mathbf{q}$ represents the discontinuity in the surface force across an interface with constant tension (see section 5.3). To remove the singularity of the kernel in (6.4.2) we write

$$I_j^{\mathrm{S}}(\mathbf{x}_0) = \int_D [g(\mathbf{x}) - g(\mathbf{x}_0)] n_i(\mathbf{x}) G_{ij}(\mathbf{x}, \mathbf{x}_0)\,\mathrm{d}S(\mathbf{x}) + g(\mathbf{x}_0) \int_D n_i(\mathbf{x}) G_{ij}(\mathbf{x}, \mathbf{x}_0)\,\mathrm{d}S(\mathbf{x}) \qquad (6.4.3)$$

The integrand of the first integral on the right-hand side of (6.4.3) is regular and thus may be computed using a standard numerical method. Equation (2.1.4) guarantees that the second integral on the right-hand side is equal to zero.

As a second example, we consider the principal value of the double-layer integral

$$I_i^{\mathrm{D}}(\mathbf{x}_0) = \int_D^{\mathscr{P}\mathscr{V}} q_j(\mathbf{x}) T_{jik}(\mathbf{x}, \mathbf{x}_0) n_k(\mathbf{x}) \, \mathrm{d}S(\mathbf{x}) \tag{6.4.4}$$

over a closed surface $D$. In this case, we write

$$I_i^{\mathrm{D}}(\mathbf{x}_0) = \int_D [q_j(\mathbf{x}) - q_j(\mathbf{x}_0)] T_{jik}(\mathbf{x}, \mathbf{x}_0) n_k(\mathbf{x}) \, \mathrm{d}S(\mathbf{x})$$

$$+ q_j(\mathbf{x}_0) \int_D^{\mathscr{P}\mathscr{V}} T_{jik}(\mathbf{x}, \mathbf{x}_0) n_k(\mathbf{x}) \, \mathrm{d}S(\mathbf{x}) \tag{6.4.5}$$

and note that the first integral on the right-hand side is regular and thus may be computed using a standard numerical method. Using (2.1.12) we find that the second integral on the right-hand side is simply equal to $-4\pi\delta_{ij}$.

It will be noted that the identities (2.1.4) and (2.1.12) are irrelevant when the density of the single-layer potential is not proportional to the normal vector or when the boundary $D$ is an open or infinite surface. In these cases we must isolate, simplify, and integrate the singularity analytically over the host boundary element. For this purpose, we introduce two surface variables $(\eta, \xi)$ over the element (see section 6.2), and rewrite the single-layer potential (6.4.2) in the equivalent form

$$I_j^{\mathrm{S}}(\mathbf{x}_0) = \int_D \left[ q_i(\mathbf{x}) G_{ij}(\mathbf{x}, \mathbf{x}_0) - \frac{q_i(\mathbf{x}_0)}{h_n(\mathbf{x})} F_{ij}(h, \xi, h_0, \xi_0) \right] \mathrm{d}S(\mathbf{x})$$

$$+ q_i(\mathbf{x}_0) \int_D F_{ij}(h, \xi, h_0, \xi_0) \, \mathrm{d}\eta \, \mathrm{d}\xi \tag{6.4.6}$$

where we recall that $\mathrm{d}S = \mathrm{h}_n \mathrm{d}\eta \mathrm{d}\xi$ (see (6.2.13)). The function $\mathbf{F}$ is designed in such a way that the first integrand on the right-hand side of (6.4.6) is regular whereas the second integral may be computed exactly in the $(\eta, \xi)$ plane (Davey & Hinduja 1988). As an example, let us consider the single-layer integral of the two-dimensional Green's function over a circular arc (Figure 6.2.2). Noting that the diagonal components of the Green's function exhibit a $-\ln \mathrm{r}$ singularity, we write

$$I_j^{\mathrm{S}}(\mathbf{x}_0) = R \int_{\theta_{i-1}}^{\theta_{i+1}} [q_i(\mathbf{x}) G_{ij}(\mathbf{x}, \mathbf{x}_0) + q_j(\mathbf{x}_0) \ln|\theta - \theta_0|] \mathrm{d}|\theta|$$

$$- q_j(\mathbf{x}_0) R \int_{\theta_{i-1}}^{\theta_{i+1}} \ln|\theta - \theta_0| \mathrm{d}|\theta| \tag{6.4.7}$$

where $R$ is the radius of the arc and $\theta_0$ is the polar angle corresponding to the point $\mathbf{x}_0$, which is presumed to lie on the arc, i.e. $\theta_0$ is between $\theta_{i-1}$

and $\theta_{i+1}$. The first integral on the right-hand side of (6.4.7) is regular and thus may be computed using a standard method. The second integral may be computed exactly as

$$\int_{\theta_{i-1}}^{\theta_{i+1}} \ln|\theta-\theta_0|\,\mathrm{d}|\theta| = |\theta_{i+1}-\theta_0|\ln|\theta_{i+1}-\theta_0| + |\theta_{i-1}-\theta_0|\ln|\theta_{i-1}-\theta_0| + |\theta_{i+1}-\theta_{i-1}| \tag{6.4.8}$$

An alternative strategy for regularizing a singular integral is to remove the singularity of the kernel by changing the variables of integration. To illustrate this strategy, let us consider a singular integral over a three-dimensional boundary element, and introduce polar coordinates $(\rho, \chi)$ in the $(\eta, \xi)$ plane so that

$$\eta-\eta_0=\rho\cos\chi \qquad \xi-\xi_0=\rho\sin\chi \tag{6.4.9}$$

We write

$$\int_D F_{ij}(h,\xi,h_0,\xi_0)\mathrm{d}\eta\,\mathrm{d}\xi = \int_D F_{ij}(\rho,\chi,\rho_0=0)\rho\,\mathrm{d}\rho\,\mathrm{d}\chi \tag{6.4.10}$$

and note that the presence of $\rho$ in the integrand on the right-hand side of (6.4.10) removes the singularity of $\mathbf{F}$ at $\rho=0$. Fairweather, Rizzo & Shippy (1979) suggested introducing two new variables $(\alpha, \beta)$ defined by the equations

$$\eta-\eta_0=(1-\beta)\alpha \qquad \xi-\xi_0=\alpha\beta \tag{6.4.11}$$

and writing

$$\int_D F_{ij}(h,\xi,h_0,\xi_0)\mathrm{d}\eta\,\mathrm{d}\xi = \int_D F_{ij}(\alpha,\beta,\alpha_0=0)\alpha\,\mathrm{d}\alpha\,\mathrm{d}\beta \tag{6.4.12}$$

The presence of $\alpha$ in the integrand on the right-hand side of (6.4.12) removes the singularity of $\mathbf{F}$ at $\alpha=0$.

**Problems**

6.4.1 Write a computer program that uses (6.4.1) to compute the area of a curved triangle defined by six nodal points.

6.4.2 Derive the equivalent of (6.4.7) for a straight segment.

## 6.5 Accuracy of boundary element methods

The accuracy of a boundary element method will depend upon the shape and distribution of the boundary elements as well as the order of approximation of the unknown function. Overall, for problems with solid boundaries, if we use linear elements (straight segments for two-dimensional or axisymmetric flow, and plane triangles for three-dimensional flow) and then, over each element, approximate the density of the single-layer potential with a constant function, and the density of the double-layer potential with a linear function, we will obtain satisfactory accuracy with a moderate number of elements. Muldowney (1989) discusses the accuracy of various boundary element collocation methods including spectral element methods.

The accuracy of the computations will decline (but fortunately, only on a local level) when the boundaries contain sharp corners, points or lines of discontinuous velocities, and three-phase contact lines. In these cases, the flow will contain singularities associated with infinite values of the velocity, the surface force, or the density of the hydrodynamic potential. Higdon (1985) and Pozrikidis (1988) found that the presence of sharp corners neither decreases the global accuracy nor discredits the overall liability of boundary element computations. Kelmanson (1983a, b), on the other hand, found that discontinuous boundary velocities may cause oscillatory solutions and slow convergence. One way to circumvent these problems is to use a high density of boundary elements in the vicinity of the singular points. Practice has shown that this is an effective remedy but not a panacea (Kelmanson 1983a, b). A more sophisticated approach is to identify the singular behavior, to decompose the unknown boundary function $\mathbf{q}$ as $\mathbf{q} = \mathbf{q}^S + \mathbf{q}^R$, where $\mathbf{q}^S$ is the known singular component, and to solve the integral equation for the regular component $\mathbf{q}^R$ (Kelmanson 1983a, b, c; Ingham & Kelmanson 1984; Chan *et al.* 1986). Implementing this strategy requires knowledge of the asymptotic behaviour of the flow in the vicinity of the singular points. This is available for a limited number of two-dimensional and axisymmetric flows (Dean & Montagnon 1949; Michael 1958; Lugt & Schwiderski 1964; Moffat 1964; Moffat & Duffy 1980; Jeffrey & Sherwood 1980; O'Neill 1983), and for several three-dimensional flows (Tokuda 1972, 1975).

Now, in order to optimize the accuracy of a boundary element computation, we must distribute the boundary elements so as to minimize a suitably defined global or local error. Experience has shown that the best results are obtained when the boundary elements are concentrated around regions of high boundary curvature and regions where the solution is expected to show strong variations (see, for instance, Ingber & Mitra 1986, Kennedy 1991). When a desired level of local or global accuracy is required, the network of the boundary elements must be refined in a global or local sense. When a three-dimensional boundary has been discretized into a set of triangles, global refinement may be performed by sub-dividing each triangle into four descendant triangles formed by connecting the middle points of the three sides of the original triangle, as illustrated in Figure 6.5.1(*a*). The logistics of refinement is considerably facilitated by numbering the nodes and the elements, and keeping tables that relate the numerical labels of the nodes and those of the associated elements and *vice versa.* It will be noted that global refinement may also be used for generating a fine from a coarse network of boundary elements.

Local refinement may be also performed via element subdivision, as shown in Figure 6.5.1(*a, b*). The method shown in Figure 6.5.1(*a*) is recommended when one angle of a triangle is larger than $\pi/2$, whereas the method illustrated in Figure 6.5.1(*b*) is recommended when the area of a triangle is excessively large.

## 6.6 Computer implementations

In the preceding sections we presented a general introduction to the basic procedures involved in a computation using a boundary element method. Details on specific issues may be found in general monographs on boundary element methods (Jaswon & Symm 1977; Banerjee & Butterfield 1981). Kelmanson (1983a, b) and Hansen (1987) discuss boundary element methods for two-dimensional flow in terms of the stream function. Higdon

*Figure 6.5.1.* Two methods for performing local refinement: (*a*) a triangle is subdivided into four descendant triangles; (*b*) a triangle and one of its neighbors are subdivided into two descendant triangles.

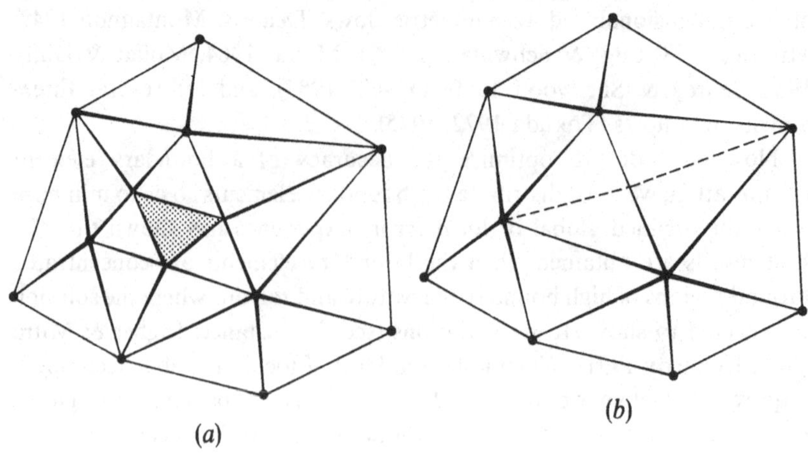

(1985) addresses the issues of sharp corners and infinite boundaries. Chan *et al.* (1986) discuss the numerical solution of Laplace's equation for steady swirling flow. Youngren & Acrivos (1975), Rallison (1981), Tran-Cong *et al.* (1989, 1990), Gavze (1990), and Kennedy (1991) present implementations for three-dimensional flow. Pozrikidis (1989b) discusses an implementation for unsteady axisymmetric flow.

In the rest of this section we present a selective compilation of computations based on boundary element methods. Our objectives are to demonstrate the diverse range of problems that may be tackled using boundary element methods and to provide references for further information on details of numerical implementations.

We begin with the simplest case of two-dimensional flow. In Figure 6.6.1 we present streamline patterns for (*a*) shear flow over a rectangular

*Figure 6.6.1.* Boundary element computations for two-dimensional flow. Streamline patterns for (*a*) shear flow over cavities and crests (Higdon 1985); (*b*) flow through a semi-infinite lattice of circular cylinders modeling a porous medium (Larson & Higdon 1987); (*c*) Couette flow in a channel confined between a plane and a sinusoidal wall (Pozrikidis 1987a); (*d*) peristaltic flow due to the propagation of waves along the flexible walls of a two-dimensional channel (Pozrikidis 1987b).

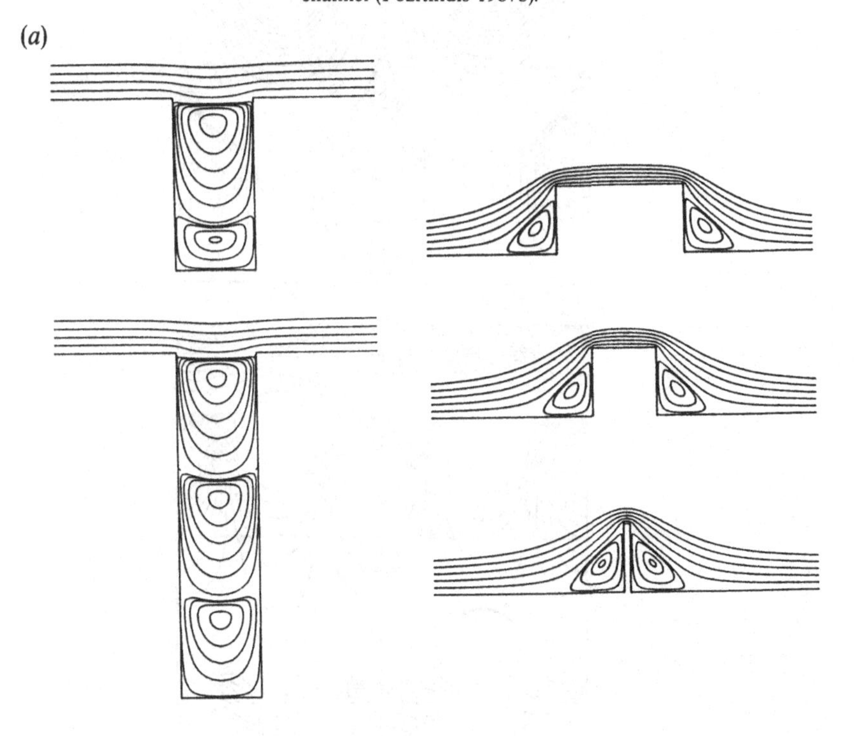

*Figure 6.6.1.* (Continued)

(*b*)

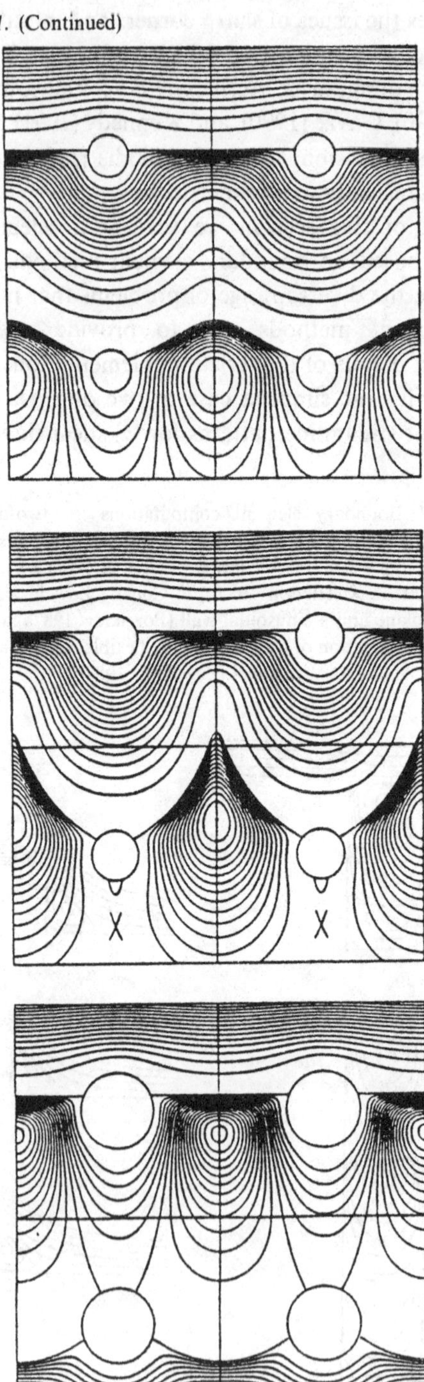

*Figure 6.6.1.* (Continued)

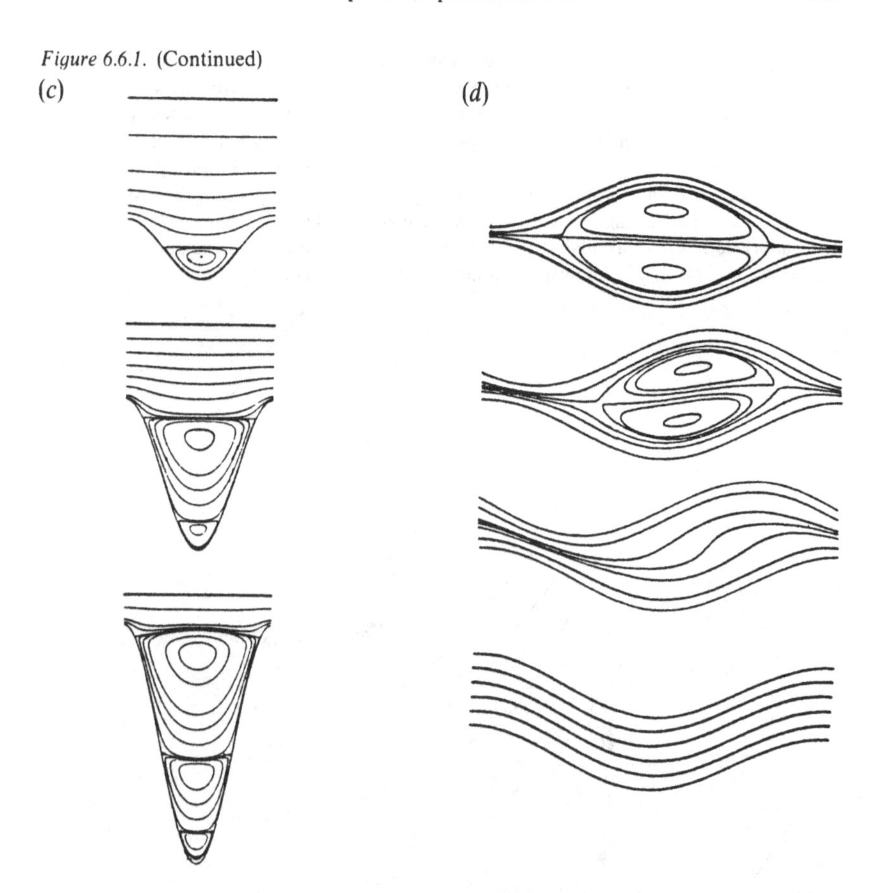


cavity (Higdon 1985), (*b*) flow through an array of elliptical cylinders modeling a porous medium (Larson & Higdon 1987), (*c*) flow in a channel confined between a plane and a sinusoidal wall (Pozrikidis 1987a), and (*d*) peristaltic flow due to the propagation of waves along the flexible walls of a channel (Pozrikidis 1987b).

In Figure 6.6.2 we present two examples of two-dimensional flow involving free surfaces and fluid interfaces. In Figure 6.6.2(*a*) we show the shape of the free surface of a liquid film flowing down an inclined wavy wall (Pozrikidis 1988), and in Figure 6.6.2(*b*) we illustrate several stages during the Rayleigh–Taylor instability of a liquid layer resting on a plane wall below another liquid of higher density (Newhouse & Pozrikidis 1990).

Proceeding to axisymmetric flow, in Figure 6.6.3 we present successive stages in (*a*) the evolution of a viscous drop settling toward a plane wall (Pozrikidis 1990b), and (*b*) the deformation of a red blood cell in a uniaxial straining flow (Pozrikidis 1990c).

*Figure 6.6.2.* Two-dimensional flow involving free surfaces and fluid interfaces; (*a*) three families of shapes of the free surface of a liquid film flowing down an inclined wavy wall for several flow rates (Pozrikidis 1988); the surface tension is constant for each family, and low in the top panel, moderate in the middle panel, and high in the bottom panel; (*b*) several stages during the Rayleigh–Taylor instability of a liquid layer resting on a plane wall below another liquid of higher density (Newhouse & Pozrikidis 1990); initially the interface has a sinusoidal shape; the bottom figure shows the instantaneous streamline pattern at an advanced stage of its evolution.

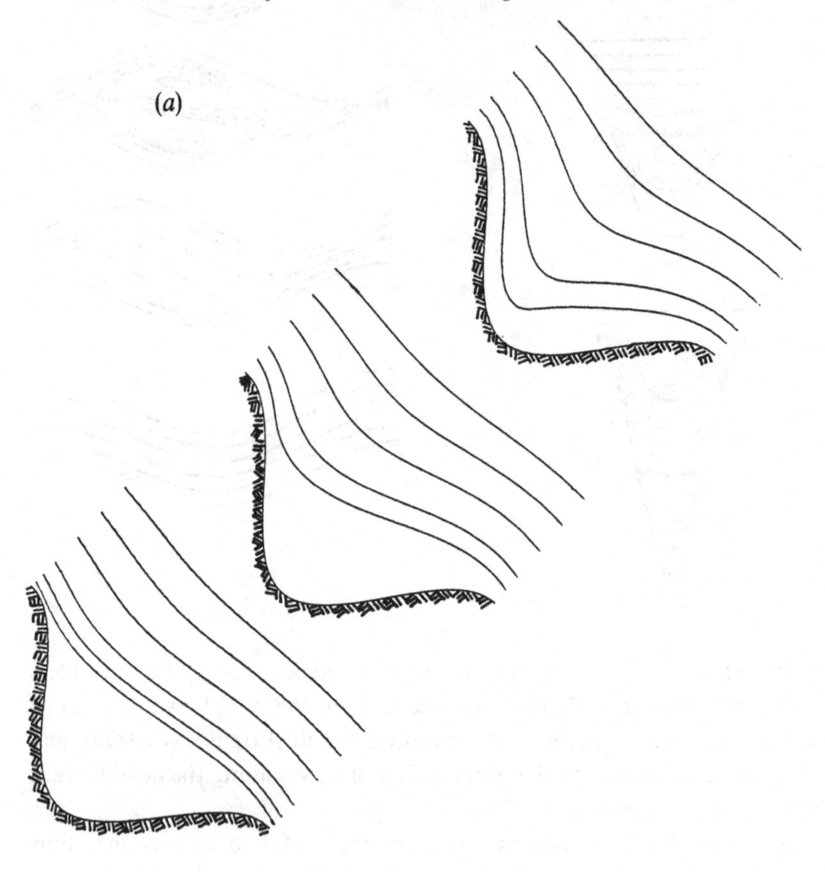

*Figure 6.6.2.* (Continued)

(*b*)

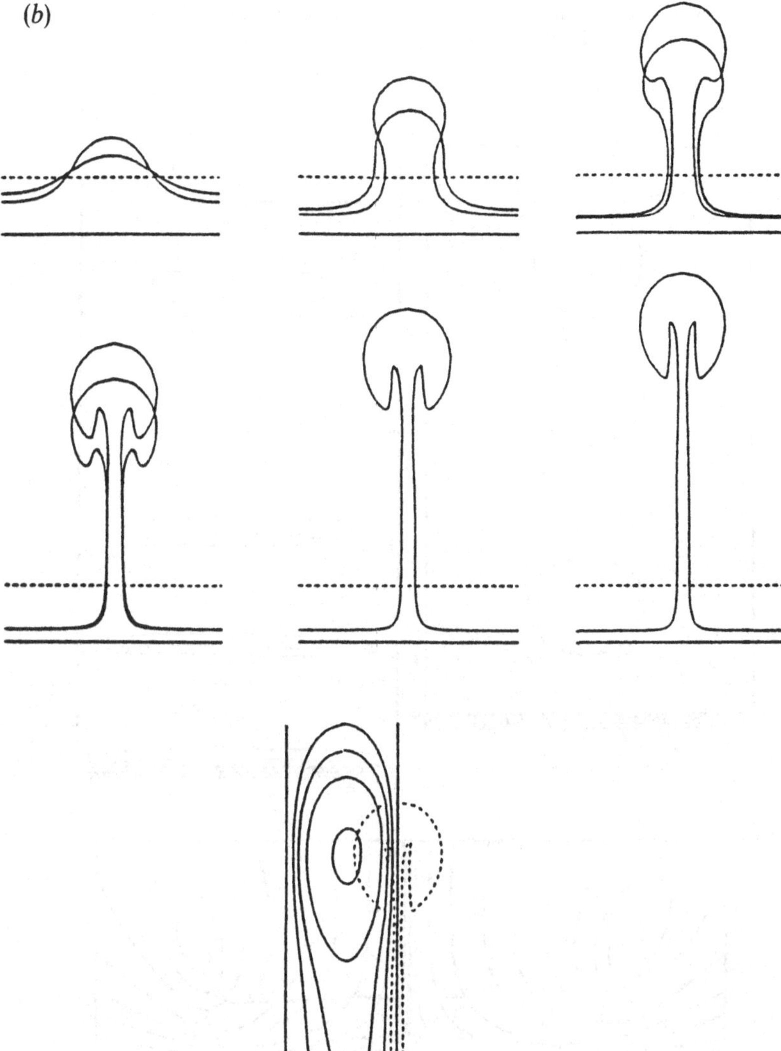

*Figure 6.6.3.* Boundary element computations for axisymmetric flow. Successive stages in the evolution of (*a*) a viscous drop falling toward a plane wall (Pozrikidis 1990b), where the bottom panel shows the instantaneous streamline pattern; (*b*) a red blood cell placed in a uniaxial extensional flow (Pozrikidis 1990c).

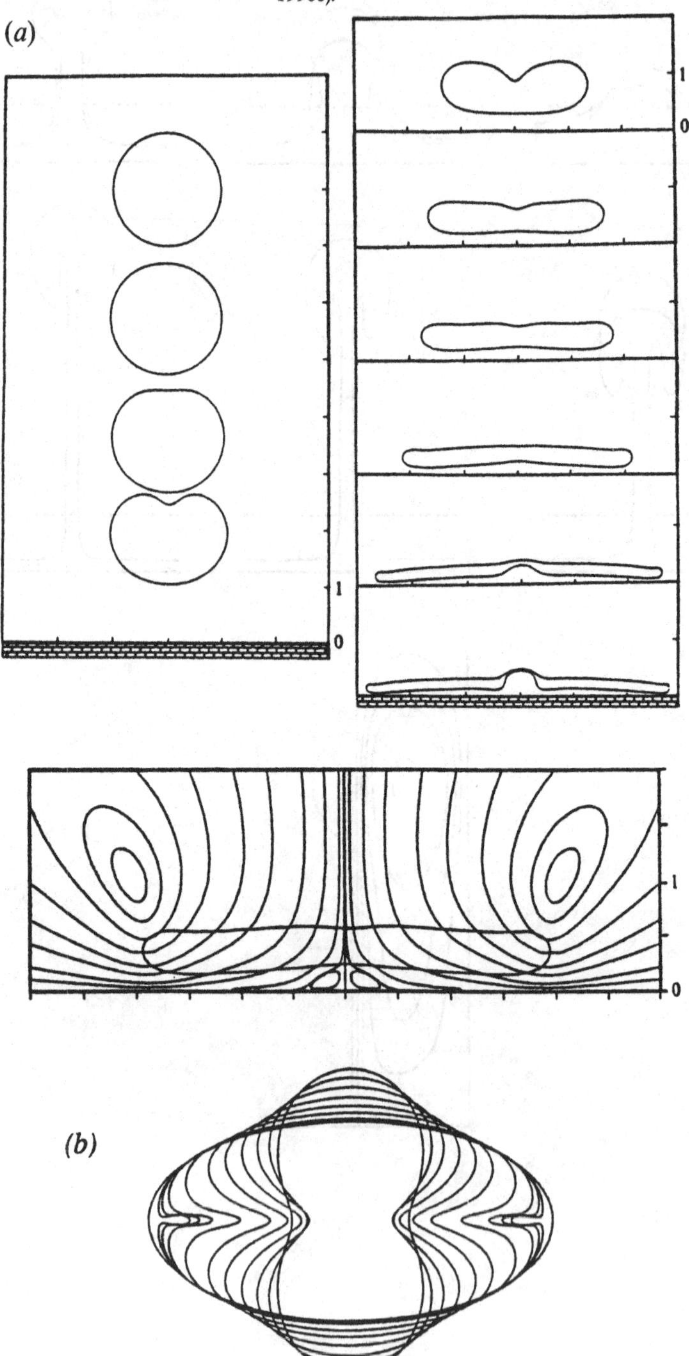

*Figure 6.6.4.* Three-dimensional flows. (*a*) The initial and advanced stages of deformation of a spherical drop evolving under the action of an incident shear flow (Kennedy *et al.* 1991); (*b*) the free surface of a liquid film flowing down an inclined plane wall over a small particle captured by the wall (Pozrikidis & Thoroddsen 1991).

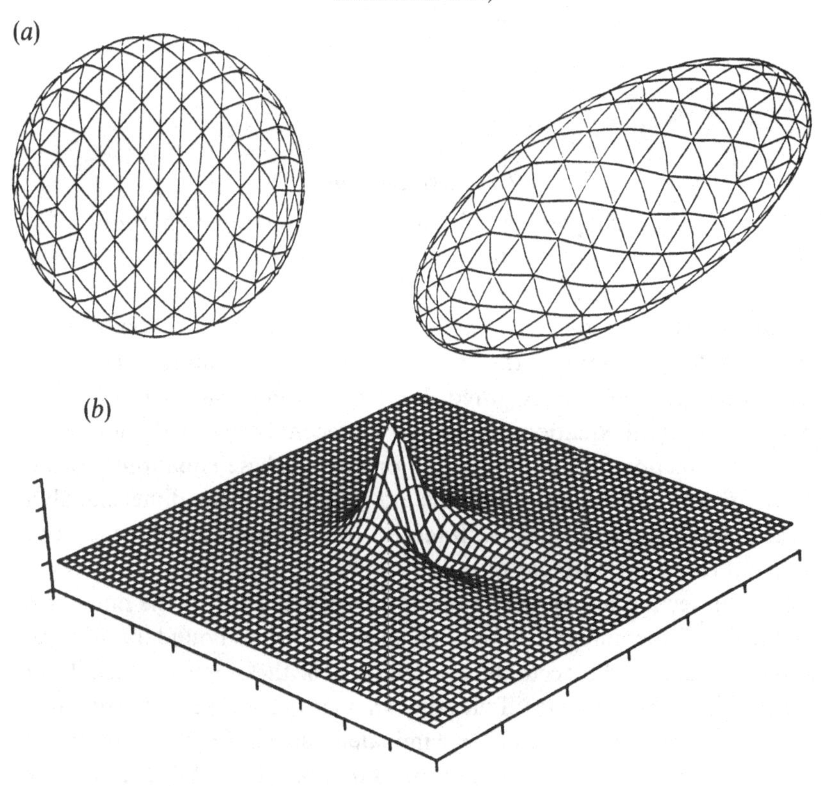


Finally, in Figure 6.6.4 we present examples of three-dimensional computations: (*a*) the initial and asymptotic shapes of a three-dimensional drop which is placed in an ambient plane shear flow (Kennedy *et al.*, 1991), and (*b*) the shape of a liquid film flowing down an inclined wall over a particle captured by the wall (Pozrikidis & Thoroddsen 1991).

# 7
# The singularity method

## 7.1 Introduction

In the preceding chapters we developed integral representations of a flow in terms of boundary distributions of Green's functions and their associated stress tensor. Physically, these representations involve boundary distributions of point forces, point sources, and point force dipoles. When we applied the representations at the boundaries of the flow and imposed the required boundary conditions, we obtained Fredholm integral equations of the first or second kind for the unknown boundary functions. Two advantages of solving these equations instead of the primary differential equations are reduction of the dimensionality of the mathematical problem with respect to the physical problem by one unit, and efficient treatment of infinite flows.

Now, in section 6.5 we saw that one important aspect of the numerical treatment of the integral equations resulting from boundary integral representations is the accurate computation of the singular single-layer and double-layer integrals. True, the singularities of the corresponding kernels may be subtracted off or eliminated using a suitable identity or changing the variables of integration, but this places burdens on the computer implementation and increases the cost of the computations. One *ad-hoc* way to circumvent this difficulty is to move the pole of the Green's function and its associated stress tensor from the boundary $D$ onto a surface $\mathscr{D}$ exterior to the flow, as illustrated in Figure 7.1.1. Of course, this is an arbitrary remedy that lacks physical foundation, but nevertheless it appears worth consideration. Furthermore, taking this action one step further, we may replace the continuous surface distribution of the singularities with line or point-wise distributions. In this manner we obtain a representation in terms of line integrals or finite collections of point forces, point sources, and point force dipoles. Finally, we may extend the menu of singularities to include not only the Green's function and its associated stress tensor, but also any other singular solutions of the governing equations of creeping flow. These extensions define a new method for computing Stokes flow, known as the *method of fundamental solutions* or simply the *singularity method* (Kupradze 1967, Webster 1975).

Having introduced the basic ideas underlying the singularity method, we now offer three comments in its favor. First, we note that in a large class of problems involving suspended particles, one is not interested in computing the precise structure of the flow, but wishes instead to evaluate certain gobal variables such as the hydrodynamic force or torque acting on a suspended particle. The singularity method promises to provide this information with relatively little computational effort, and without direct reference to the detailed structure of the flow. This is to be contrasted with the boundary integral method, which relies on the accurate computation of boundary functions, such as the boundary surface force or the density of a hydrodynamic potential, before it can produce any global variable of the flow. Second, as we saw in section 2.5, singularity representations allow us to derive the generalized Faxen relations in a direct fashion, and with very little analytical effort. Third, the singularity method may be combined effectively with the boundary integral method yielding a compound method that maintains many of the advantages of its constituents. An example of a compound method has been discussed already in section 4.7 in connection with the double-layer representation of an external flow.

To place the singularity method into perspective, it is helpful to summarize what follows in this chapter. In section 7.2 we discuss the singularities of Stokes flow. For convenience, we classify the singularities

*Figure 7.1.1.* In the singularity method, the domain of distribution $\mathscr{D}$ of the fundamental solutions resides outside the domain of flow. (*a*) Internal flow; (*b*) external flow.

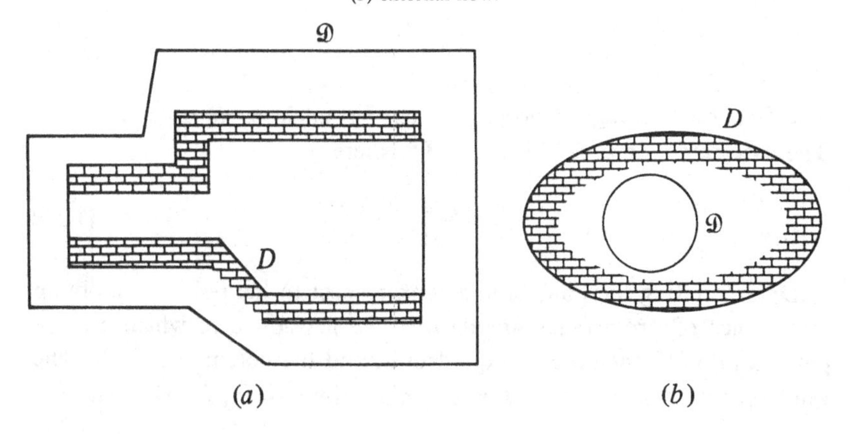

into three categories according to the topology of the domain of flow. The first category includes the free-space singularities for infinite unbounded flow, the second category includes singularities for bounded flow, and the third category includes singularities for internal flow. The velocity field associated with all singularities is required to be regular throughout the domain of flow except at a single point, namely the pole. The velocity field associated with the singularities of internal flow is allowed to be unbounded at infinity. Finally, the velocity field associated with the singularities of bounded or internal flow is required to vanish over the boundaries of the flow. In section 7.3 we present examples of exact and asymptotic singularity representations, and in section 7.4 we discuss numerical methods for computing singularity representations. To conclude, in section 7.5 we discuss extensions of the singularity method for unsteady Stokes flow.

## 7.2 The singularities of Stokes flow

### *Free-space singularities*

First, we consider singularities for infinite unbounded flow. To derive these singularities, we find it convenient to identify the pressure with a spherical harmonic function and then to compute the associated flow.

Beginning with the simplest choice, we set $P = c$ where $c$ is a constant, and use the Stokes equation to find that the velocity is a harmonic function, i.e. $\nabla^2 \mathbf{u} = 0$. Clearly, any irrotational flow $\mathbf{u} = \nabla \Phi$, where $\Phi$ is the potential function, constitutes an acceptable Stokes flow. Selecting, in particular $\Phi = -1/(4\pi r)$, where $r = |\hat{\mathbf{x}}|$, $\hat{\mathbf{x}} = \mathbf{x} - \mathbf{x}_0$, we obtain the point source with pole at the point $\mathbf{x}_0$. The associated velocity field is $\mathbf{u} = \mathscr{s}\boldsymbol{\Sigma}$ where $\mathscr{s}$ is a constant and

$$\Sigma_i = \frac{\hat{x}_i}{r^3} \tag{7.2.1}$$

The flow rate through a closed surface that encloses $\mathbf{x}_0$ is equal to $4\pi\mathscr{s}$. The associated stress field is $\boldsymbol{\sigma} = \mu\mathscr{s}\mathbf{T}^{\Sigma}$ where

$$T_{ik}^{\Sigma} = 2\frac{\delta_{ik}}{r^3} - 6\frac{\hat{x}_i\hat{x}_k}{r^5} \tag{7.2.2}$$

Differentiating the point source with respect to the pole $\mathbf{x}_0$ we obtain a sequence of irrotational singularities the first three of which are the potential dipole, the potential quadruple, and the potential octuple. The velocity due to a potential dipole is given by $u_i = D_{ij}\mathscr{d}_j$, where $\mathscr{d}$ is a constant, and

$$D_{ij} = \frac{\partial \Sigma_i}{\partial x_{0,j}} = -\frac{\delta_{ij}}{r^3} + 3\frac{\hat{x}_i \hat{x}_j}{r^5} \tag{7.2.3}$$

The associated stress field is $\sigma_{ik} = \mu T^{\mathrm{D}}_{ijk} d_j$, where

$$T^{\mathrm{D}}_{ijk} = 6\frac{\delta_{ij}\hat{x}_k + \delta_{ik}\hat{x}_j + \delta_{jk}\hat{x}_i}{r^5} - 30\frac{\hat{x}_i \hat{x}_j \hat{x}_k}{r^7} \tag{7.2.4}$$

The velocity due to a potential quadruple is $u_i = Q_{ijl} q_{jl}$, where $q$ is a constant, and

$$Q_{ijl} = \frac{\partial D_{ij}}{\partial x_{0,l}} = -3\frac{\delta_{ij}\hat{x}_l + \delta_{il}\hat{x}_j + \delta_{jl}\hat{x}_i}{r^5} + 15\frac{\hat{x}_i \hat{x}_j \hat{x}_l}{r^7} \tag{7.2.5}$$

The associated stress field is $\sigma_{ik} = \mu T^{\mathrm{Q}}_{ijlk} q_{jl}$, where

$$T^{\mathrm{Q}}_{ijlk} = \frac{\partial T^{\mathrm{D}}_{ijk}}{\partial x_{0,l}} = -\frac{6}{r^5}(\delta_{ij}\delta_{kl} + \delta_{ik}\delta_{jl} + \delta_{jk}\delta_{il}) + \frac{30}{r^7}(\delta_{ij}\hat{x}_k + \delta_{ik}\hat{x}_j + \delta_{jk}\hat{x}_i)\hat{x}_l$$
$$+ \frac{30}{r^7}(\delta_{il}\hat{x}_j\hat{x}_k + \delta_{jl}\hat{x}_i\hat{x}_k + \delta_{kl}\hat{x}_i\hat{x}_j) - 210\frac{\hat{x}_i \hat{x}_j \hat{x}_k \hat{x}_l}{r^9} \tag{7.2.6}$$

Continuing to differentiate, we derive the potential octuple and higher order multipoles of the point source.

Next, we turn our attention to rotational singularities with non-constant pressure and non-vanishing vorticity. First, we set the pressure equal to the singular harmonic function

$$P = 2\mu\beta\left(\frac{1}{r}\right) = \mu\beta\nabla^2 r \tag{7.2.7}$$

where $\beta$ is a constant. To compute the corresponding velocity field we substitute (7.2.7) into the Stokes equation obtaining

$$\nabla^2(\mathbf{u} - \beta\nabla r) = 0 \tag{7.2.8}$$

It is not possible to compute a velocity field $\mathbf{u}$ that satisfies both (7.2.8) and the continuity equation. The pressure stresses due to (7.2.7) are too strong to be balanced by viscous stresses, and the choice (7.2.7) must be dismissed.

All partial derivatives of the pressure (7.2.7) with respect to $\mathbf{x}_0$ constitute acceptable pressure fields. Taking the first derivatives with respect to $x_0$, $y_0$, and $z_0$ individually, and multiplying the results by three arbitrary constants constituting the vector $\mathbf{g}$, we find $P = \mu \mathbf{p}^{\mathrm{G}} \cdot \mathbf{g}$, where

$$p^{\mathrm{G}}_j = 2\frac{\hat{x}_j}{r^3} \tag{7.2.9}$$

To compute the corresponding velocity we substitute this pressure field into the Stokes equation obtaining

$$\nabla^2 u_i = 2\left(\frac{\delta_{ij}}{r^3} - 3\frac{\hat{x}_i\hat{x}_j}{r^5}\right)g_j \tag{7.2.10}$$

The functional form of (7.2.10) suggests a solution in the form

$$u_i = \left(a\frac{\delta_{ij}}{r} + b\frac{\hat{x}_i\hat{x}_j}{r^3}\right)g_j \tag{7.2.11}$$

where $a$ and $b$ are constants. Substituting (7.2.11) into the continuity equation will show that $a = b$; furthermore requiring that (7.2.9) and (7.2.11) satisfy the Stokes equation will show that $a = b = 1$. Clearly, the emerging singularity is the free-space Green's function representing the flow due to a point force of strength $8\pi\mu g$ acting in an infinite fluid. Conforming with standard notation, we cast (7.2.11) in the equivalent form $u_i = G_{ij} g_j$, where

$$G_{ij} = \frac{\delta_{ij}}{r} + \frac{\hat{x}_i\hat{x}_j}{r^3} \tag{7.2.12}$$

is the Stokeslet. Considering (2.2.11) we find that the stress field associated with the Stokeslet is given by $\sigma_{ik} = \mu T^{G}_{ijk} g_j$, where

$$T^{G}_{ijk} = -6\frac{\hat{x}_i\hat{x}_j\hat{x}_k}{r^5} \tag{7.2.13}$$

Continuing our search, we take the first derivatives of (7.2.9) with respect to $\mathbf{x}_0$ to compute the pressure field $P = \mu p^{GD}_{jl} d_{jl}$, where $\mathbf{d}$ is a constant matrix, and

$$p^{GD}_{jl} = 2\frac{\partial}{\partial x_{0,l}}\left(\frac{\hat{x}_j}{r^3}\right) = -2\frac{\delta_{jl}}{r^3} + 6\frac{\hat{x}_j\hat{x}_l}{r^5} \tag{7.2.14}$$

The superscript GD stands for Green's function doublet. To compute the associated velocity we differentiate the Stokeslet with respect to $\mathbf{x}_0$ finding $u_i = G^{D}_{ijl} d_{jl}$, where

$$G^{D}_{ijl} = \frac{\partial G_{ij}}{\partial x_{0,l}} = \frac{1}{r^3}(\delta_{ij}\hat{x}_l - \delta_{il}\hat{x}_j - \delta_{jl}\hat{x}_i) + 3\frac{\hat{x}_i\hat{x}_j\hat{x}_l}{r^5} \tag{7.2.15}$$

is the Stokeslet doublet. Physically, the Stokeslet doublet represents the velocity due to a point force dipole. The associated stress field is given by $\sigma_{ik} = \mu T^{GD}_{ijlk} d_{jl}$, where

$$T^{GD}_{ijkl} = -6\frac{\partial}{\partial x_{0,l}}\left(\frac{\hat{x}_i\hat{x}_j\hat{x}_k}{r^5}\right) = \frac{6}{r^5}(\delta_{il}\hat{x}_j\hat{x}_k + \delta_{jl}\hat{x}_i\hat{x}_k + \delta_{kl}\hat{x}_i\hat{x}_j) - 30\frac{\hat{x}_i\hat{x}_j\hat{x}_k\hat{x}_l}{r^7} \tag{7.2.16}$$

Now, we decompose the coefficient of the Stokeslet doublet $\mathbf{d}$ into a symmetric component $\mathbf{s} = \frac{1}{2}(\mathbf{d} + \mathbf{d}^{\mathrm{T}})$ and an antisymmetric component $\mathbf{r} = \frac{1}{2}(\mathbf{d} - \mathbf{d}^{\mathrm{T}})$, where the superscript T stands for transpose, and write the velocity due to the Stokeslet doublet in the equivalent form

$$u_i = G^{\text{D-S}}_{ijl} s_{jl} + G^{\text{D-A}}_{ijl} r_{jl} \tag{7.2.17}$$

where the superscripts -S and -A denote the symmetric and antisymmetric part with respect to the indices $j$ and $l$, respectively. Inspecting (7.2.15) we find

$$G^{\text{D-S}}_{ijl} = -\delta_{jl}\frac{\hat{x}_i}{r^3} + 3\frac{\hat{x}_i\hat{x}_j\hat{x}_l}{r^5} \qquad G^{\text{D-A}}_{ijl} = \frac{\delta_{ij}\hat{x}_l - \delta_{il}\hat{x}_j}{r^3} \tag{7.2.18}$$

Furthermore, exploiting the antisymmetry of $\mathbf{r}$, we write

$$r_{jl} = -\tfrac{1}{2}\varepsilon_{jlm}\mathscr{L}_m \tag{7.2.19}$$

where $\mathscr{L}$ is defined as

$$\mathscr{L}_m = -\varepsilon_{mjl} r_{jl} \tag{7.2.20}$$

Using (7.2.19) we cast (7.2.17) in the equivalent form

$$u_i = G^{\text{D-S}}_{ijl} s_{jl} - \tfrac{1}{2} G^{\text{D-A}}_{ijl}\varepsilon_{jlm}\mathscr{L}_m \equiv G^{\text{D-S}}_{ijl} s_{jl} + G^{\text{C}}_{im}\mathscr{L}_m \tag{7.2.21}$$

where we have introduced a new singularity

$$G^{\text{C}}_{im} \equiv -\tfrac{1}{2}\varepsilon_{jlm} G^{\text{D-A}}_{ijl} = -\tfrac{1}{2}\varepsilon_{jlm} G^{\text{D}}_{ijl} = \tfrac{1}{2}\varepsilon_{mlj}\frac{\partial G_{ij}}{\partial x_{0,l}} = \varepsilon_{iml}\frac{\hat{x}_l}{r^3} \tag{7.2.22}$$

termed the *couplet* or *rotlet*. The pressure field associated with the couplet is constant, whereas the stress field is given by $\sigma_{ik} = \mu T^{\text{C}}_{imk}\mathscr{L}_m$, where

$$T^{\text{C}}_{imk} = -\tfrac{1}{2}\varepsilon_{jlm} T^{\text{D}}_{ijlk} = 3\frac{\varepsilon_{ijm}\hat{x}_k + \varepsilon_{kjm}\hat{x}_i}{r^5}\hat{x}_j \tag{7.2.23}$$

Inspecting the symmetric component of the Stokeslet doublet given in (7.2.18), we recognize a point source, and a residual singularity called the stresslet, i.e.

$$G^{\text{D-S}}_{ijl} \equiv \tfrac{1}{2}(G^{\text{D}}_{ijl} + G^{\text{D}}_{ilj}) = -\delta_{jl}\Sigma_i + G^{\text{STR}}_{ijl} \tag{7.2.24}$$

where $\boldsymbol{\Sigma}$ is the point source. The flow due to a stresslet is $u_i = G^{\text{STR}}_{ijl} s_{jl}$, where

$$G^{\text{STR}}_{ijl} = 3\frac{\hat{x}_i\hat{x}_j\hat{x}_l}{r^5} \tag{7.2.25}$$

It will be noted that apart from a proportionality constant, the stresslet is identical to the stress tensor associated with the Stokeslet. The pressure field corresponding to the stresslet is identical to that of the Stokeslet doublet, i.e. $P = \mu p^{\text{GD}}_{jl} s_{jl}$. The associated stress field is given by $\sigma_{ik} = \mu T^{\text{STR}}_{ijlk} s_{jl}$, where

$$T^{\text{STR}}_{ijlk} = -\delta_{ik} p^{\text{SD}}_{jl} + \frac{\partial G^{\text{STR}}_{ijl}}{\partial x_k} + \frac{\partial G^{\text{STR}}_{kjl}}{\partial x_i} = \tfrac{1}{2}(T^{\text{SD}}_{ijlk} + T^{\text{SD}}_{iljk}) + \delta_{lj} T^{\Sigma}_{ik}$$
$$= \delta_{ik}\delta_{jl}\frac{2}{r^3} + \frac{3}{r^5}(\delta_{ij}\hat{x}_k\hat{x}_l + \delta_{il}\hat{x}_k\hat{x}_j + \delta_{kj}\hat{x}_i\hat{x}_l + \delta_{kl}\hat{x}_i\hat{x}_j) - 30\frac{\hat{x}_i\hat{x}_j\hat{x}_k\hat{x}_l}{r^7} \tag{7.2.26}$$

where we recall that $\mathbf{T}^{\Sigma}$ is the stress tensor corresponding to the point source.

Differentiating the Stokeslet doublet we obtain the Stokeslet quadruple. The associated velocity field is given by $u_i = G^{Q}_{ijlm} q_{jlm}$, where $\mathbf{q}$ is a constant, and

$$G^{Q}_{ijlm} = \frac{\partial^2 G_{ij}}{\partial x_{0,l} \partial x_{0,m}}$$
$$= \frac{1}{r^3}(\delta_{il}\delta_{jm} + \delta_{im}\delta_{jl} - \delta_{ij}\delta_{lm})$$
$$-\frac{3}{r^5}(\delta_{lm}\hat{x}_i\hat{x}_j + \delta_{jm}\hat{x}_i\hat{x}_l + \delta_{jl}\hat{x}_i\hat{x}_m + \delta_{im}\hat{x}_j\hat{x}_l + \delta_{il}\hat{x}_j\hat{x}_m - \delta_{ij}\hat{x}_l\hat{x}_m) + 15\frac{\hat{x}_i\hat{x}_j\hat{x}_l\hat{x}_m}{r^7} \tag{7.2.27}$$

whereas the associated pressure field is given by $P = \mu p^{GQ}_{jlm} q_{jlm}$, where

$$p^{GQ}_{jlm} = \frac{\partial p^{GD}_{jl}}{\partial x_{0,m}} = -\frac{6}{r^5}(\delta_{jl}\hat{x}_m + \delta_{jm}\hat{x}_l + \delta_{lm}\hat{x}_j) + 30\frac{\hat{x}_j\hat{x}_l\hat{x}_m}{r^7} \tag{7.2.28}$$

It is important to note that

$$G^{Q}_{ijll} = -2\left(-\frac{\delta_{ij}}{r^3} + 3\frac{\hat{x}_i\hat{x}_j}{r^5}\right) \tag{7.2.29}$$

and to recognize that the term in the parentheses on the right-hand side of (7.2.29) is the potential dipole $\mathbf{D}$. This allows us to write

$$D_{ij} = -\tfrac{1}{2}G^{Q}_{ijll} = -\tfrac{1}{2}\nabla_0^2 G_{ij} \tag{7.2.30}$$

Using (7.2.30) we can express all potential singularities, except for the point force, in terms of derivatives of the Laplacian of the Green's function.

### *Singularities of bounded flow*

The singularities of bounded flow are required to be regular throughout the domain of the flow except at the pole $\mathbf{x}_0$, to decay to zero at infinity, and to vanish over the boundaries of the flow.

To expedite the computation of the singularities of bounded flow, it is helpful to recall that the flow due to a point source is identical to the pressure field associated with the Green's function multiplied by $-\frac{1}{2}$ (see section 3.2) and also that high-order singularities may be constructed from low-order singularities by differentiating with respect to the pole $\mathbf{x}_0$. A word of caution is here in order: differentiation with respect to the pole $\mathbf{x}_0$ may not be equivalent to differentiation with respect to the field point $\mathbf{x}$ followed by sign inversion, for the flow due to the singularities may depend not only on the difference $\mathbf{x} - \mathbf{x}_0$ but also explicitly on $\mathbf{x}_0$.

To present an example of a set of singularities for bounded flow, we consider a semi-infinite domain of flow bounded by a plane wall. The velocity associated with the singularities is required to vanish over the plane wall at $x = w$ and to decay to zero at infinity. Using (3.3.10) we find that the point source is given by

$$\mathbf{\Sigma}^{\mathrm{W}}(\mathbf{x}, \mathbf{x}_0) = -\tfrac{1}{2}\mathbf{p}^{\mathrm{W}}(\mathbf{x}_0, \mathbf{x}) \tag{7.2.31}$$

where the pressure vector $\mathbf{p}^{\mathrm{W}}$ was given in (3.3.10) and the superscript W stands for wall. Evaluating the expression on the right-hand side of (7.2.31) we obtain

$$\Sigma_i^{\mathrm{W}} = \frac{\hat{x}_i}{r^3} + \frac{\hat{X}_i}{R^3} + 2\hat{X}_i\left(-\frac{1}{R^3} + 3\frac{\hat{X}_1^2}{R^5}\right) - 2h_0\left(-\frac{\delta_{i1}}{R^3} + 3\frac{\hat{X}_1\hat{X}_i}{R^5}\right) \tag{7.2.32}$$

where $\hat{\mathbf{X}} = \mathbf{x} - \mathbf{x}_0^{\mathrm{IM}}$, $R = |\hat{\mathbf{X}}|$, $\mathbf{x}_0^{\mathrm{IM}} = (2w - x_0, y_0, z_0)$ is the image of the pole with respect to the wall, $h_0 = x_0 - w$ is the distance of the pole from the wall, and the 1-dimension corresponds to the $x$-axis. Blake & Chwang (1974) derived (7.2.32) in an alternative fashion, using the method of Fourier transforms. Inspecting the right-hand side of (7.2.32) we identify the individual terms as a primary point source, an image point source, an image Stokeslet doublet, and an image potential dipole. The last two singularities are oriented perpendicular to the wall. Thus, we write

$$\Sigma_i^{\mathrm{W}}(\mathbf{x}, \mathbf{x}_0) = \Sigma_i(\hat{\mathbf{x}}) + \Sigma_i(\hat{\mathbf{X}}) + 2G_{i11}^{\mathrm{D}}(\hat{\mathbf{X}}) - 2h_0 D_{i1}(\hat{\mathbf{X}}) \tag{7.2.33}$$

Because of the presence of the point force dipole, the flow due to the point source has non-vanishing vorticity and non-constant pressure. Using (7.2.14), we find that the pressure associated with (7.2.23) is given by $P = \mu P^{\mathrm{SW}}$ where

$$P^{\mathrm{SW}} = -\frac{4}{R^3} + 12\frac{\hat{X}_1^2}{R^5} \tag{7.2.34}$$

To compute the point source dipole we differentiate the point source with respect to the pole $\mathbf{x}_0$, obtaining

$$D_{ij}^{\mathrm{W}}(\mathbf{x}, \mathbf{x}_0) = D_{ij}(\hat{\mathbf{x}}) \pm [D_{ij}(\hat{\mathbf{X}}) + 2G_{i1j}^{\mathrm{Q}}(\hat{\mathbf{X}}) - 2h_0 Q_{i1j}(\hat{\mathbf{X}})] - 2\delta_{j1} D_{i1}(\hat{\mathbf{X}}) \tag{7.2.35}$$

where the plus sign is for $j = 2, 3$, and the minus sign is for $j = 1$. The corresponding pressure is due exclusively to the image Stokeslet quadruple. Thus,

$$p_j^{\mathrm{DW}} = \frac{\partial P^{\mathrm{SW}}}{\partial x_{0,j}} = \pm\left(-24\delta_{j1}\frac{\hat{X}_1}{R^5} - 12\frac{\hat{x}_j}{R^5} + 60\frac{\hat{X}_j\hat{X}_1^2}{R^7}\right) \tag{7.2.36}$$

in agreement with (7.2.28).

The point force above a plane wall was discussed in detail in section 3.3. Singularities expressing multipoles of the point force may be computed by straightforward differentiation with respect to the pole. The couplet

$$G_{im}^{\mathrm{CW}} \equiv \tfrac{1}{2}\varepsilon_{mlj}\frac{\partial G_{ij}^{\mathrm{W}}}{\partial x_{0,l}} \tag{7.2.37}$$

where $\mathbf{G}^{\mathrm{W}}$ is the Green's function given in (3.3.7), is discussed in detail by Blake & Chwang (1974).

Nigam & Srinivasan (1975) discuss the point force, the rotlet, the point source, and the source doublet in an infinite flow that is bounded internally by a solid sphere. The orientation of these singularities is chosen so as to produce axisymmetric flow. In all cases, the image system consists of isolated free-space singularities at the inverse point of the pole, or distributions of free-space singularities extending from the centre of the sphere up to the inverse point of the pole. Hackborn *et al.* (1986) and Hackborn (1990) discuss the flow due to a couplet located inside a spherical container and between two parallel plane walls.

### *Singularities of internal flow*

According to our previous classification, the singularities of internal flow are required to be regular throughout the domain of flow but are allowed to become singular at infinity.

To derive the singularities of internal flow we follow the procedure outlined in the beginning of this section for the free-space singularities. Thus, recalling that the pressure is a harmonic function, we set $P = 10\mu\hat{\mathbf{x}}\cdot\mathbf{e}$, where $\mathbf{e}$ is a constant vector. Substituting this expression into the Stokes equation we obtain $\nabla^2\mathbf{u} = 10\,\mathbf{e}$ and find $\mathbf{u} = \mathbf{E}\cdot\mathbf{e}$ where

$$E_{ij} = 2r^2\delta_{ij} - \hat{x}_i\hat{x}_j \tag{7.2.38}$$

is a new singularity called the Stokeson (Chwang & Wu 1975). The stress field associated with the Stokeson is given by $\sigma_{ik} = \mu T_{ijk}^{\mathrm{E}} e_j$, where

$$T_{ijk}^{\mathrm{E}} = 3(-4\delta_{ik}\hat{x}_j + \delta_{ij}\hat{x}_k + \delta_{kj}\hat{x}_i) \tag{7.2.39}$$

All derivatives of the Stokeson with respect to the pole constitute legitimate singularities of internal flow. Differentiating the Stokeson once we obtain the Stokeson dipole with associated pressure $P = -10\mu f_{ii}$, velocity $u_j = E_{ijl}^{\mathrm{D}} f_{jl}$, and vorticity $\omega_i = 5\varepsilon_{ilk} f_{lk}$, where

$$E_{ijl}^{\mathrm{D}} \equiv \frac{\partial E_{ij}}{\partial x_{0,l}} = -4\delta_{ij}\hat{x}_l + \delta_{il}\hat{x}_j + \delta_{jl}\hat{x}_i \tag{7.2.40}$$

and $\mathbf{f}$ is a constant matrix. The symmetric part of the Stokeson dipole is the stresson, whereas the antisymmetric part is the roton. The stresson represents a linear purely straining flow with vanishing vorticity and constant pressure, whereas the roton represents rigid body rotation.

### *The contribution of the singularities to the global properties of a flow*

Having introduced the explicit forms of the singularities of Stokes flow, we proceed now to identify their contribution to certain global properties of the flow, i.e. the flow rate through, the force and torque exerted on, and the coefficient of the stresslet associated with a closed surface.

Inspecting the functional form of the singularities we find that the flow rate $Q$ through any closed surface that encloses the pole of a singularity and no boundaries is equal to zero, except for the point source for which

$$Q = 4\pi \mathscr{s}, \tag{7.2.41}$$

and the stresslet for which

$$Q = 4\pi s_{ii} \tag{7.2.42}$$

The force exerted on any surface that encloses the pole of a singularity is equal to zero, except for the Stokeslet for which

$$\mathbf{F} = -8\pi\mu\mathbf{g} \tag{7.2.43}$$

The torque with respect to an arbitrary point $\mathbf{x}_1$ exerted on any surface that encloses the pole of a singularity is equal to zero, except for the Stokeslet doublet for which

$$L_i = 8\pi\mu\varepsilon_{ijl}d_{jl} \tag{7.2.44}$$

the couplet for which

$$\mathbf{L} = -8\pi\mu\mathscr{L} \tag{7.2.45}$$

and the Stokeslet for which

$$\mathbf{L} = 8\pi\mu(\mathbf{x}_1 - \mathbf{x}_0) \times \mathbf{g} \tag{7.2.46}$$

(see problem 2.2.3.)

The coefficient of the stresslet $\mathscr{S}$ is relevant only for the point force dipole and the stresslet for which

$$\mathscr{S} = -8\pi\mu\mathbf{s} \tag{7.2.47}$$

It should be noted that $\mathscr{S}$ has the same value on any two reducible surfaces that enclose the pole of a stresslet (problem 7.2.3).

### *Two-dimensional flow*

The whole set of singularities for three-dimensional flow may be translated verbatim to two-dimensional flow. The point source may be identified with the pressure field of a Green's function of infinite flow, whereas derivative singularities may be produced by straightforward differentiation with respect to the pole.

Several singularities representing the flow due to a point force have already been discussed in section 3.5. Ranger (1977) discusses the flow in the exterior of an elliptical cylinder due to a point force located and directed along the minor axis of the cylinder. Dorrepaal *et al.* (1984) and Avudainayagam & Jothiram (1987) discuss singularities for a flow bounded internally by a circular cylinder. Hackborn (1990) discusses the flow due to a couplet located between two parallel plane walls.

The two-dimensional couplet merits special consideration due to its ability for producing flow with involved streamline patterns. The stream function due to the free-space couplet is given by

$$\Psi = -\ln|\mathbf{x} - \mathbf{x}_0| \tag{7.2.48}$$

The stream function associated with a couplet located at a distance $h_0$ above a plane wall located at $y = w$ is given by

$$\Psi = -\ln|\hat{\mathbf{x}}| + \ln|\hat{\mathbf{X}}| - 2\frac{y(y + h_0)}{|\hat{\mathbf{X}}|^2} \tag{7.2.49}$$

where $\hat{\mathbf{x}} = \mathbf{x} - \mathbf{x}_0, \hat{\mathbf{X}} = \mathbf{x} - \mathbf{x}_0^{\text{IM}}$, and $\mathbf{x}_0^{\text{IM}}$ is the image of the pole of the couplet with respect to the wall (Ranger 1980). The stream function associated with a couplet located at a radial position $r_0$ and polar angle $\theta_0$ inside a circular cylinder is given by

$$\Psi = \ln|\hat{\mathbf{x}}| + \ln|\hat{\mathbf{X}}| - \frac{1}{2}\left(\frac{r^2 - a^2}{|\hat{\mathbf{X}}|^2}\right)\left(\frac{a^2}{r_0^2} - \frac{r^2}{a^2}\right) \tag{7.2.50}$$

where $a$ is the radius of the cylinder, $\hat{\theta} = \theta - \theta_0$, $\hat{\mathbf{x}} = \mathbf{x} - \mathbf{x}_0, \hat{\mathbf{X}} = \mathbf{x} - \mathbf{x}_0^{\text{IM}}$, and $\mathbf{x}_0^{\text{IM}}$ is the image point with respect to the cylinder ($\mathbf{x}_0^{\text{IM}}$ is located outside the cylinder at $r = a^2/r_0$ and $\theta_0$ (Ranger 1980)). The stream function associated with an array of $n$ couplets placed at a radial coordinate value $r_0$ at even angular separations inside a circular cylinder is given by

$$\Psi = -\frac{1}{2}\ln\left|\frac{r^{2n} - 2r^n r_0^n \cos(n\hat{\theta}) + r_0^{2n}}{c^{2n}r^{2n} - 2r^n r_0^n \cos(n\hat{\theta}) + a^{2n}}\right|$$
$$-\frac{n}{2}\left(\frac{r^2}{a^2} - 1\right)\left(\frac{a^{2n} - c^{2n}r^{2n}}{c^{2n}r^{2n} - 2r^n r_0^n \cos(n\hat{\theta}) + a^{2n}}\right) \tag{7.2.51}$$

where $c = r_0/a$, and one of the couplets is located at $\theta_0$ (Smith 1987a). The case of the single couplet is recovered by setting $n = 1$.

## Problems

7.2.1 Show that the first moment of the surface force exerted on a spherical surface centered at the stresslet is

$$\int_{\text{sphere}} x_j \sigma_{ik} n_k \, dS = -8\pi\mu(\tfrac{3}{5}s_{ij} + \tfrac{7}{5}\delta_{ij}s_{ll})$$

Using this equation show that the torque acting on a surface enclosing the pole of the stresslet is equal to zero.

7.2.2 Show that the first moment of the surface force acting on a spherical surface centered at the couplet is

$$\int_{\text{sphere}} x_j \sigma_{ik} n_k \, \mathrm{d}S = 4\pi\mu\varepsilon_{ijm}\mathscr{L}_m$$

Using this equation show that the torque acting on a surface enclosing the pole of the couplet is $\mathbf{L} = -8\pi\mu\mathscr{L}$.

7.2.3 Show that the coefficient of the stresslet

$$\mathscr{S}_{ik} = \frac{1}{2}\int_D [\hat{x}_k f_i + \hat{x}_i f_k - \delta_{ik}\tfrac{2}{3}\hat{x}_l f_l - 2\mu(u_k n_i + u_i n_k)] \, \mathrm{d}S$$

has the same values on two reducible surfaces that do not enclose any singularities.

7.2.4 Explain why the flow due to a singularity of bounded flow must cease when the pole of the singularity is placed on a solid surface bounding the flow.

7.2.5 Assess the asymptotic behaviour of the flow due to a point source that is located above a plane wall, far from the wall.

## 7.3 Singularity representations

Following the general ideas discussed in section 7.1, we implement the singularity method by expressing a flow in terms of a discrete or continuous distribution of singularities as

$$\begin{aligned} u_i(\mathbf{x}) = &\int_{\mathscr{D}_\Sigma} \Sigma_i(\mathbf{x}, \mathbf{x}_0)\, \mathscr{s} \, \mathrm{d}S(\mathbf{x}_0) + \int_{\mathscr{D}_D} D_{ij}(\mathbf{x}, \mathbf{x}_0)\, \mathscr{d}_j \, \mathrm{d}S(\mathbf{x}_0) \\ &+ \int_{\mathscr{D}_Q} Q_{ijl}(\mathbf{x}, \mathbf{x}_0)\, \mathscr{q}_{jl} \, \mathrm{d}S(\mathbf{x}_0) + \cdots + \int_{\mathscr{D}_G} G_{ij}(\mathbf{x}, \mathbf{x}_0)\, g_j \, \mathrm{d}S(\mathbf{x}_0) \\ &+ \int_{\mathscr{D}_{GD}} G^{\mathrm{D}}_{ijl}(\mathbf{x}, \mathbf{x}_0)\, d_{jl} \, \mathrm{d}S(\mathbf{x}_0) + \cdots \end{aligned} \qquad (7.3.1)$$

where the domains of the distributions $\mathscr{D}_\Sigma, \mathscr{D}_D, \ldots$ are located in the exterior of the flow. The densities of the distributions are either continuous functions or one-dimensional or two-dimensional delta functions. Clearly, the flow expressed by (7.3.1) satisfies the equations of Stokes flow throughout the domain as well as on the boundaries of the flow. The first series on the right-hand side of (7.3.1) contains the point source, the point source dipole, the point source quadrupole, etc., whereas the second series contains the point force (Green's function), the point force doublet, etc. Recalling that the point source dipole, as well as higher-order singularities that are derivatives of the point source, may be written in terms of the Laplacian or derivatives of the Laplacian of the Green's function, we deduce that apart from the point source, all singularities in the first series may be incorporated into the second series. For clarity and convenience, however, we prefer to maintain the explicit form (7.3.1). The problem is reduced to selecting the individual domains of distribution and to computing the densities of the distributions in order to satisfy the prescribed boundary conditions for the velocity or surface force over the boundary $D$.

In section 7.1 we argued that the singularity method allows an easy computation of several global properties of the flow. Indeed, using the general properties of the singularities discussed in the previous section, we find that the total flow rate $Q$ through a closed surface that encloses the domain of distribution of the point sources and stresslets is

$$Q = 4\pi \int_{\mathscr{D}} (\mathscr{s} + s_{ii})\, dS \tag{7.3.2}$$

The force $\mathbf{F}$ exerted on a closed surface that encloses the domain of distribution of the Green's functions is

$$\mathbf{F} = -8\pi\mu \int_{\mathscr{D}} \mathbf{g}\, dS \tag{7.3.3}$$

The torque and the coefficient of the stresslet are proportional to the total density of the couplets, stresslets, and point force dipoles, but in addition, they depend on the specific form of the distribution of Green's functions.

To illustrate the general principles and the computational methodology associated with the singularity method, in the rest of this section we shall discuss several examples of exact and approximate singularity representations.

### *A translating solid sphere*

First, we consider the flow produced by the translation of a solid sphere. Inspecting the functional form of the various singularities presented in section 7.2, we decide to represent the flow in terms of a Stokeslet and a potential dipole with poles at the center of the sphere. Thus, we set

$$u_i(\mathbf{x}) = G_{ij}(\mathbf{x}, \mathbf{x}_0) g_j + D_{ij}(\mathbf{x}, \mathbf{x}_0) \mathscr{d}_j \tag{7.3.4}$$

where $\mathbf{x}_0$ is the centre of the sphere. Introducing the explicit forms of the singularities we obtain

$$u_i(\mathbf{x}) = \left( \frac{\delta_{ij}}{r} + \frac{\hat{x}_i \hat{x}_j}{r^3} \right) g_j + \left( -\frac{\delta_{ij}}{r^3} + 3\frac{\hat{x}_i \hat{x}_j}{r^5} \right) \mathscr{d}_j \tag{7.3.5}$$

Applying the boundary condition $\mathbf{u} = \mathbf{U}$ at $r = a$, where $a$ is the radius of the sphere and $\mathbf{U}$ is the velocity of translation, we obtain two equations for the coefficients of the singularities, namely

$$\mathbf{g}a^2 - \boldsymbol{d} = \mathbf{U}a^3, \qquad \mathbf{g}a^2 + 3\boldsymbol{d} = 0 \tag{7.3.6}$$

Solving this system we obtain

$$\mathbf{g} = \tfrac{3}{4}a\mathbf{U} \qquad \boldsymbol{d} = -\tfrac{1}{4}a^3\mathbf{U} \tag{7.3.7}$$

The surface force exerted on the sphere is

$$f_i = \mu(T^{G}_{ijk}g_j + T^{D}_{ijk}d_j)n_k = -\frac{3}{2}\frac{\mu}{a}U_i \tag{7.3.8}$$

namely, a constant. To calculate the force exerted on the sphere we may either integrate the surface force over the surface of the sphere, or use (7.3.3). Either way we obtain

$$\mathbf{F} = \int_{\text{sphere}} \mathbf{f}\, dS = -8\pi\mu\mathbf{g} = -6\pi\mu a\mathbf{U} \tag{7.3.9}$$

in agreement with the Stokes law. The torque exerted on the sphere with respect to its center is equal to zero.

To derive the Faxen relation for the force on the sphere, it is convenient to express (7.3.4) exclusively in terms of the Green's function. Using (7.3.7) and (7.2.30) we obtain

$$u_i(\mathbf{x}) = \tfrac{1}{8}U_j a(6 + a^2\nabla_0^2)G_{ij}(\mathbf{x}, \mathbf{x}_0) \tag{7.3.10}$$

*A sphere in linear flow*

Next, we consider an infinite linear flow $\mathbf{u}^\infty = \mathbf{A}\cdot\mathbf{x}$ past a stationary solid sphere. After inspecting the functional form of the various singularities presented in section 7.2, we decide to represent the disturbance flow $\mathbf{u}'$ due to the sphere in terms of a Stokeslet doublet and a potential quadrupole each with a pole at the center of the sphere. Thus, we write

$$u_i'(\mathbf{x}) = G^{D}_{ijl}(\mathbf{x}, \mathbf{x}_0)d_{jl} + Q_{ijl}(\mathbf{x}, \mathbf{x}_0)q_{jl} \tag{7.3.11}$$

or explicitly,

$$u_i'(\mathbf{x}) = \left(\frac{\delta_{ij}\hat{x}_l - \delta_{il}\hat{x}_j - \delta_{jl}\hat{x}_i}{r^3} + 3\frac{\hat{x}_i\hat{x}_j\hat{x}_l}{r^5}\right)d_{jl} + \left(-3\frac{\delta_{ij}\hat{x}_l + \delta_{il}\hat{x}_j + \delta_{jl}\hat{x}_i}{r^5} + 15\frac{\hat{x}_i\hat{x}_j\hat{x}_l}{r^7}\right)q_{jl} \tag{7.3.12}$$

In order to satisfy the condition of zero velocity over the surface of the sphere we require $\mathbf{u}' = -\mathbf{A}\cdot\mathbf{x}$ at $r = a$ obtaining

$$\mathbf{d} = -\frac{5}{a^2}\mathbf{q} \tag{7.3.13}$$

and

$$\mathbf{A} = \frac{2}{a^5}[4\mathbf{\varphi} - \mathbf{\varphi}^{\mathrm{T}} - \mathbf{I}\,\mathrm{Trace}(\mathbf{\varphi})] \tag{7.3.14}$$

We note that Trace$(\mathbf{A}) = 0$ and this suggests that Trace$(\mathbf{\varphi}) = 0$. To solve for $\mathbf{\varphi}$, we split both $\mathbf{A}$ and $\mathbf{q}$ into their symmetric and antisymmetric components, setting $\mathbf{\varphi} = \mathbf{\varphi}^{\mathrm{S}} + \mathbf{\varphi}^{\mathrm{A}}$, and $\mathbf{A} = \mathbf{A}^{\mathrm{S}} + \mathbf{A}^{\mathrm{A}}$. Substituting into (7.3.14) we obtain

$$\mathbf{A}^{\mathrm{S}} + \mathbf{A}^{\mathrm{A}} = \frac{2}{a^5}(3\mathbf{\varphi}^{\mathrm{S}} + 5\mathbf{\varphi}^{\mathrm{A}}) \tag{7.3.15}$$

and conclude that

$$\mathbf{\varphi}^{\mathrm{S}} = \frac{a^5}{6}\mathbf{A}^{\mathrm{S}}, \qquad \mathbf{\varphi}^{\mathrm{A}} = \frac{a^5}{10}\mathbf{A}^{\mathrm{A}} \tag{7.3.16}$$

and hence,

$$\mathbf{\varphi} = \frac{a^5}{30}(4\mathbf{A} + \mathbf{A}^{\mathrm{T}}) \tag{7.3.17}$$

Using (7.3.13) we calculate the coefficient of the Stokeslet dipole as

$$\mathbf{d} = -\frac{a^3}{6}(4\mathbf{A} + \mathbf{A}^{\mathrm{T}}) \tag{7.3.18}$$

The symmetric and antisymmetric components of $\mathbf{d}$ are

$$\mathbf{s} = -\tfrac{5}{12}a^3(\mathbf{A} + \mathbf{A}^{\mathrm{T}}) \qquad \mathbf{r} = -\tfrac{1}{4}a^3(\mathbf{A} - \mathbf{A}^{\mathrm{T}}) \tag{7.3.19}$$

To compute the coefficient of the couplet inherent in the Stokeslet dipole, we use (7.2.20), obtaining

$$\mathscr{L}_m = \tfrac{1}{2}a^3\varepsilon_{mjl}A_{jl} \tag{7.3.20}$$

The disturbance surface force exerted on the sphere is given by

$$f'_i = \mu(T^{\mathrm{GD}}_{ijlk}d_{jl} + T^{\mathrm{Q}}_{ijlk}\varphi_{jl})n_k = 3\frac{\mu}{a}A_{ij}\hat{x}_j \tag{7.3.21}$$

Due to the absence of a point force, the total force exerted on the sphere is equal to zero. Using (7.2.45) and the coefficient of the couplet given in (7.3.20), we find that the torque exerted on the sphere is equal to

$$L_m = -8\pi\mu\mathscr{L}_m = -4\pi\mu a^3\varepsilon_{mjl}A_{jl} \tag{7.3.22}$$

Furthermore, using (7.2.47) we find that the coefficient of the stresslet is

$$\mathscr{S} = -8\pi\mu\mathbf{s} = \tfrac{10}{3}\pi\mu a^3(\mathbf{A} + \mathbf{A}^{\mathrm{T}}) \tag{7.3.23}$$

Restricting our attention to a purely straining incident flow for which $\mathbf{A}$

is a symmetric matrix, we obtain

$$\mathscr{q} = \frac{a^5}{6}\mathbf{E} \qquad \mathbf{d} = \mathbf{s} = -\tfrac{5}{6}a^3\mathbf{E} \qquad \mathbf{r} = \mathscr{L} = 0 \qquad \mathscr{S} = \tfrac{20}{3}\pi\mu a^3\mathbf{E} \tag{7.3.24}$$

where in order to emphasize that $\mathbf{A}$ is symmetric, we have set $\mathbf{A} = \mathbf{E}$.

To derive the Faxen relation for the stresslet, it will be helpful to cast (7.3.11) in an alternative form involving only the Green's function. Using the definitions of the Stokeslet doublet and potential quadrupole, we obtain

$$u_i'(\mathbf{x}) = (d_{jl} - \tfrac{1}{2}\mathscr{q}_{jl}\nabla_0^2)\frac{\partial}{\partial x_{0,l}}G_{ij}(\mathbf{x}, \mathbf{x}_0) \tag{7.3.25}$$

For a purely straining incident flow, $\mathbf{A} = \mathbf{E}$, we obtain

$$u_i'(\mathbf{x}) = -E_{jl}\frac{1}{6}(5a^3 + a^5\nabla_0^2)\frac{\partial}{\partial x_{0,l}}G_{ij}(\mathbf{x}, \mathbf{x}_0) \tag{7.3.26}$$

***Flow due to a rotating sphere***

The velocity on the surface of a solid sphere that rotates with angular velocity $\mathbf{\Omega}$ is identical to the disturbance velocity due to a sphere immersed in a linear flow $\mathbf{A}\cdot\mathbf{x}$, where

$$A_{ik} = -\varepsilon_{ijk}\Omega_j \tag{7.3.27}$$

Using (7.3.17) and (7.3.18) and noting that $\mathbf{A}$ is antisymmetric, i.e. $\mathbf{A} = -\mathbf{A}^T$, we obtain $\mathscr{q} = (1/10)a^5\mathbf{A}$ and $\mathbf{d} = -(1/2)a^3\mathbf{A}$. Furthermore, noting that $\mathscr{q}$ is antisymmetric but $Q_{ijl}$ is symmetric with respect to the last two indices, we deduce that the quadrupole makes a vanishing contribution to the flow. Reverting to (7.3.11) and using (7.2.22) we obtain

$$u_i = -a^3\tfrac{1}{2}G_{ijl}^{D}A_{jl} = a^3\tfrac{1}{2}G_{ijl}^{D}\varepsilon_{jml}\Omega_m = a^3 G_{im}^{C}\Omega_m = a^3\varepsilon_{iml}\Omega_m\frac{\hat{x}_l}{r^3} \tag{7.3.28}$$

Equation (7.3.28) suggests that the flow may be represented merely in terms of a couplet with strength $a^3\mathbf{\Omega}$ located at the center of the sphere. Using either (7.2.45), or (7.2.22) and (7.2.27), we find that the torque acting on the sphere is

$$\mathbf{L} = -8\pi\mu a^3\mathbf{\Omega} \tag{7.3.29}$$

Now, in order to derive the Faxen relation for the torque, it will be convenient to cast (7.3.28) in terms of the Green's function. Using (7.2.22) we obtain

$$u_i(\mathbf{x}) = \tfrac{1}{2}a^3\Omega_k\varepsilon_{klj}\frac{\partial G_{ij}}{\partial x_{0,l}}(\mathbf{x}, \mathbf{x}_0) \tag{7.3.30}$$

### *Flow due to the translation or rotation of a prolate spheroid*

Chwang & Wu (1974) showed that the flow due to a prolate spheroid rotating about its major axis may be represented in terms of a distribution of couplets over the focal length of the spheroid as

$$u_i(\mathbf{x}) = \omega_1 \frac{1}{\left(\dfrac{2e}{1-e^2}\right) - \ln\left(\dfrac{1+e}{1-e}\right)} \int_{-c}^{c} (c^2 - x_0^2) G_{i1}^{C}(\mathbf{x}, \mathbf{x}_0)\, dx_0 \qquad (7.3.31)$$

where $e$ is the eccentricity of the spheroid, defined as $e = c/a$, $0 < e < 1$, where $c$ is the focal length of the spheroid, defined by $c^2 = a^2 - b^2$, and $a$ and $b$ are the major and minor axes of the spheroid (see problem 2.5.4). Furthermore, Chwang and Wu (1975) showed that the flow produced by the translation of a prolate spheroid may be represented in terms of a distribution of Stokeslets and potential dipoles over the focal length of the spheroid with constant and parabolic densities respectively:

$$u_i(\mathbf{x}) = V_k a_{kj} \int_{-c}^{c} \left[ G_{ij}(\mathbf{x}, \mathbf{x}_0) - \left(\frac{1-e^2}{2e^2}\right)(c^2 - x_0^2) D_{ij}(\mathbf{x}, \mathbf{x}_0) \right] dx_0 \qquad (7.3.32)$$

where $\mathbf{a}$ is a diagonal matrix with

$$a_{11} = \frac{e^2}{-2e + (1+e^2)\ln\left(\dfrac{1+e}{1-e}\right)} \qquad a_{22} = a_{33} = \frac{-2e^2}{-2e + (1-3e^2)\ln\left(\dfrac{1+e}{1-e}\right)} \qquad (7.3.33)$$

and the 1-axis is aligned with the major axis of the spheroid. It will be noted that both the Stokeslets and the potential dipoles in (7.3.32) are oriented in the direction of translation. As $e \to 0, a_{11}, a_{22}, a_{33} \to (3/8)e$, yielding the results for the sphere (see problem 2.5.3). Expressing the couplet and the potential dipole in (7.3.31) and (7.3.32) in terms of the Green's function, we obtain singularity representations in forms that are suitable for producing the Faxen relations for the force and torque (see problems 2.5.3 and 2.5.4).

### *A translating liquid drop*

Next, we consider the flow due to a translating spherical liquid drop. One necessary assumption of our analysis will be that the surface tension is high enough for the drop to maintain its spherical shape during the motion. Cursory inspection of the menu of the available singularities results in the following selections:

$$u_i^{\mathrm{Ext}}(\mathbf{x}) = G_{ij}(\mathbf{x}, \mathbf{x}_0) g_j + D_{ij}(\mathbf{x}, \mathbf{x}_0) d_j \qquad u_i^{\mathrm{Int}}(\mathbf{x}) = c_i + E_{ij}(\mathbf{x}, \mathbf{x}_0) e_j \qquad (7.3.34)$$

where $\mathbf{G}, \mathbf{D}$, and $\mathbf{E}$ are the Stokeslet, potential dipole, and Stokeson respectively, and $\mathbf{c}$ is a constant.

Continuity of the velocity at the surface of the drop requires

$$\mathbf{g}a^2 - \mathscr{d} - \mathbf{c}a^3 - \mathbf{e}2a^5 = 0 \qquad \mathbf{g}a^2 + 3\mathscr{d} + \mathbf{e}a^5 = 0 \tag{7.3.35}$$

We note that in a frame of reference moving with the drop, the component of the velocity normal to the surface of the drop must be equal to zero, i.e. $(\mathbf{u}^{\text{Int}} - \mathbf{U})\cdot\mathbf{n} = 0$ where $\mathbf{U}$ is the velocity of translation of the drop, and using (7.3.34), we obtain

$$\mathbf{c} + a^2\mathbf{e} = \mathbf{U} \tag{7.3.36}$$

Furthermore, using the formulae of section 7.2, we obtain

$$f_k^{\text{Ext,Sh}} = \mu_1\left[-6\frac{\hat{x}_i\hat{x}_j}{a^4}g_j + \left(6\frac{\delta_{ij}}{a^4} - 18\frac{\hat{x}_i\hat{x}_j}{a^6}\right)\mathscr{d}_j\right](\delta_{ik} - n_in_k) \tag{7.3.37}$$

$$f_k^{\text{Int,Sh}} = \mu_2\left(3a\delta_{ij} - 9\frac{\hat{x}_i\hat{x}_j}{a}\right)e_j(\delta_{ik} - n_in_k) \tag{7.3.38}$$

where $\mu_1$ and $\mu_2$ are the viscosities of the external and internal fluid respectively and the superscript Sh denotes the shearing component of the surface force. It will be noted that the projection operator $\mathbf{I} - \mathbf{nn}$ extracts the tangential component of the surface force from the total surface force. Requiring that the shear stress is continuous across the surface of the drop, we set the right-hand sides of (7.2.37) and (7.2.38) equal and obtain

$$\mathscr{d} = \tfrac{1}{2}\lambda a^5\mathbf{e} \tag{7.3.39}$$

where $\lambda = \mu_2/\mu_1$. Solving the system (7.2.35), (7.2.36), and (7.2.39) for the coefficients of the singularities we find

$$\begin{aligned} \mathbf{g} &= \tfrac{1}{4}\left(\frac{3\lambda + 2}{\lambda + 1}\right)a\mathbf{U} \qquad & \mathscr{d} &= -\tfrac{1}{4}\left(\frac{\lambda}{\lambda + 1}\right)a^3\mathbf{U} \\ \mathbf{c} &= \tfrac{1}{2}\left(\frac{2\lambda + 3}{\lambda + 1}\right)\mathbf{U} \qquad & \mathbf{e} &= -\tfrac{1}{2}\left(\frac{\lambda}{\lambda + 1}\right)\frac{1}{a^2}\mathbf{U} \end{aligned} \tag{7.3.40}$$

It will be noted that in the limit as $\lambda \to \infty$, $\mathbf{g}$ and $\mathscr{d}$ assume the values for the solid sphere given previously in (7.3.7). The drag force exerted on the drop is

$$\mathbf{F} = -8\pi\mu\mathbf{g} = -2\pi\mu a\left(\frac{3\lambda + 2}{\lambda + 1}\right)\mathbf{U} \tag{7.3.41}$$

To derive the Faxen relation for the force, it is necessary to have an expression for the external flow in terms of the Green's function. Substituting (7.3.40) into (7.3.34) and expressing the potential dipole in terms of the Green's function we obtain

$$u_i^{\mathrm{Ext}}(\mathbf{x}) = U_j a \tfrac{1}{8}\left[2\left(\frac{3\lambda+2}{\lambda+1}\right) + a^2\left(\frac{\lambda}{\lambda+1}\right)\nabla_0^2\right] G_{ij}(\mathbf{x}, \mathbf{x}_0) \qquad (7.3.42)$$

In the limit as $\lambda \to \infty$, (7.3.42) reduces to (7.3.10) and we recover the results for the solid sphere.

### *Further singularity representations*

Chwang & Wu (1974, 1975), Chwang (1975), and Huang & Chwang (1986) have developed singularity representations for several families of flows involving axisymmetric bodies including the following: flow due to the axial rotation of dumbbell-shaped and simply-connected or doubly-connected oblate bodies; streaming, linear, and quadratic flow past prolate spheroids; flow between two confocal prolate spheroids in axial rotation; flow between two concentric spheres in relative translation. Kim (1986) and Kim & Arunachalam (1987) developed singularity representations for an arbitrary flow past a prolate or oblate ellipsoid, and demonstrated that the natural domain of the singularity distributions is the focal ellipse of the ellipsoid. Furthermore, they showed that the order of singularities necessary to represent the flow increases with the order of the incident flow.

### *Thin and slender bodies*

The simplified boundary integral equations (2.3.37) and (2.3.30) guarantee that a flow due to a moving rigid body or a flow past a stationary or moving rigid body may be represented exactly in terms of a distribution of Green's functions over the surface of the body. If the body has infinitesimal thickness, two facing sides of the surface of the body will collapse into a single surface, and the corresponding distributions will join to form a unified distribution. As an example, we note that the flow produced by the translation of a circular disk of infinitesimal thickness may be represented exactly in terms of a distribution of Stokeslets over the surface of the disk as

$$u_i(\mathbf{x}) = \frac{1}{\pi^2}\int_D \mathcal{S}_{ij}(\mathbf{x}_0, \mathbf{x}) \frac{c_j}{(a^2 - \sigma_0^2)^{1/2}} \, \mathrm{d}A(\mathbf{x}_0) \qquad (7.3.43)$$

where $a$ is the radius and $D$ is the area of the disk, $c_1 = U_1, c_2 = (2/3)U_2$, $c_3 = (2/3)U_3$, $\mathbf{U}$ is the velocity of translation, and the 1-axis is directed along the axis of the disk. It will be noted that the Stokeslets are oriented in the direction of motion, and furthermore, that the density of the distribution becomes singular at $\sigma = a$ reflecting the singular behaviour of the surface force at the rim.

Now, it appears reasonable to surmise that a flow past a smooth body of small but finite thickness might be represented in terms of distributions of singularities over the asymptotic surface that would arise if the body had infinitesimal thickness. In this perspective, the singularity distribution for a body of infinitesimal thickness is seen to provide the asymptotic behaviour of the singularity distribution, in the limit as a small parameter that characterizes the thickness of the body tends to zero. To support this argument, we note that the flow due to the axial translation of a thin disk may be represented in terms of a distribution of Stokeslets and potential dipoles both oriented in the direction of motion over the mid-plane of the disk (Barshinger & Geer 1984). The domain of the distribution $\mathscr{D}$ is a circular disk whose radius is smaller than that of the actual disk. It is interesting to note that the reduction of the domain of distribution (with respect to that of a disk of infinitesimal thickness) is necessary in order to eliminate the singular behaviour of the surface force around the rim.

Considering next slender bodies, and following the above arguments, we find it reasonable to represent the flow in terms of line distributions of singularities over the central axes of the bodies. Enforcing the required boundary conditions we obtain a system of Fredholm integral equation of the first kind for the strength of the singularities. In certain simple cases, it will be possible to find the solution by simple inspection. Burgers (1938, p. 122), for instance, showed that the flow produced by the translation of a straight rod may be described in terms of a uniform distribution of Stokeslets whose density is proportional to the slenderness parameter $1/\ln(L/a)$; $L$ is the length of the rod, and $a$ is the effective radius of the cross-section of the rod. Formal asymptotic analyses for bodies with non-axisymmetric cross-sectional shapes, flexible bodies, bodies whose volume changes in time, and bodies executing general types of motion have shown that, in general, distributions of point forces, potential dipoles, couplets, and point sources will be necessary to represent an arbitrary flow (Cox 1970, Tillett 1970, Batchelor 1970b, Geer 1976, Keller & Rubinow 1976).

Slender body theory has found important applications in the field of biofluiddynamics, and in particular, in studies of the flow due to flagella motions. Hancock (1953) and Gray & Hancock (1955) represented the flow due to a moving flagella in terms of distributions of point forces and potential dipoles. In the simplest version of their theory, the resistance coefficient theory, they assumed that the strength of the singularities is proportional to the local velocity of translation. A comprehensive account and extensions of their theory are presented by Lighthill (1975, Chapter 3; see also Higdon 1979).

**Problems**

7.3.1 Derive the singularity representation of the flow due to the radial expansion of a spherical air bubble. Your results should be given in terms of the pressure inside the bubble.

7.3.2 Derive a singularity representation of the disturbance flow produced by a spherical air bubble placed in a purely straining linear ambient flow. Based on this representation derive the Faxen relation for the stresslet.

7.3.3 Verify the validity of (7.3.22) by computing the torque as an integral of the surface force over the sphere.

7.3.4 Using (7.3.43) show that the force exerted on a translating circular disk of infinitesimal thickness is given by $\mathbf{F} = -16\mu a(U_1, \frac{2}{3}U_2, \frac{2}{3}U_3)$.

7.3.5 Develop an approximate singularity representation for the flow due to the rolling of a slender torus around its central axis (Chwang & Hwang 1990).

## 7.4 Numerical methods

Exact singularity representations are feasible only for a limited number of flows: those that are bounded by spherical and spheroidal surfaces. To obtain representations for arbitrary flows we must resort to numerical methods. The general strategy is to express the flow in terms of discrete or continuous singularity distributions and then to compute the densities of the distributions, and possibly the location of the poles, so as to satisfy the required boundary conditions in some approximate sense.

In a pioneering work, Burgers (1938, p. 120) represented the disturbance flow due to a sphere in an incident streaming flow simply by a point force. He computed the strength of the point force by requiring that the mean velocity over the surface of the sphere is equal to zero, and found that his approximate solution produces a drag force on the sphere identical to that given by Stokes' law. In retrospect, this perfect agreement can be understood by recalling that the disturbance flow may be expressed exactly with a Stokeslet and a potential dipole, and noting that the average velocity of the dipole over the surface of the sphere is equal to zero. Burgers performed a similar computation for a sphere that is held still in a paraboloidal flow, and found that the drag force is in perfect agreement with the exact solution given by the Faxen relation. Dąbroś (1985) extended the computations of Burgers to include coaxial arrays of spheres. In his least accurate calculations, for touching spheres, his numerical results differed from the exact solution by less than 9%. These results certainly advocate the use of approximate singularity representations for computing global properties of a flow.

Now, to develop a formal procedure for computing approximate singularity representations, we introduce a positive functional expressing the difference between the required boundary conditions and the corresponding boundary values due to the singularity representation. For instance, when the problem requires $\mathbf{u} = \mathscr{u}$ over the boundary $D$, we introduce the positive functional $d(\mathbf{u} - \mathscr{u})$, whereas when the problem requires that the surface force $\mathbf{f}$ over the boundary $D$ is equal to $\mathscr{f}$, we introduce the alternative functional $\delta(\mathbf{f} - \mathscr{f})$. For simplicity, we consider problems involving boundary conditions for the velocity. The general strategy is to express $\mathbf{u}$ in terms of a singularity expansion and then to minimize $d(\mathbf{u} - \mathscr{u})$ with respect to the strength of the singularities and, possibly, the location of the poles. A popular choice for $d$ is

$$d(\mathbf{u} - \mathscr{u}) = \sum_{m=1}^{M} [\mathbf{u}(\mathbf{x}_m) - \mathscr{u}(\mathbf{x}_m)] \cdot [\mathbf{u}(\mathbf{x}_m) - \mathscr{u}(\mathbf{x}_m)] \tag{7.4.1}$$

where $\{\mathbf{x}_m, m = 1, \ldots, M\}$, is a collection of collocation points over the boundary $D$, yielding the least-squares boundary collocation method. It will be noted that when the poles of the singularities are fixed, minimizing (7.4.1) produces a quadratically non-linear problem with respect to the strengths of the singularities. When the force or torque acting on the boundary $D$ is required to have prescribed values, the strengths of the singularities must satisfy an additional linear constraint such as (7.3.3).

The least-squares boundary collocation method has found important applications in a variety of numerical studies of flows involving suspended particles. Dąbroś (1985) and Malysa *et al.* (1986), for instance, computed singularity representations for several types of flow past spherical and ellipsoidal particles, as well as flow due to the motion of a sphere in the proximity of another sphere attached to a plane wall. Their representations included point forces and point sources with fixed poles, and their results compared satisfactorily with known analytical solutions. In certain cases, instead of using singularity expansions, it is more convenient to use alternative expansions in terms of spherical harmonic functions (Lamb 1932, section 335). Brady & Bossis (1988) and Weinbaum *et al.* (1990) review various applications of boundary collocation methods, whereas Kim & Karrila 1991, Chapter 13) present comprehensive tests of accuracy. Burgess & Mahajerin (1984) present applications of the least-squares boundary collocation method for two-dimensional internal flow. Nitsche

& Brenner (1990) use the singularity method to compute the flow within axisymmetric channels.

All the above computations assume singularities with fixed poles. Allowing the poles to be mobile, i.e. minimizing the functional with respect to the coordinates of the poles, produces a highly non-linear problem but adds new degrees of freedom, which is likely to furnish more accurate results with fewer singularities (Mathon & Johnston 1977; Karageorghis & Fairweather 1987, 1988, 1989). The detailed numerical investigation of singularity representations with mobile poles emerges as an interesting topic for future research.

### Problems

7.4.1 Following Burgers (1938), represent the flow produced by the translation of a sphere in the form $\mathbf{u}(\mathbf{x}) = \mathscr{S}(\mathbf{x}, \mathbf{x}_0)\cdot\mathbf{g}$ where $\mathscr{S}$ is the Stokeslet and $\mathbf{x}_0$ is the center of the sphere. Then, by requiring that the average velocity on the surface of the sphere is equal to the velocity of translation, compute the coefficient $\mathbf{g}$. Based on the computed value of $\mathbf{g}$, calculate the drag force exerted on the sphere, and compare it with that given by Stokes' law. Perform a similar computation for paraboloidal flow past a sphere $\mathbf{u}^{\mathrm{P}} = u[(y^2 + z^2)/a^2, 0, 0]$, where $a$ is the radius of the sphere, and compare your result for the force on the sphere with that given by the Faxen relation.

7.4.2 Based on the singularity solution of Burgers (1938) discussed in problem 7.4.1, derive an approximate Faxen relation for the force on a sphere.

## 7.5 Unsteady flow

The whole apparatus of the singularity method for steady flow may be translated to unsteady flow. While the general ideas remain the same, the specific formulae for the singularities become considerably more involved and the computation of singularity representations becomes considerably more cumbersome.

To present some examples of singularities for infinite flow, we note that the point source is identical to (7.2.1). The corresponding pressure field, however, is not constant but equal to $\lambda^2/r$. The Stokeslet was discussed in section 2.7. The symmetric Stokeslet quadrupole (see (7.2.30)) and Stokeson are given by

$$-\tfrac{1}{2}\nabla_0^2\mathscr{S}_{ij} = -\frac{\delta_{ij}}{r^3}\mathrm{e}^{-R}(1 + R + R^2) + 3\frac{\hat{x}_i\hat{x}_j}{r^3}\mathrm{e}^{-R}\left(1 + R + \frac{R^2}{2}\right) \quad (7.5.1)$$

and

$$E_{ij} = 2\delta_{ij} r^2 Q(R) + \frac{\hat{x}_i \hat{x}_j}{r} J(R) \tag{7.5.2}$$

where $\mathscr{S}$ is the unsteady Stokeslet, $R = \lambda r$, and

$$Q(R) = \frac{15}{R^5}[R(R^2 + 6)\cosh R - 3(R^2 + 2)\sinh R] \tag{7.5.3}$$

$$J(R) = -10 - \frac{15}{R^5}[R(R^2 + 18)\cosh R - (7R^2 + 18)\sinh R] \tag{7.5.4}$$

(Pozrikidis 1989a). The associated pressure fields are identical to those of the corresponding singularities of steady flow. It will be noted that the symmetric Stokeslet quadrupole is not identical to the point source doublet, as it is in the case of steady flow.

One exceptional feature of the singularity method for unsteady flow is that it is capable of producing the force $\mathbf{F}$ exerted on a particle that executes translational oscillations solely in terms of the strength of the unsteady Stokeslets $\mathbf{g}$ necessary to represent the flow (Pozrikidis 1989a). Specifically, $\mathbf{F} = -8\pi\mathbf{g} + \lambda^2 V_p \mathbf{V}$, where $V_p$ is the volume of the particle, and $\mathbf{V}$ is the amplitude of the oscillations (problem 7.5.4).

Exact singularity representations of unsteady flow are limited to those for flow produced by the translational or rotational oscillations of a solid or liquid sphere, and for the disturbance flow due to a sphere embedded in a linear ambient flow (Pozrikidis 1989a; Kim & Karrila 1991, Chapter 6). It should be noted that since linear flow $\mathbf{u} = \mathbf{A} \cdot \mathbf{x}$ is not an exact solution to the equations of unsteady Stokes flow, the solution for a sphere in a linear ambient flow must be viewed as the response of the sphere to the linear term in the Taylor series about the center of the sphere of a general unsteady incident flow.

Pozrikidis (1989a) computed approximate singularity representations of the flow due to the translational oscillations of a prolate spheroid, in terms of distributions of Green's functions and symmetric Stokeslet quadruples over the focal length of the spheroid. The densities of the distributions range between those for steady flow at low frequencies, and those for inviscid potential flow at high frequencies.

Singularities for two-dimensional unsteady flow are discussed by Smith (1987b).

**Problems**

7.5.1 Verify that in the limit of low frequencies (7.5.1) and (7.5.2) reduce to the corresponding singularities for steady flow.

7.5.2 Compute the singularity representation of the flow produced by the translational oscillations of a solid sphere, and derive the corresponding Faxen relation for the force (Pozrikidis 1989a).

7.5.3 Show that the unsteady couplet is given by

$$\mathscr{S}^{C}_{ij} = \varepsilon_{ijk}\frac{\hat{x}_k}{r^3}e^{-R}(R+1)$$

Compute the singularity representations of the flow due to the rotary oscillations of a sphere and derive the corresponding Faxen relation for the torque (Pozrikidis 1989a).

7.5.4 Show that the force on a solid particle that oscillates with velocity $\mathbf{V}$ is equal to $\mathbf{F} = -8\pi\mathbf{g} + \lambda^2 V_p \mathbf{V}$, where $\mathbf{g}$ is the total strength of the Stokeslets necessary to represent the flow.

# *Answers and keys*

## *Chapter 1*

### Section 1.1

*1.1.1.* For simplicity, we shall assume that there is no body force acting on the fluid. In an accelerating frame of reference attached to the body, the Navier–Stokes equation takes the form of (1.1.2) with $\mathbf{b} = -\mathrm{d}\mathbf{V}/\mathrm{dt}$, where $\mathbf{V}$ is the velocity of the body. If $U$ is the characteristic velocity of the fluid in a stationary frame of reference, then $U - V$ is the characteristic velocity of the fluid in the moving frame of reference. There are two available length scales: $D$, associated with the large-scale motion of the fluid, and $D_B$, the characteristic size of the body. In the absence of temporal forcing of the ambient flow, the characteristic time scale of the flow in the vicinity of the body $T$ is found by noting that

$$\frac{\partial|\mathbf{u}|}{\partial t} \approx \frac{\partial(U-V)}{\partial t} = \frac{\partial U}{\partial x_i}\frac{\mathrm{d}x_i}{\mathrm{d}t} - \frac{\mathrm{d}V}{\mathrm{d}t} \approx \frac{U}{D}V - \dot{V}$$

where $\dot{V} = \mathrm{d}V/\mathrm{d}t$. Setting $\partial|\mathbf{u}|/\partial t = (U-V)/T$, and combining with the above equation we find

$$T = \frac{U-V}{VU/D - \dot{V}}$$

It will be noted that when the body is stationary $T$ becomes infinite, whereas when the body moves with the velocity of the fluid $T = D/U$.

In a frame of reference attached to the body, the frequency parameter, Reynolds number, and Froude number of the flow in the vicinity of the body are given by

$$\beta = \frac{D_B^2}{\nu T} = Re\frac{D_B}{D}\frac{(UV - \dot{V}D)}{(U-V)^2} \qquad Re = \frac{(U-V)D_B}{\nu}, \qquad Fr = \frac{(U-V)^2}{\dot{V}D_B}$$

Clearly, the Reynolds number is small when the body travels with a velocity that is nearly the local velocity of the fluid, i.e. $U \approx V$.

**Section 1.2**

*1.2.1.* Applying the divergence theorem we obtain

$$\int_D \sigma_{ik} n_k x_j \, dS = \int_V \frac{\partial}{\partial x_k}(\sigma_{ik} x_j) \, dS = \int_V \sigma_{ij} dV + \int_V \frac{\partial \sigma_{ik}}{\partial x_k} x_j dV$$

Note that for a regular Stokes flow the last integral is equal to zero.

*1.2.2.* Using the divergence theorem and the continuity equation we obtain

$$\int_D \left( u_i' \frac{\partial u_k}{\partial x_i} - u_i \frac{\partial u_k'}{\partial x_i} \right) n_k \, dS = \int_V \frac{\partial}{\partial x_k} \left( u_i' \frac{\partial u_k}{\partial x_i} - u_i \frac{\partial u_k'}{\partial x_i} \right) dV$$

$$= \int_V \left( \frac{\partial u_i'}{\partial x_k} \frac{\partial u_k}{\partial x_i} - \frac{\partial u_i}{\partial x_k} \frac{\partial u_k'}{\partial x_i} \right) dV = 0$$

*1.2.3.* Substituting $z = x + iy$ and $z^* = x - iy$ we find

$$\chi = \mu \tfrac{1}{3}[x(x^2 + 3y^2 - 6) + 2iy(y^2 - 3)]$$

Recalling that $\chi = \Phi - 2i\mu\Psi$ we conclude

$$\Phi = \mu \tfrac{1}{3} x(x^2 + 3y^2 - 6), \quad \Psi = -\tfrac{1}{3} y(y^2 - 3)$$

Differentiating $\Psi$ and $\Phi$ and using (1.2.11) and (1.2.17) we compute

$$u = 1 - y^2, \qquad v = 0, \qquad P = -2\mu x$$

*1.2.4.* The differential change of the ratio $u_\phi/\sigma$ along a line that lies in an azimuthal plane is

$$d\left(\frac{u_\phi}{\sigma}\right) = \frac{1}{\sigma}\left(\frac{\partial u_\phi}{\partial x}\right) dx + \frac{\partial}{\partial \sigma}\left(\frac{u_\phi}{\sigma}\right) d\sigma = \frac{1}{\mu\sigma}(\sigma_{x\phi} dx + \sigma_{\sigma\phi} d\sigma)$$

$$= \frac{1}{\sigma^3}\left(\frac{\partial \chi}{\partial \sigma} dx - \frac{\partial \chi}{\partial x} d\sigma\right) = -\frac{1}{\sigma^3} \nabla\chi \cdot \mathbf{n} \, dl = -\frac{1}{\sigma^3} \frac{\partial \chi}{\partial n} dl$$

If along this line $u_\phi = \Omega\sigma$, then $d(u_\phi/\sigma) = 0$ and $\partial\chi/\partial n = 0$.

*1.2.5.* Using the definition $\Phi = \chi \cos 2\phi/\sigma^2$, and (1.2.32) it is a straightforward task to show that $\Phi$ satisfies the three-dimensional Laplace equation

$$\frac{\partial^2 \Phi}{\partial \sigma^2} + \frac{1}{\sigma} \frac{\partial \Phi}{\partial \sigma} + \frac{1}{\sigma^2} \frac{\partial^2 \Phi}{\partial \phi^2} + \frac{\partial^2 \Phi}{\partial x^2} = 0$$

**Section 1.3**

*1.3.1.* We resolve the angular velocity of the sphere into a component perpendicular to the wall and a component parallel to the wall. Because of the linearity of Stokes flow, we may consider the motions associated with the two components independently. Considering first rotation around an axis perpendicular tò the wall, we invoke axisymmetry and reversibility to find that the components of the force parallel and normal to the wall must be equal to zero. Considering next rotation around an axis parallel to the wall we invoke reversibility to find that the component of the force normal to the wall must be equal to zero.

**Section 1.4**

*1.4.1.* To prove this relation we consider (1.4.1) and (1.4.2), where in (1.4.2) we use $\mu'$ in place of $\mu$. Then, we multiply (1.4.1) by $\mu'$ and (1.4.2) by $\mu$, and subtract the resulting expressions.

*1.4.2.* The simplest way to prove this relation is by subtracting the left-hand side of the identity discussed in problem 1.2.2 from the left-hand side of (1.4.4).

*1.4.3.* To prove (2.5.39), we expand $\mathbf{u}$ in a Taylor series about the centre of the sphere obtaining

$$\mathbf{u}(\mathbf{x})=\mathbf{u}(\mathbf{x}_0)+\frac{\partial \mathbf{u}}{\partial x_i}(\mathbf{x}_0)\hat{x}_i+\frac{1}{2!}\frac{\partial^2 \mathbf{u}}{\partial x_i \partial x_j}(\mathbf{x}_0)\hat{x}_i\hat{x}_j + \frac{1}{3!}\frac{\partial^3 \mathbf{u}}{\partial x_i \partial x_j \partial x_k}(\mathbf{x}_0)\hat{x}_i\hat{x}_j\hat{x}_k+\cdots \tag{1}$$

where $\hat{\mathbf{x}}=\mathbf{x}-\mathbf{x}_0$. Integrating this equation over the surface of the sphere we find

$$\int_{\text{sphere}} \mathbf{u}\,\mathrm{d}S = 4\pi a^2\mathbf{u}(\mathbf{x}_0)+\frac{\partial \mathbf{u}}{\partial x_i}(\mathbf{x}_0)A_i+\frac{1}{2!}\frac{\partial^2 \mathbf{u}}{\partial x_i \partial x_j}(\mathbf{x}_0)A_{ij} + \frac{1}{3!}\frac{\partial^3 \mathbf{u}}{\partial x_i \partial x_j \partial x_k}(\mathbf{x}_0)A_{ijk}+\cdots \tag{2}$$

where

$$A_{ijkl\ldots}=\int_{\text{sphere}} \hat{x}_i\hat{x}_j\hat{x}_k\hat{x}_l\cdots \mathrm{d}S \tag{3}$$

Writing $\hat{\mathbf{x}}=a\mathbf{n}$ and using the divergence theorem we compute

$$A_{ijkl\ldots}=a\int_{\text{sphere}} n_i\hat{x}_j\hat{x}_k\hat{x}_l\cdots \mathrm{d}S = a\int_{\text{sphere}}\frac{\partial}{\partial \hat{x}_i}(\hat{x}_j\hat{x}_k\hat{x}_l\cdots)\,\mathrm{d}V$$
$$=\delta_{ij}a\int_{\text{sphere}} \hat{x}_k\hat{x}_l\cdots \mathrm{d}V+\delta_{ik}a\int_{\text{sphere}}\hat{x}_j\hat{x}_l\cdots \mathrm{d}V+\cdots \tag{4}$$

from which we deduce that $A_i=0$, $A_{ij}=\delta_{ij}4\pi a^4/3$, $A_{ijk}=0$. Furthermore, by induction, we find that all $A_{ijk}$ with an odd number of indices are equal to zero. Next, we compute

$$\int_{\text{sphere}} \hat{x}_k\hat{x}_l\cdots \mathrm{d}V=\int_0^a\left(\frac{r}{a}\right)^2\frac{r}{a}\frac{r}{a}\cdots \mathrm{d}r\,A_{kl\ldots} \tag{5}$$

Combining (4) and (5) we find that $A_{ijkl...}$ (more than three indices) may be expressed as a sum of terms each of which contains the product of at least two Kronecker deltas with distinct indices, for instance $\delta_{ij}\delta_{kl}$. This suggests that the partial derivatives of $\mathbf{u}$ of degree four or higher in the expansion (2) may be expressed in terms of $\nabla^4\mathbf{u}$ and its derivatives, which are equal to zero (see (1.2.5)). Thus, only the first and third terms on the right-hand side of (2) survive. Dividing (2) by $4\pi a^2$ yields (2.5.39). To prove (2.5.40) we follow a similar procedure.

*1.4.4.* Direct substitution into (1.4.15) shows that the force is oriented in the streamwise direction, and furthermore,

$$F_x = \frac{3}{2}\frac{\mu}{a^3}U\int_{S_P}(x^2+y^2)\,dS = \tfrac{3}{2}\mu U a\int_0^\pi \sin^3\theta\,d\theta = 4\pi\mu U a$$

*1.4.5.* We apply the reciprocal identity (1.4.5) for the flows $\mathbf{u}$ and $\mathbf{u}'$ that are produced when the particle translates with velocities $\mathbf{V}$ and $\mathbf{V}'$, respectively. The domain $D$ consists of the surface of the particle and a surface of large radius enclosing the particle. Following the arguments of section 1.4 we neglect the integrals at infinity, and apply the boundary conditions $\mathbf{u}=\mathbf{V}$ and $\mathbf{u}'=\mathbf{V}'$ on the surface of the particle to find $\mathbf{V}\cdot\mathbf{F}'=\mathbf{V}'\cdot\mathbf{F}$. Using the definition (1.4.18) we obtain $\mathbf{V}\cdot(\mathbf{X}\cdot\mathbf{V}')=\mathbf{V}'\cdot(\mathbf{X}\cdot\mathbf{V})$ which is equivalent to $\mathbf{V}\cdot[(\mathbf{X}-\mathbf{X}^{TR})\cdot\mathbf{V}']=0$, where the superscript TR indicates the transpose. We note that $\mathbf{V}$ and $\mathbf{V}'$ are arbitrary and this dictates that $\mathbf{X}=\mathbf{X}^{TR}=0$ and thus that $\mathbf{X}$ is symmetric.

To prove that $\mathbf{Y}$ is symmetric we proceed along the same lines, using two flows $\mathbf{u}$ and $\mathbf{u}'$ that are produced when the particle rotates with angular velocities $\mathbf{\Omega}$ and $\mathbf{\Omega}'$.

To prove that $\mathbf{P}$ is the transpose of $\mathbf{P}'$, we use the flows $\mathbf{u}$ and $\mathbf{u}'$ that are produced when the particle translates with velocity $\mathbf{V}$ or rotates with angular velocity $\mathbf{\Omega}$. We require that on the surface of the particle $\mathbf{u}=\mathbf{V}$ and $\mathbf{u}'=\mathbf{\Omega}\times\mathbf{x}$, respectively. Applying (1.4.5) we find

$$\mathbf{V}\cdot\mathbf{F}^R=\mathbf{\Omega}\cdot\int_{S_P}\mathbf{x}\times\mathbf{f}^T\,dS$$

The integral on the right-hand side of this equation is simply the torque $\mathbf{L}^T$. Using the definitions (1.4.18) and (1.4.19) we obtain $\mathbf{V}\cdot(\mathbf{P}\cdot\mathbf{\Omega})=\mathbf{\Omega}\cdot(\mathbf{P}'\cdot\mathbf{V})$ or $\mathbf{V}\cdot[(\mathbf{P}-\mathbf{P}'^{TR})\cdot\mathbf{\Omega}]=0$ which indicates that $\mathbf{P}'$ is the transpose of $\mathbf{P}$.

*1.4.6.* The proof is based on the arguments that led to (1.4.10) and (1.4.12), with the additional remark that the integrals over the surface of the tube vanish because the velocity there is equal to zero.

**Section 1.5**

*1.5.1.* Considering first the diagonal components of $\mathbf{e}$ we require $\partial u/\partial x = \partial v/\partial y = \partial w/\partial z = 0$ to find

$$u = f(y, z), \qquad v = g(x, z), \qquad w = q(x, y) \tag{1}$$

where $f, g, q$ are three regular functions. Considering next the off-diagonal components of $\mathbf{e}$ and requiring that $\partial u/\partial y = -\partial v/\partial x$, $\partial u/\partial z = -\partial w/\partial x$, and $\partial v/\partial z = -\partial w/\partial y$, we obtain

$$\frac{\partial f(y,z)}{\partial y} = -\frac{\partial g(x,z)}{\partial x}, \qquad \frac{\partial f(y,z)}{\partial z} = -\frac{\partial q(x,y)}{\partial x}, \qquad \frac{\partial g(x,z)}{\partial z} = -\frac{\partial q(x,y)}{\partial y} \tag{2}$$

The first two of these equations imply

$$f(y,z) = y\alpha(z) + \beta(z) \qquad g(x,z) = -x\alpha(z) + \gamma(z) \tag{3}$$

$$f(y,z) = z\delta(y) + \varepsilon(y) \qquad q(x,y) = -x\delta(y) + \zeta(y) \tag{4}$$

Substituting the second equations of (3) and (4) into the third equation of (2) we obtain

$$-x\alpha'(z) + \gamma'(z) = x\delta'(y) - \zeta'(y)$$

which implies that

$$\alpha(z) = az + b, \quad \delta(y) = -ay + c, \quad \gamma(z) = dz + e, \quad \zeta(y) = -dy + e$$

Substituting the expressions for $\alpha(z)$ and $\delta(y)$ into the first equations of (3) and (4), we find that $a = 0$, and

$$f(y,z) = by + cz + U \tag{5}$$

Working similarly with $g(x,z)$ and $q(x,y)$ we find that the velocity may be expressed in the form $\mathbf{u} = \mathbf{U} + \mathbf{\Omega} \times \mathbf{x}$, which is the definition of rigid body motion.

*1.5.2.* Considering the mechanical energy balance for incompressible Newtonian flow, we find that (1.5.3) is also valid for flow produced by the steady motion of a rigid body at finite Reynolds numbers. This equation implies that the lower the energy dissipation, the weaker the drag force or torque opposing the translation or rotation of a body.

**Section 1.6**

*1.6.1.* The proof parallels that for steady flow outlined at the beginning of section 1.4.

*1.6.2.* The proof parallels that of (1.5.1).

*1.6.3.* The proof follows the arguments presented immediately after (1.6.10).

## Chapter 2

### Section 2.2

*2.2.1.* For convenience, we set the origin at the pole $\mathbf{x}_0$. Taking the three-dimensional Fourier transform of (2.1.3) and (2.1.9) we obtain

$$k_i \hat{G}_{ij} = 0 \tag{4}$$

$$ik_l \hat{p}_j + |\mathbf{k}|^2 \hat{G}_{lj} = \frac{4}{(2\pi)^{1/2}} \delta_{lj} \tag{5}$$

where in (2.1.9) we have changed the index $k$ to $l$. Multiplying (5) by $k_l$ and using (4) we derive the second equation in (1). Substituting this equation into (5) we derive the first equation in (1).

To carry out the inverse Fourier transforms we will need to use the integrals

$$\int_{\substack{\text{all}\\ \text{space}}} \frac{1}{|\mathbf{k}|^2} \exp(i\mathbf{k}\cdot\mathbf{x})\, d\mathbf{k} = 2\pi^2 \frac{1}{|\mathbf{x}|} \qquad \int_{\substack{\text{all}\\ \text{space}}} \frac{1}{|\mathbf{k}|^4} \exp(i\mathbf{k}\cdot\mathbf{x})\, d\mathbf{k} = -\pi^2 |\mathbf{x}| \tag{6}$$

which may be established most readily by performing the required spatial integrations in spherical polar coordinates. It is instructive to observe that the first equation in (6) may be derived from the second one by operating with $\nabla_x^2$, where the subscript $x$ indicates differentiation with respect to $\mathbf{x}$. Inverting the Fourier transforms given in (1) we obtain

$$\begin{aligned}
\mathbf{G}(\mathbf{x}) &= \frac{1}{\pi^2} \int_{\substack{\text{all}\\ \text{space}}} \frac{1}{|\mathbf{k}|^2} \left( \mathbf{I} - \frac{\mathbf{kk}}{|\mathbf{k}|^2} \right) \exp(i\mathbf{k}\cdot\mathbf{x})\, d\mathbf{k} \\
&= \frac{2}{|\mathbf{x}|} \mathbf{I} + \frac{1}{\pi^2} \nabla_x \nabla_x \int_{\substack{\text{all}\\ \text{space}}} \frac{1}{|\mathbf{k}|^4} \exp(i\mathbf{k}\cdot\mathbf{x})\, d\mathbf{k} \\
&= \frac{2}{|\mathbf{x}|} \mathbf{I} - \nabla_x \nabla_x |\mathbf{x}| = \frac{1}{|\mathbf{x}|} \mathbf{I} + \frac{\mathbf{xx}}{|\mathbf{x}|^3}
\end{aligned}$$

$$\begin{aligned}
\mathbf{p}(\mathbf{x}) &= -\frac{i}{\pi^2} \int_{\substack{\text{all}\\ \text{space}}} \frac{\mathbf{k}}{|\mathbf{k}|^2} \exp(i\mathbf{k}\cdot\mathbf{x})\, d\mathbf{k} \\
&= -\frac{1}{\pi^2} \nabla_x \int_{\substack{\text{all}\\ \text{space}}} \frac{1}{|\mathbf{k}|^2} \exp(i\mathbf{k}\cdot\mathbf{x})\, d\mathbf{k} = -2\nabla_x \frac{1}{|\mathbf{x}|} = 2\frac{\mathbf{x}}{|\mathbf{x}|^2}
\end{aligned}$$

*2.2.2.* The stress field associated with the flow (2.2.12) is

$$\sigma_{ik}(\mathbf{x}_0, \mathbf{x}) = \mu \left[ -\delta_{ik} \Pi_{jl}(\hat{\mathbf{x}}) - \frac{\partial T_{jil}(\hat{\mathbf{x}})}{\partial \hat{x}_k} - \frac{\partial T_{jkl}(\hat{\mathbf{x}})}{\partial \hat{x}_i} \right] q_{jl}$$

Substituting the expressions (2.2.11) and (2.2.13), and carrying out the necessary differentiations, we find the equation given in the statement of the problem.

*2.2.3.* The torque exerted on any surface that encloses the pole of a point force must be the same. For simplicity, we consider a spherical surface of radius $a$ centered at the pole $\mathbf{x}_0$, and compute

$$L_i(\mathbf{x}_0) = \varepsilon_{ijk} \int_{\text{sphere}} \hat{x}_j f_k \, dS = -\frac{3}{4\pi a^4} g_l \varepsilon_{ijk} \int_{\text{sphere}} \hat{x}_j \hat{x}_k \hat{x}_l \, dS = 0$$

where $\hat{\mathbf{x}} = \mathbf{x} - \mathbf{x}_0$ and $\mathbf{g}$ is the strength of the point force. The torque with respect to another point $\mathbf{x}_1$ is

$$\mathbf{L}(\mathbf{x}_1) = \int_{\text{sphere}} (\mathbf{x} - \mathbf{x}_1) \times \mathbf{f} \, dS = \mathbf{L}(\mathbf{x}_0) + (\mathbf{x}_0 - \mathbf{x}_1) \times \int_{\text{sphere}} \mathbf{f} \, dS = (\mathbf{x}_1 - \mathbf{x}_0) \times \mathbf{g}$$

**Section 2.3**

*2.3.1.* This identity arises by applying the reciprocal identity (1.4.3) for the flow $\mathbf{u}$ and the flow produced by a point force, and using (2.1.10). It will be noted that the derivative of the Green's function on the right-hand side of (1.4.3) will produce a singular term which, however, is weaker than the delta function and thus is neglected.

To prove (2.3.11) we integrate the equation given in the statement of the problem over a selected control volume $V$, use the divergence theorem to convert the volume integral of the left-hand side into a surface integral over the boundaries of $V$, and exploit the properties of the delta function to simplify the right-hand side.

*2.3.2.* The only difference from the procedure discussed in section 2.3 is that now $S_\varepsilon$ is the surface of a hemisphere enclosing the pole. The corresponding integral (2.3.9) may readily be computed, using the divergence theorem, to be $\delta_{ij} 2\pi\varepsilon^4/3$.

*2.3.3.* The procedure leads to (2.3.8) and (2.3.9) where $S_\varepsilon$ is the portion of the surface of a sphere of radius $\varepsilon$ that is centered at $\mathbf{x}_0$ and is confined by the two sides of the wedge. Thus, we find

$$c_{ij}(\mathbf{x}_0) = \frac{3}{4\pi\varepsilon^4} \int_{S_\varepsilon} \hat{x}_i \hat{x}_j \, dS(\mathbf{x}) \tag{1}$$

Note that $\mathbf{c}$ is a symmetric matrix. To compute $\mathbf{c}$ for a point along the corner of a two-dimensional wedge we introduce a Cartesian coordinate system in which the $x$-axis is directed along the wedge and the $y$-axis lies in the lower side of the wedge. Furthermore, we introduce the associated spherical polar coordinate system $(r, \theta, \phi)$, where the $y$-axis corresponds to $\phi = 0$. The surface $S_\varepsilon$ is defined by $r = \varepsilon$, $0 \leqslant \theta \leqslant \pi$, $0 \leqslant \phi \leqslant \alpha$, where $\alpha$ is the angle of the wedge. We write $x = \varepsilon \cos\theta$, $y = \varepsilon \sin\theta \cos\phi$, $z = \varepsilon \sin\theta \sin\phi$, $dS = \varepsilon^2 \sin\theta \, d\theta d\phi$. Substituting these expressions into (1) and carrying out the required integrations we find

$$c_{xx} = \frac{\alpha}{2\pi}, \quad c_{yy} = \frac{\alpha}{2\pi}\left(1 + \frac{\sin 2\alpha}{2\alpha}\right), \quad c_{zz} = \frac{\alpha}{2\pi}\left(1 - \frac{\sin 2\alpha}{2\alpha}\right)$$

$$c_{xy} = c_{xz} = 0, \quad c_{yz} = \frac{1}{4\pi}(1 - \cos 2\alpha)$$

*2.3.4.* The pressure $P$ of a linear flow $\mathbf{u} = \mathbf{A}\cdot\mathbf{x}$ is constant and the stress field is given by $\boldsymbol{\sigma} = -P\mathbf{I} + \mu(\mathbf{A} + \mathbf{A}^T)$ (where $\mathbf{A}$ is a traceless constant matrix). Substituting these values into (2.3.11) and using (2.1.4) we obtain

$$A_{jl}\mathbf{x}_{0,l} = \frac{1}{8\pi}(A_{il} + A_{li})\int_D G_{ij}(\mathbf{x}, \mathbf{x}_0)n_l(\mathbf{x})\,\mathrm{d}S(\mathbf{x}) - \frac{1}{8\pi}A_{il}\int_D x_l T_{ijk}(\mathbf{x}, \mathbf{x}_0)n_k(\mathbf{x})\mathrm{d}S(\mathbf{x}) \tag{1}$$

where the normal vector $\mathbf{n}$ is directed outwards from the control volume bounded by $D$. Using the above equation we deduce

$$\int_D [G_{ij}(\mathbf{x}, \mathbf{x}_0)n_l(\mathbf{x}) + G_{lj}(\mathbf{x}, \mathbf{x}_0)n_i(\mathbf{x})]\mathrm{d}S(\mathbf{x}) - \int_D x_l T_{ijk}(\mathbf{x}, \mathbf{x}_0)n_k(\mathbf{x})\mathrm{d}S(\mathbf{x}) = 8\pi\delta_{ij}\mathbf{x}_{0,l} \tag{2}$$

Starting with (2.3.4) and (2.3.13) and repeating the above procedure we derive two similar expressions for a point $\mathbf{x}_0$ located outside or on the boundary of $D$; these expressions are identical to (2) except that in the first case the right-hand side is equal to zero, and in the second case the coefficient $8\pi$ in the right-hand side should be replaced by $4\pi$, and the double-layer integral should be interpreted in the principal value sense. Working in a similar manner, we derive analogous expressions for parabolic flow.

*2.3.5.* We simply apply (2.3.11) with $D$ being the total (upper and lower) surface of the paper, note that the velocities on the two sides of the paper are the same whereas the normal vectors are opposite, and reduce (2.3.11) to the simplified form

$$u_j(\mathbf{x}_0) = -\frac{1}{8\pi\mu}\int_{S^+} \Delta f_i(\mathbf{x})G_{ij}(\mathbf{x}, \mathbf{x}_0)\mathrm{d}S(\mathbf{x})$$

where $\Delta\mathbf{f} = \mathbf{f}^+ - \mathbf{f}^-$, and the superscripts $^+$ and $^-$ indicate the upper and the lower surface of the paper respectively. We note that the density of the single-layer integral is equal to the difference in the surface force on either side of the paper.

*2.3.6.* Noting that $\mathbf{G}(\mathbf{x}, \mathbf{x}_0) = 0$ when $\mathbf{x}$ is on $D$, we reduce (2.3.11) to the simple form

$$u_j(\mathbf{x}_0) = \frac{1}{8\pi}\int_D u_i(\mathbf{x})T_{ijk}(\mathbf{x}, \mathbf{x}_0)n_k(\mathbf{x})\mathrm{d}S(\mathbf{x})$$

expressing the flow in terms of a double-layer potential whose density is proportional to the boundary velocity. Note that (2.3.12) is not applicable when the Green's function is required to vanish on $D$.

*2.3.7.* To prove (1) we write the boundary integral equation (2.3.11) for the disturbance flow $\mathbf{u}^{\mathrm{D}}$ due to the drop at a point $\mathbf{x}_0$ outside the drop, obtaining

$$u_j^{\mathrm{D}}(\mathbf{x}_0) = -\frac{1}{8\pi\mu_1}\int_{D^+} f_i^{\mathrm{D}}(\mathbf{x})G_{ij}(\mathbf{x},\mathbf{x}_0)\,\mathrm{d}S(\mathbf{x}) + \frac{1}{8\pi}\int_D u_i^{\mathrm{D}}(\mathbf{x})T_{ijk}(\mathbf{x},\mathbf{x}_0)n_k(\mathbf{x})\mathrm{d}S(\mathbf{x}) \tag{4}$$

Next, we consider the incident flow $\mathbf{u}^\infty$, integrate (2.3.3) over the volume of the drop, and apply the divergence theorem to derive an equation similar to (2.3.4):

$$-\frac{1}{8\pi\mu_1}\int_D f_i^\infty(\mathbf{x})G_{ij}(\mathbf{x},\mathbf{x}_0)\,\mathrm{d}S(\mathbf{x}) + \frac{1}{8\pi}\int_D u_i^\infty(\mathbf{x})T_{ijk}(\mathbf{x},\mathbf{x}_0)n_k(\mathbf{x})\,\mathrm{d}S(\mathbf{x}) = 0 \tag{5}$$

Adding the left-hand side of (5) to the right-hand side of (4) and recalling that $\mathbf{u} = \mathbf{u}^{\mathrm{D}} + \mathbf{u}^\infty$ is the total flow, we obtain (1). Equation (2) for the internal flow arises directly from (2.3.11); the minus sign on the right-hand side is due to our standard convention that the normal vector is directed outside the drop. Finally, equation (3) arises from (1) by taking the limit as $\mathbf{x}_0$ approaches $D$ and using (2.3.12).

*2.3.8.* This becomes evident by considering (2.3.11) and using (2.3.18) and (2.3.19). Alternatively, we may argue that the single-layer potential is continuous as $\mathbf{x}_0$ crosses the boundary and thus (2.3.27) and (2.3.30) remain valid over the boundary.

*2.3.9.* To prove (2) we apply (2.3.27) for the flow produced by a sphere that translates with velocity $\mathbf{V}$ in an infinite fluid. Using (1.4.6) we find that the corresponding distribution of the surface force is $\mathbf{f} = -3\mu\mathbf{V}/2a$, where $\mathbf{x}_c$ is the centre of the sphere. Substituting this expression into (2.3.27), applying the resulting equation at a point $\mathbf{x}_0$ that lies on the surface of the sphere, and imposing the boundary condition $\mathbf{u} = \mathbf{V}$ we obtain (2). Integrating (1) over the surface of the sphere and using (2) we obtain

$$\int_{S_P} u_j^\infty(\mathbf{x}_0)\mathrm{d}S(\mathbf{x}_0) = \frac{1}{8\pi\mu}\int_{S_P} f_i(\mathbf{x})\int_{S_P}\mathcal{S}_{ij}(\mathbf{x}-\mathbf{x}_0)\mathrm{d}S(\mathbf{x}_0)\mathrm{d}S(\mathbf{x})$$
$$= \frac{2}{3}\frac{a}{\mu}\int_{S_P} f_i(\mathbf{x})\delta_{ij}\mathrm{d}S(\mathbf{x}) = \frac{2}{3}\frac{a}{\mu}F_j$$

from which the first equation of (1.4.15) follows.

*2.3.10.* To prove (1) we apply (2.3.27) for the flow produced by a sphere that rotates with angular velocity $\mathbf{\Omega}$ in an infinite fluid. Using (1.4.11) and (1.4.14) we find that the corresponding distribution of the surface force is $\mathbf{f} = (-3\mu/a)\mathbf{\Omega} \times (\mathbf{x} - \mathbf{x}_c)$, where $\mathbf{x}_c$ is the center of the sphere. Substituting this expression into (2.3.27), applying the resulting equation at a point $\mathbf{x}_0$ that lies on the surface of the sphere, and imposing the boundary condition $\mathbf{u} = \mathbf{\Omega} \times \mathbf{x}$ we obtain (1). The rest of the proof proceeds as in problem 2.3.9.

**Section 2.4**

*2.4.1.* Essentially, we are asked to compute the limiting form of the free-space axisymmetric Green's function $\mathbf{M}$ in the limit as $\sigma \to 0$. The computations are simplified by considering the primary expression (2.4.5). Using the asymptotic expansions

$$I_{10} \approx 2\pi \frac{1}{r} \quad I_{11} \approx \pi \frac{\sigma_0 \sigma}{r^3},$$

$$I_{30} \approx 2\pi \frac{1}{r^3} \quad I_{31} \approx 3\pi \frac{\sigma_0 \sigma}{r^5} \quad I_{32} \approx \pi \frac{1}{r^3}$$

which are accurate to the leading order in $\sigma$, and where $r^2 = (x_0 - x)^2 + \sigma_0^2$, and substituting into (2.4.5) we obtain

$$\begin{bmatrix} M_{xx} \\ M_{\sigma x} \end{bmatrix} = 2\pi\sigma \begin{bmatrix} \dfrac{1}{r} + \dfrac{(x_0 - x)^2}{r^3} \\ \dfrac{(x_0 - x)\sigma_0}{r^3} \end{bmatrix} \quad \begin{bmatrix} M_{x\sigma} \\ M_{\sigma\sigma} \end{bmatrix} = \pi\sigma^2 \frac{\sigma_0^2 - 2(x_0 - x)^2}{r^5} \begin{bmatrix} x_0 - x \\ \sigma_0 \end{bmatrix}$$

Clearly, the first column of $\mathbf{M}$ represents the flow produced by a Stokeslet with strength $2\pi\sigma$ oriented in the $x$-direction. The second column of $\mathbf{M}$ represents the flow $\mathbf{u}^\sigma$ produced by a Stokeslet doublet of strength $-\pi\sigma^2$ oriented along the $x$-axis. Specifically,

$$u_i^\sigma(\mathbf{x}_0 - \mathbf{x}) = -\pi\sigma^2 \frac{\partial}{\partial x} \mathcal{S}_{ix}(\mathbf{x}_0 - \mathbf{x})$$

*2.4.2.* We consider (2.3.27), and expand the velocity on the surface of the body in a Fourier series similar to (2.4.24). Thus, we derive (2.4.25) with a minus sign on the right-hand side. Realizing that $\mathscr{u}_{\phi 0} = u_\phi$ and $\mathscr{f}_{\phi 0} = f_\phi$, whereas all other $\mathscr{u}_{\alpha n}$ and $\mathscr{f}_{\alpha n}$ are equal to zero, we obtain

$$\int_C Q_{\phi\phi 0}(\hat{\mathbf{x}}, \sigma, \sigma_0) f_\phi \sigma \, dl = -8\pi\mu u_\phi(\mathbf{x}_0)$$

From (2.4.21) we read $Q_{\phi\phi 0} = 2I_{11}$ and substitute into the above equation to obtain the desired result.

*2.4.3.* This becomes evident by noting that $f_{\beta 0} = f_\beta$, whereas all other $f_{\beta n}$ are equal to zero, substituting into (2.4.19), and realizing that the top left $2 \times 2$ block of $Q_{\alpha\beta 0}$ is the matrix $\mathbf{M}$ given in (2.4.5).
*2.4.4.* The derivation is straightforward, yet tedious. It involves using (2.3.27) and expanding the boundary velocity $\mathbf{u} = \mathbf{V} + \boldsymbol{\Omega} \times \mathbf{x}$ into a Fourier series with respect to $\phi$. Note that this series will involve *only* the constant and the first harmonic components. The final result is a system of nine real integral equations.
2.4.5. This is the system (2.4.28) where $C$ is the trace of the projection and the wall in an azimuthal plane. If we use a Green's function that vanishes on the wall this reduces $C$ to the trace of the projection alone.

**Section 2.5**

*2.5.1.* This becomes evident by writing $\mathscr{S}$ and $\mathscr{R}$ over the surface of the particle $S_P$, and applying the divergence theorem to convert the surface integral over $S_P$ into a surface integral over $D$ plus a volume integral over the intervening volume of fluid. The integrand of the volume integral may be shown to vanish yielding the desired result.
*2.5.2.* To prove (2.5.32), we apply the reciprocal identity (1.4.5) for the disturbance flow $\mathbf{u}^D$ produced when the particle is held still in the purely straining flow $\mathbf{u} = \mathbf{E}\cdot\mathbf{x}$, and the disturbance flow $\mathbf{v}^D$ produced when the particle is held still in the incident flow $\mathbf{u}^\infty$. Neglecting the integrals at infinity, using the boundary conditions $\mathbf{u}^D = -\mathbf{E}\cdot\mathbf{x}$ and $\mathbf{v}^D = -\mathbf{u}^\infty$ on the surface of the particle, and noting that $\mathbf{E}$ is a traceless symmetric matrix we obtain

$$E_{ij}\mathscr{S}^D_{ij} = \int_{S_P} u_i^\infty f_i^D \mathrm{d}S \tag{1}$$

where $\mathbf{f}^D$ is the surface force associated with the flow $\mathbf{u}^D$, and $\mathscr{S}^D$ is the coefficient of the stresslet associated with the flow $\mathbf{v}^D$.

Next, we write the boundary integral equation for the flow $\mathbf{u}^\infty$ at a point $\mathbf{x}_0$ in the particle. Operating on both sides by $E_{lk}\mathscr{L}^0_{lkj}\langle\rangle$, using (2.5.31) and the boundary condition $\mathbf{u}^D = -\mathbf{E}\cdot\mathbf{x}$ on $S_P$, we obtain

$$\int_{S_P} u_i^\infty f_i^D \mathrm{d}S = -8\pi\mu E_{lk}\mathscr{L}^0_{lkj}\langle u_j^\infty(\mathbf{x}_0)\rangle - E_{ij}\int_{S_P} x_j f_i^\infty \mathrm{d}S \tag{2}$$

Combining (1) and (2) yields

$$E_{ij}\mathscr{S}_{ij} = -8\pi\mu E_{lk}\mathscr{L}^0_{lkj}\langle u_j^\infty(\mathbf{x}_0)\rangle \tag{3}$$

which implies the Faxen relation (2.5.32).

*2.5.3.* Consistency with the asymptotic limit of the sphere (2.5.36) may be verified by performing a straightforward asymptotic expansion. The force exerted on a spheroid held stationary in a uniform flow $\mathbf{u}^\infty = \mathbf{V}$ is $\mathbf{F} = 16\pi\mu c\,\mathbf{a}\cdot\mathbf{V}$. Note that as the spheroid becomes a sphere this expression agrees with the Stokes law $\mathbf{F} = 6\pi\mu a\mathbf{V}$. The force exerted on a spheroid held stationary in a paraboloidal flow is

$$F_1 = \tfrac{32}{3}\pi\mu a U a_{11} \mathrm{e}(1-\mathrm{e}^2), \quad F_2 = \tfrac{16}{3}\pi\mu a V a_{22} \mathrm{e}(2-\mathrm{e}^2)$$

*2.5.4.* Consistency with the asymptotic limit, which is a sphere, (2.5.37) may be verified by performing a straightforward asymptotic expansion. The torque on a spheroid that is held stationary in the linear flow $\mathbf{u}^\infty = [0, kz, 0]$ is

$$M_1 = -\tfrac{16}{3}\pi\mu k\, a^3 \frac{\mathrm{e}^3}{\dfrac{2\mathrm{e}}{1-\mathrm{e}^2} - \ln\dfrac{1+\mathrm{e}}{1-\mathrm{e}}}$$

*2.5.5.* The proof follows by simple substitution.

*2.5.6.* We assume that the distance between the particles is much larger than the size of the particles, and consider the boundary integral equation (2.3.27), where $D$ is the surface of the two particles, and $\mathbf{G}$ is the Stokeslet $\mathscr{S}$. Following the steps that led us to (2.5.49), we arrive at the equation

$$\begin{bmatrix}\mathbf{F}^{(1)}\\ \mathbf{L}^{(1)}\end{bmatrix} = -\mu\begin{bmatrix}\mathbf{X}^{(1)}\\ \mathbf{P}'^{(1)}\end{bmatrix}\cdot\left[\mathbf{V}^{(1)} + \frac{1}{8\pi\mu}\mathbf{F}^{(2)}\cdot\mathscr{S}(\mathbf{x}_1-\mathbf{x}_2)\right] - \mu\begin{bmatrix}\mathbf{P}^{(1)}\\ \mathbf{Y}^{(1)}\end{bmatrix}\cdot\boldsymbol{\Omega}^{(1)}$$

as well as to an identical equation with transposed superscripts. The superscripts (1) and (2) indicate the first and second particle, respectively, $\mathbf{x}_1$ and $\mathbf{x}_2$ are two points within the first and the second particle, and $\mathbf{V}$ and $\boldsymbol{\Omega}$ are the velocities of translation and of rotation. These equations provide us with a system of linear relations between the forces and torques exerted on the particles. If we require that the $x$-axis passes through the centers $\mathbf{x}_1$ and $\mathbf{x}_2$, we shall find that $\mathscr{S}(\mathbf{x}_1-\mathbf{x}_2)$ is a diagonal matrix with $\mathscr{S}_{11} = 2/R$, $\mathscr{S}_{22} = 1/R$, $\mathscr{S}_{33} = 1/R$, where $R = |\mathbf{x}_1 - \mathbf{x}_2|$. In the case of $N$ particles we find

$$\begin{bmatrix}\mathbf{F}^{(n)}\\ \mathbf{L}^{(n)}\end{bmatrix} = -\mu\begin{bmatrix}\mathbf{X}^{(n)}\\ \mathbf{P}'^{(n)}\end{bmatrix}\cdot\left[\mathbf{V}^{(n)} + \frac{1}{8\pi\mu}\sum_{\substack{m=1\\ m\neq n}}^{N}\mathbf{F}^{(m)}\cdot\mathscr{S}(\mathbf{x}_n-\mathbf{x}_m)\right] - \mu\begin{bmatrix}\mathbf{P}^{(n)}\\ \mathbf{Y}^{(n)}\end{bmatrix}\cdot\boldsymbol{\Omega}^{(n)}$$

*2.5.7.* This follows from the symmetry property of the Green's functions expressed by (2.3.14).

**Section 2.6**

*2.6.1.* The two-dimensional Fourier transform of a function and its inverse are given in equations (2) and (3) of problem 2.2.1 where $\mathbf{k}$ is the two-dimensional wave number vector, and the coefficients $1/(2\pi)^{3/2}$ in front of both integrals must be replaced by $1/2\pi$. Taking the two-dimensional complex Fourier transforms of the two-dimensional Stokes equation and of the continuity equation and solving for the transformed velocity and pressure we obtain

$$\hat{G}_{ij} = \frac{2}{|\mathbf{k}|^2}\left(\delta_{ij} - \frac{k_i k_j}{|\mathbf{k}|^2}\right) \qquad \hat{p}_j = -\frac{2ik_j}{|\mathbf{k}|^2}$$

To carry out the inverse Fourier transforms we follow a procedure similar to that outlined in the solution of problem 2.2.1. Note that it will be necessary to use the integrals

$$\int_{\substack{\text{all}\\ \text{space}}} \frac{1}{|\mathbf{k}|^2}\exp(i\mathbf{k}\cdot\mathbf{x})d\mathbf{k} = -2\pi \ln|\mathbf{x}|, \int_{\substack{\text{all}\\ \text{space}}} \frac{1}{|\mathbf{k}|^4}\exp(i\mathbf{k}\cdot\mathbf{x})d\mathbf{k} = \frac{\pi}{2}|\mathbf{x}|^2(\ln|\mathbf{x}| - 1)$$

The first of these integrals may be derived from the second one by operating with the two-dimensional Laplacian with respect to $\mathbf{x}$. The second integral may be computed using the fundamental solution of the biharmonic equation given in (2.6.28).

*2.6.2.* See the discussion preceding (2.3.18).

*2.6.3.* The proof follows the arguments outlined in problem 4.3.1.

*2.6.4.* The proof parallels that of problem 2.3.3.

*2.6.5.* The Green's function $G(\mathbf{x}, \mathbf{x}_0)$ of the two-dimensional biharmonic equation is a solution of the equation $\nabla^4 G = \delta(\mathbf{x} - \mathbf{x}_0)$ or equivalently, the Poisson equation $\nabla^2 G = (1/2\pi)\ln r$, where $r = |\mathbf{x} - \mathbf{x}_0|$, and $\delta$ is the two-dimensional delta function. The solution to the last equation may be carried out most effectively using spherical polar coordinates and stipulating that $G$ is a function of $r$ only.

*2.6.6.* The second Green's identity states that for any two functions $\phi_1$ and $\phi_2$,

$$\int_D (\phi_1 \nabla^2 \phi_2 - \phi_2 \nabla^2 \phi_1) dA = -\int_{\mathscr{C}} (\phi_1 \nabla \phi_2 - \phi_2 \nabla \phi_1)\cdot \mathbf{n} dl$$

where the unit normal vector $\mathbf{n}$ points into the control area $D$. Applying this identity twice, the first time for $\phi_1 = \psi$, $\phi_2 = \nabla^2\chi$, and the second time for $\phi_1 = \chi$, $\phi_2 = \nabla^2\psi$, and subtracting the resulting expressions yields (2.6.31). Following the procedure outlined in the text yields (2.6.32).

*2.6.7.* This may be proven by simple evaluation.

*2.6.8.* This is evident from the Cauchy–Reimann equations (1.2.15).

*2.6.9.* This may be proven by simple substitution into the biharmonic equation.

**Section 2.7**

*2.7.1.* This may be shown by direct evaluation using $B(0) = 1$.

*2.7.2.* This problem requires a straightforward inversion of the Laplace transform.

*2.7.3.* A key step in the derivation of the generalized Faxen relations is the use of the boundary integral equation (2.5.27), which is also applicable for unsteady Stokes flow.

*2.7.4.* This problem requires us to eliminate the double-layer potential from the general boundary integral equation (2.3.11). For this purpose, we consider uniform unsteady Stokes flow $\mathbf{u} = \mathbf{V}$, within the volume occupied by the body, with associated pressure $P = -\lambda^2 \mathbf{V} \cdot \mathbf{x}$, stress tensor $\boldsymbol{\sigma} = \lambda^2 \mathbf{I} \mathbf{V} \cdot \mathbf{x}$, and surface force $\mathbf{f} = \lambda^2 \mathbf{n} \mathbf{V} \cdot \mathbf{x}$. Selecting a point $\mathbf{x}_0$ outside the body and applying (2.3.4) (in dimensionless form), we obtain the identity

$$\lambda^2 V_k \int_{S_P} G_{ij}(\mathbf{x}, \mathbf{x}_0) x_k n_i(\mathbf{x}) \mathrm{d}S(\mathbf{x}) = V_i \int_{S_P} T_{ijk}(\mathbf{x}, \mathbf{x}_0) n_k(\mathbf{x}) \mathrm{d}S(\mathbf{x}) \qquad (2)$$

Next, we consider the boundary integral equation (2.3.11), impose the boundary condition $\mathbf{u} = \mathbf{V}$ on the surface of the body and use (2) to derive (1).

*2.7.5.* To derive this Green's function we follow a procedure similar to that for steady Stokes flow discussed in section 2.6. Thus, we find that $\mathbf{G}$ is given by (2.2.4) where the function $H$ satisfies the equation

$$(\nabla^2 - \lambda^2) H = \frac{1}{2\pi} \ln r \qquad (1)$$

One may readily verify that

$$H = -\frac{1}{2\pi\lambda^2} (\ln r + 2\pi E) \qquad (2)$$

is a solution to (1), where $E$ is the free-space fundamental solution of the two-dimensional Helmholtz operator, i.e.

$$(\nabla^2 - \lambda^2) E + \delta(\mathbf{x} - \mathbf{x}_0) = 0 \qquad (3)$$

where $\delta$ is the two-dimensional delta function.

In the case of a Laplace transform $\lambda^2$ is real and positive (see section 1.6), and we obtain

$$E = \frac{1}{2\pi} K_0(\lambda r) \qquad (4)$$

where $K_0$ is a modified Bessel function (Stakgold 1968, Vol. II, p. 265; Abramowitz & Stegun 1972, p. 374). Substituting (4) into (2) and then into (2.2.4), and using the identities

$$\frac{dK_0(x)}{dx} = -K_1(x) \qquad \frac{dK_1(x)}{dx} = -K_0(x) - \frac{K_1(x)}{x} \tag{5}$$

we compute the velocity field in the form (2.6.1). The Green's function is

$$G_{ij} = -\delta_{ij}A(R) + \frac{\hat{x}_i\hat{x}_j}{r^2}B(R) \tag{6}$$

where $R = \lambda r$, and the functions $A(R)$ and $B(R)$ are given by

$$A(R) = 2\left[\frac{1}{R^2} - \frac{K_1(R)}{R} - K_0(R)\right] \quad B(R) = 2\left[\frac{2}{R^2} - K_0(R) - 2\frac{K_1(R)}{R}\right] \tag{7}$$

where $K_1$ is a modified Bessel function. Abramowitz & Stegun (1972, p. 378) provide accurate polynomial approximations for $K_0$ and $K_1$. In the limit of small $R$,

$$K_0(R) \approx -\ln\frac{R}{2} - 0.57721 + \cdots \quad K_1(R) \approx \frac{R}{2}\ln\frac{R}{2} + \frac{1}{R} + 0.03861R + \cdots \tag{8}$$

Substituting these asymptotic expansions into (7) we find that as $R \to 0$, $A(R) \to \ln R$ and $B(R) \to 1$, recovering the steady two-dimensional Stokeslet.

The pressure vector corresponding to (6) is given by (2.6.19). Using (2.1.8) we compute the stress tensor

$$T_{ijk} = 2\delta_{ik}\frac{x_j}{r^2}[B(R) - 1] + \frac{\delta_{ij}x_k + \delta_{kj}x_i}{r^2}C(R) - 4\frac{x_i x_j x_k}{r^4}D(R) \tag{9}$$

where

$$C(R) = \frac{8}{R^2} - 4K_0(R) - 2\left(R + \frac{4}{R}\right)K_1(R)$$

$$D(R) = \frac{8}{R^2} - 4K_0(R) - \left(R + \frac{8}{R}\right)K_1(R) \tag{10}$$

Using the expansions (8) one may verify that as $R \to 0$, $C(R) \to 0$ and $D(R) \to 1$, recovering the results for the steady two-dimensional Stokeslet.

In the case of the Fourier transform or oscillatory flow, $\lambda^2 = -i\omega L^2/\nu$ is an imaginary number (see section 1.6), and we obtain

$$E = \frac{i}{4}H_0^{(1)}(e^{i3\pi/4}|\lambda|r) = \frac{1}{2\pi}[ker_0(|\lambda|r) + i\,kei_0(|\lambda|r)] \tag{11}$$

where $H_0^{(1)}$ is a Hankel function and $ker_0$, $kei_0$ are Kelvin functions (Stakgold 1968, Vol. II, p. 265; Abramowitz & Stegun 1972, p. 379).

Substituting (11) into (2) we obtain

$$H = -\frac{1}{2\pi\lambda^2}[\ln r + \mathrm{ker}_0(|\lambda| r) + \mathrm{i}\, kei_0(|\lambda| r)] \qquad (12)$$

Furthermore, substituting into (2.2.4), and using the formulae

$$ker'_0(x) = \frac{1}{(2)^{1/2}}[ker_1(x) + kei_1(x)] \qquad kei'_0(x) = \frac{1}{(2)^{1/2}}[-ker_1(x) + kei_1(x)]$$

$$ker'_1(x) = -\frac{1}{x}ker_1(x) - \frac{1}{(2)^{1/2}}[ker_0(x) + kei_0(x)]$$

$$kei'_1(x) = -\frac{1}{x}kei_0(x) - \frac{1}{(2)^{1/2}}[kei_0(x) - ker_0(x)]$$

provided by Abramowitz & Stegun, we express the Green's function in terms of $ker_0$ and $kei_0$. These functions may be computed accurately using polynomial approximations (Abramowitz & Stegun 1972, p. 384).

*2.7.6.* Using the defining expression (2.1.8), equation (2.7.12), and the symmetry property (2.3.14) we find

$$T_{ijk}(\mathbf{x}, \mathbf{x}_0) = 2\delta_{ik}\frac{x_{0,j} - x_j}{r^3} + \left[\frac{\partial G_{ji}}{\partial x_k} + \frac{\partial G_{jk}}{\partial x_i}\right](\mathbf{x}_0, \mathbf{x})$$

which shows that $\mathbf{T}$ is composed of a point source and two point force dipoles. Noting that the pressure field due to a point source is $\lambda^2/r$ and using (2.7.12) we obtain the desired result.

### Section 2.8

*2.8.1.* In this problem it is proper to use a Green's function that vanishes over the tube. Decomposing this Green's function into the free-space component given in (2.8.6) and a regular complementary component, substituting into (2.8.5), and imposing the boundary condition $u_\phi = W\sigma$, we derive the integral equation

$$W\sigma_0 = -\frac{1}{4\pi\mu}\int_C [I_{11}(\hat{x}, \sigma, \sigma_0) + Q(\hat{x}, \sigma, \sigma_0)]f_\phi \sigma \,\mathrm{d}l \qquad (1)$$

where $Q$ expresses the effect of the complementary component. Note that physically, $u_\phi = -(I_{11} + Q)/4\pi\mu$ expresses the flow produced by a ring of point forces of density per unit length $f_\phi$ oriented in the azimuthal direction with axis at the center line of the tube. Regularity requires that the velocity vanish along the axis of rotation. Also, as the radius of the ring $\sigma_0$ vanishes, the flow $-I_{11}/4\pi$ produced by the primary component ceases, and $Q$ must vanish. These two observations suggest that we may write $Q(\hat{x}, \sigma, \sigma_0) = \sigma\sigma_0 H(\hat{x}, \sigma, \sigma_0)$, where $H$ is a regular function.

Now, if the tube is large, the value of the function $H$ over the body is approximately constant and has a small value $\alpha$. Substituting $Q = \sigma\sigma_0\alpha$ into (1) and using (2.8.10) we obtain

$$W\left(1+\frac{\alpha}{8\pi^2\mu W}L\right)\sigma_0 = -\frac{1}{4\pi\mu}\int_C I_{11}(\hat{x},\sigma,\sigma_0)f_\phi\sigma\,\mathrm{d}l \tag{2}$$

which implies the equation given in the statement of the problem.

*2.8.2.* Combining (2.8.9) and (2.8.10) we find that the torque exerted on the disk is given by

$$L = -16\mu Wa\int_0^a \frac{\sigma^3}{(a^2-\sigma^2)^{1/2}}\mathrm{d}\sigma = -\frac{32}{3}\mu Wa^3$$

*2.8.3.* Expanding the kernel (2.8.14) in a Taylor series with respect to $\lambda$ we find

$$K = I_{11} + \frac{1}{2}\lambda^2\int_0^{2\pi}\cos\phi(\hat{x}^2+\sigma^2+\sigma_0^2-2\sigma\sigma_0\cos\phi)^{1/2}\mathrm{d}\phi + \cdots$$

Correspondingly, we expand $f$ in the series $f = f^{(0)} + \lambda^2 f^{(2)} + \cdots$. Substituting into (2.8.13) and imposing the boundary condition $u_\phi = \sigma$, we derive an integral equation for $f^{(2)}$, namely

$$2\int_C I_{11}f^{(2)}\sigma\,\mathrm{d}l(\mathbf{x})$$
$$= -\int_C f^{(0)}\int_0^{2\pi}\cos\phi(\hat{x}^2+\sigma^2+\sigma_0^2-2\sigma\sigma_0\cos\phi)^{1/2}\mathrm{d}\phi\,\sigma\,\mathrm{d}l(\mathbf{x})$$

## *Chapter 3*

### Section 3.1

*3.1.1.* These identities arise directly from (2.1.3), (2.1.4) and the symmetry property (3.1.3).

*3.1.2.* We consider an incident flow $\mathbf{u}^\infty$ past a stationary particle. Applying (2.3.30) at the surface of the particle $S_P$, requiring that the velocity at the surface of the particle vanishes, dot multiplying the resulting equation by the surface force $\mathbf{f}^T$ exerted on the particle when it translates with velocity $\mathbf{V}$, and integrating the product over the surface of the particle, we obtain

$$\int_{S_P} u_j^\infty(\mathbf{x}_0)f_j^T(\mathbf{x}_0)\,\mathrm{d}S(\mathbf{x}_0) = \frac{1}{8\pi\mu}\int_{S_P} f_i(\mathbf{x})\int_{S_P} G_{ij}(\mathbf{x},\mathbf{x}_0)f_j^T(\mathbf{x}_0)\,\mathrm{d}S(\mathbf{x}_0)\,\mathrm{d}S(\mathbf{x}) \tag{1}$$

Furthermore, using (3.1.3) and (2.3.27) we find that when $\mathbf{x}_0$ is on $S_P$,

$$\int_{S_P} G_{ij}(\mathbf{x},\mathbf{x}_0)f_j^T(\mathbf{x}_0)\mathrm{d}S(\mathbf{x}_0) = \int_{S_P} f_j^T(\mathbf{x}_0)G_{ji}(\mathbf{x}_0,\mathbf{x})\,\mathrm{d}S(\mathbf{x}_0) = -8\pi\mu V_i \tag{2}$$

Combining (1) and (2), and using the definition (1.4.6) we derive (1.4.10). Equation (1.4.12) may be derived in a similar way.

*3.1.3.* The torque with respect to the point $\mathbf{x}_1$ exerted on the union of the external boundaries of an internal flow owing to a point force of strength $\mathbf{g}$ located at $\mathbf{x}_0$ is $\mathbf{L} = -(\mathbf{x}_1 - \mathbf{x}_0) \times \mathbf{g}$ (see problem 2.2.3, and note that the minus sign arises because the normal vector is directed into the volume enclosed by the boundaries). Comparing this equation with (3.1.14) we find $Y_{ij} = -\varepsilon_{ikj}(x_{1,k} - x_{0,k})$. Substituting this expression into (3.1.15) yields $u_j = -\Omega_i \varepsilon_{ikj}(x_{1,k} - x_{0,k}) = \varepsilon_{jik}\Omega_i(x_{0,k} - x_{1,k})$ i.e. rigid body motion. Thus, we conclude that if the external boundary of an internal flow rotates as a rigid body, the flow must express rigid body motion.

*3.1.4.* Combining (3.1.13) with (7.3.5) we find

$$X_{ij}(\mathbf{x}) = \frac{3a}{4}\left(\frac{\delta_{ij}}{r} + \frac{x_i x_j}{r^3}\right) - \frac{a^3}{4}\left(-\frac{\delta_{ij}}{r^3} + 3\frac{x_i x_j}{r^5}\right) \tag{1}$$

where $a$ is the radius of the sphere, and the origin is set at the centre of the sphere. Next, we decompose $\mathbf{g} = \mathbf{g}_r + \mathbf{g}_t$ where $\mathbf{g}_r$ and $\mathbf{g}_t$ are the radial and transverse components of $\mathbf{g}$, given by $\mathbf{g}_r = \mathbf{x}(\mathbf{x}\cdot\mathbf{g})/r^2$ and $\mathbf{g}_t = \mathbf{g} - \mathbf{g}_r$, respectively. Using (1) and (3.1.12) we obtain

$$\mathbf{F}' = \frac{1}{2}\left(\frac{3}{R} - \frac{1}{R^3}\right)\mathbf{g}_r + \frac{1}{4}\left(\frac{3}{R} + \frac{1}{R^3}\right)\mathbf{g}_t \tag{2}$$

where $R = r/a$. Similarly, combining (7.3.28) with (3.1.15) we find

$$Y_{ij}(\mathbf{x}) = \varepsilon_{jil} a^3 \frac{x_l}{r^3} \tag{3}$$

and using (3) and (3.1.14) we obtain

$$\mathbf{L}' = \frac{1}{R^3}\mathbf{x} \times \mathbf{g} \tag{4}$$

*3.1.5.* We consider the disturbance flow $\mathbf{u}^D$ due to a solid boundary $S_B$ immersed in a purely straining incident flow $\mathbf{u}^\infty = \mathbf{E}\cdot\mathbf{x}$. Following the discussion in the text we find that the coefficient of the stresslet associated with the flow produced by a point force, $\mathscr{S}'$, obeys the relationship

$$E_{ij}\mathscr{S}'_{ij} = -u_i^D(\mathbf{x}_0)g_i \tag{1}$$

where $\mathbf{g}$ is the strength of the point force (see (2.5.4)). Furthermore, exploiting the linear dependence of $\mathscr{S}'$ on $\mathbf{g}$ we write $\mathscr{S}'_{ij} = -Z_{ijk}(\mathbf{x}_0)g_k$ where $Z_{ijk}(\mathbf{x}_0)$ is a resistance matrix, and find

$$u_k^D(\mathbf{x}_0) = E_{ij}Z_{ijk}(\mathbf{x}_0) \tag{2}$$

The precise value of $Z_{ijk}(\mathbf{x}_0)$ for a sphere may be deduced using the solution (7.3.12) as indicated in problem 3.1.4.

**Section 3.2**

*3.2.1.* That this flow is consistent with the equations of creeping motion was already proven in the discussion following (2.5.12). The source dipole character of this solution becomes evident by examining the behaviour close to the pole.

**Section 3.3**

*3.3.1.* We select a volume of fluid bounded by the plane wall and a hemispherical cap of large radius enclosing the pole of the point force. The force exerted on the union of the wall and the hemispherical cap must be equal to $-\mathbf{g}$, where $\mathbf{g}$ is the strength of the point force (see section 2.2). We note that the velocity at infinity decays faster than $1/r$, which indicates that as the radius of the cap tends to infinity, the corresponding force vanishes. Switching the direction of the normal vector into the fluid, we find that the force acting on the wall must be equal to $\mathbf{g}$.
*3.3.2.* Within the fluid we draw a closed surface that is bounded by the wall and a large hemispherical cap enclosing the protrusion. The force acting on this closed surface must be equal to zero. We note that the velocity at infinity decays faster than $1/r$, which means that as the radius of the hemispherical cap tends to infinity, the force acting on this surface and hence, on the wall, must vanish.
*3.3.3.* Using (3.3.26) we find

$$\text{(a)}\quad \mathbf{F} = \frac{1}{2}\left(-2 + \frac{3}{R} - \frac{1}{R^3}\right)\mathbf{g}_r + \frac{1}{4}\left(-4 + \frac{3}{R} + \frac{1}{R^3}\right)\mathbf{g}_t$$

$$\text{(b)}\quad \mathbf{F} = \frac{1}{2}\left(\frac{3}{R} - \frac{1}{R^3}\right)\mathbf{g}_r + \frac{1}{4}\left(\frac{3}{R} + \frac{1}{R^3}\right)\mathbf{g}_t,$$

$$\text{(c)}\quad \mathbf{F} = -\mathbf{g}$$

where $R = r/a$, $r$ is the distance of the point force from the center of the sphere, and $a$ is the radius of the sphere.
*3.3.4.* Far from the sphere the flow behaves as if it were induced by a point force of strength $-\mathbf{F}$, where $\mathbf{F}$ is the answer to part (a) of problem 3.3.3. Thus

$$u_i(\mathbf{x}) = -\frac{1}{8\pi\mu}\mathcal{S}_{ij}(\mathbf{x}, \mathbf{x}_0)F_j$$

where $\mathcal{S}$ is the Stokeslet.

**Section 3.4**

*3.4.1.* This Green's function may be derived in a straightfoward manner by integrating the three-dimensional Green's function for a flow bounded by a plane wall as discussed in section 3.4. Details are given in the cited references.

*3.4.2.* To prove this identity we write

$$H(N) = \exp(\mathrm{i}\hat{x}t)(z^{-N} + \cdots + z^{N}) = \exp(\mathrm{i}\hat{x}t)\frac{z^{-N} - z^{N+1}}{1 - z}$$

where $z = \exp(\mathrm{i}lt)$, resolve $z$ into its real and imaginary parts, substitute into the fraction on the right-hand side of the above equation, and simplify the resulting expressions. Note that the real part of $H(N)$ is a sum of cosines, whereas the imaginary part is a sum of sines. Setting the numerator of the fraction on the right-hand side of (3.4.15) equal to zero we obtain $t = 2\pi m/l$ and $t = 2\pi m/[l(2N + 1)]$ where $m$ is an integer. The first choice is dismissed, for it makes the denominator of the fraction in (3.4.15) equal to zero. Setting the real and imaginary parts of $\exp(\mathrm{i}\hat{x}t)$ equal to zero we find $t = (m + 1/2)\pi/\hat{x}$ and $t = m\pi/\hat{x}$ respectively.

**Section 3.5**

*3.5.1.* Modifying the procedure outlined in the beginning of section 3.1 we derive the equivalent of (3.1.6) for two-dimensional flow. Integrating this equation as explained in the text we arrive at (3.1.7). In the case of infinite flow we must ensure that the integral

$$\int_{C_\infty} [G_{ik}(\mathbf{x}, \mathbf{x}_2)T_{ilj}(\mathbf{x}, \mathbf{x}_1) - G_{il}(\mathbf{x}, \mathbf{x}_1)T_{ikj}(\mathbf{x}, \mathbf{x}_2)]n_j(\mathbf{x})\,\mathrm{d}l(\mathbf{x})$$

over a closed contour $C_\infty$ of large radius enclosing the poles $\mathbf{x}_1, \mathbf{x}_2$ and the body tends to zero as the radius of $C_\infty$ tends to infinity. Scaling arguments indicate that the limiting value of the above integral will not change if we collapse the two poles. Thus, we consider the modified integral

$$\int_{C_\infty} [G_{ik}(\mathbf{x}, \mathbf{x}_1)T_{ilj}(\mathbf{x}, \mathbf{x}_1) - G_{il}(\mathbf{x}, \mathbf{x}_1)T_{ikj}(\mathbf{x}, \mathbf{x}_1)]n_j(\mathbf{x})\,\mathrm{d}l(\mathbf{x})$$

Applying the reciprocal theorem for the flows (3.1.4) with $\mathbf{x}_2$ identical to $\mathbf{x}_1$ we find that the value of this integral will not change if we replace $C_\infty$ by a circle of small radius enclosing $\mathbf{x}_1$. Noting that close to the pole the Green's function behaves like the Stokeslet, and substituting the exact expressions for **G** and **T** using (2.6.17) and (2.6.20) we find that this integral is equal to zero.

*3.5.2.* This may be shown by direct evaluation.

*3.5.3.* Writing out (3.5.4) we find

$$p_i^{\mathrm{W}}(\mathbf{x}_0) = -2\frac{\hat{X}_i}{|\hat{X}|^2} - 2\frac{\hat{Y}_i}{|\hat{\mathbf{Y}}|^2} + 4\frac{\partial \mathcal{S}_{i2}}{\partial \hat{Y}_2}(\hat{\mathbf{Y}}) + 4(w-y)\frac{\partial}{\partial Y_2}\left(\frac{\hat{\mathbf{Y}}}{|\hat{\mathbf{Y}}|^2}\right)$$

where

$$\hat{\mathbf{X}} = \mathbf{x}_0 - \mathbf{x}, \quad \hat{\mathbf{Y}} = \mathbf{x}_0 - \mathbf{x}^{\mathrm{IM}}, \quad \mathbf{x}^{\mathrm{IM}} = (x, 2y - w)$$

and $\mathcal{S}$ is the two-dimensional Stokeslet. Clearly, $\mathbf{p}^{\mathrm{W}}(\mathbf{x}_0)$ represents the flow due to a collection of singularities including a primary point source and an image system composed of a point source, a Stokeslet doublet, and a potential dipole. One may show by direct evaluation that $\mathbf{p}^{\mathrm{W}}(\mathbf{x}_0)$ indeed, vanishes when $\mathbf{x}_0$ is on the wall, i.e. when $x_0 = w$.

*3.5.4.* This may be shown merely by expanding the function $A$ and its derivatives in a Taylor series with respect to $\hat{\mathbf{x}}$.

*3.5.5.* Without loss of generality, we shall assume that $\omega$ is a positive number. Thus, we extend $x$ into the complex $z$-plane, and define a contour $C$ composed of the $x$-axis, and a semicircle of large radius enclosing the first two quadrants. Combining (3.5.33) and (3.5.36) we obtain

$$\mathscr{f}(\omega, h) = \int_{-\infty}^{\infty} f(x, h)\exp(\mathrm{i}\omega x)\,\mathrm{d}x = \frac{\pi}{2H}\sin(\pi h/2H)$$
$$\times \int_C \frac{z\exp(\mathrm{i}\omega z)}{\cosh(\pi z/2H) - \cos\pi h/2H)}\,\mathrm{d}z$$

To compute the contour integral on the right-hand side of the above equation we use the theory of residues. The poles of the integrand are $z_n = iy_n$ where $y_n = -h + 4nH$ and $n$ is an integer. Applying the standard theory of residues yields

$$\mathscr{f}(\omega, h) = 2\pi\mathrm{i}\sin(\pi h/2H) \sum_{\substack{n=-\infty \\ y_n > 0}}^{\infty} \frac{y_n \exp(-\omega y_n)}{\sin(\pi y_n/2H)}$$
$$= 2\pi\mathrm{i}\sin(\pi h/2H) \sum_{n=-\infty}^{\infty} y_n \exp(-\omega|y_n|)$$

Summing the last series yields (3.5.36).

**Section 3.6**

*3.6.1.* This requires a straightforward asymptotic expansion.

## *Chapter 4*

**Section 4.1**

*4.1.1.* To prove (4.1.5) we multiply (4.1.4) from the right with the normal vector, integrate over $S$, and use (2.1.12). Similarly for (4.1.6).

*4.1.2.* We consider equation (2) of problem 2.3.9. This equation implies that the velocity on the surface of a sphere produced by a distribution of Stokeslets over the surface of the sphere of constant density $\mathbf{q}$ is $\mathbf{u} = 16\pi a\mathbf{q}/3$. Owing to the uniqueness of Stokes flow, the velocity field in the interior of the sphere will have the same value.

*4.1.3.* We note that the singularity of the integrand of $\mathbf{J}$ is one order less than that of $\mathbf{T}$, and deduce that $\mathbf{J}$ represents a continuous function. Using (2.1.12) we obtain $\mathbf{f}^+ = \mathbf{J} + 8\mu\pi\mathbf{q}$ and $\mathbf{f}^- = \mathbf{J}$, from which (2) follows immediately.

*4.1.4.* This is a consequence of the fact that the single-layer operator is self-adjoint. We compute

$$(\mathbf{L}^S\mathbf{q}, \mathbf{q}) = \beta \int_D \mathbf{q}(\mathbf{x})\cdot\mathbf{q}^*(\mathbf{x})\,\mathrm{d}S(\mathbf{x}) \qquad (\mathbf{q}, \mathbf{L}^S\mathbf{q}) = \beta^* \int_D \mathbf{q}(\mathbf{x})\cdot\mathbf{q}^*(\mathbf{x})\,\mathrm{d}S(\mathbf{x})$$

Using (4.1.11) we find that $\beta = \beta^*$ concluding that $\beta$ must be real.

**Section 4.2**

*4.2.1.* See problem 4.1.2.

*4.2.2.* That (4.2.4) is a general solution of (4.1.7) may be shown readily with the help of (3.2.7). To show that (4.2.3) is a general solution of (4.1.8) we must prove that each $\mathbf{q}'^{(j)}$ satisfies (4.4.3) with $\beta = -1$. Now, using (2.3.27) we find that the surface force on the external surface of $S_B$ is equal to

$$f_i^+(\mathbf{x}_0) = -\frac{1}{8\pi} n_k(\mathbf{x}_0) \int_{S_B^+} f_j(\mathbf{x}_0) T_{ijk}(\mathbf{x}_0, \mathbf{x})\,\mathrm{d}S(\mathbf{x})$$

where $\mathbf{x}_0$ is on $S_B$. Using (4.1.7) to convert the integral on the right-hand side into the corresponding principal value integral, we arrive at the required identity.

**Section 4.3**

*4.3.1.* The proof follows the arguments discussed in the solution of problem 4.1.3.

*4.3.2.* This becomes evident by using the results of problem 2.2.2.

*4.3.3.* The proof merely requires a substitution.

*4.3.4.* We note that close to the pole the kernel of the double-layer potential resembles that of the Stokeslet. Thus,

$$T_{ijk}(\mathbf{x}, \mathbf{x}_0) n_k(\mathbf{x}) \approx -6 \frac{\hat{x}_i \hat{x}_j}{|\hat{\mathbf{x}}|^5} \hat{\mathbf{x}}\cdot\mathbf{n}$$

where $\hat{\mathbf{x}} = \mathbf{x} - \mathbf{x}_0$. The key is to realize that close to the pole $\hat{\mathbf{x}}$ tends to become perpendicular to $\mathbf{n}$. Thus, if the domain of distribution is a

Lyapunov surface, the dot product $\hat{\mathbf{x}}\cdot\mathbf{n}$ tends to zero at a rate that is faster than linear, and the kernel of the double-layer potential tends to become infinite at a rate which is slower than quadratic.

**Section 4.4**

*4.4.1.* Assume that $\mathbf{q}$ is the solution of the integral equation (4.4.1) and that $\mathbf{q}^{(n)}$ is the $n$th approximation to the solution. We observe that the error $\mathbf{e}^{(n)} = \mathbf{q}^{(n)} - \mathbf{q}$ satisfies the evolution equation

$$e_i^{(n+1)}(\mathbf{x}_0) = \frac{\alpha}{4\pi}\int_D^{\mathscr{PV}} e_j^{(n)}(\mathbf{x})\,T_{jik}(\mathbf{x},\mathbf{x}_0)n_k(\mathbf{x})\,\mathrm{d}S(\mathbf{x})$$

This equation defines a mapping that will converge to the fixed point zero for an arbitrary initial choice of $\mathbf{e}^{(0)}$ as long as the Neumann series is convergent.

**Section 4.5**

*4.5.1.* We are required to prove that

$$n_i(\mathbf{x}_0) = \frac{1}{4\pi}\int_{\text{sphere}}^{\mathscr{PV}} n_j(\mathbf{x})\,T_{jik}(\mathbf{x},\mathbf{x}_0)n_k(\mathbf{x})\,\mathrm{d}S(\mathbf{x})$$

This identity follows immediately by applying the boundary integral equation (2.3.11) for the flow exterior to a sphere due to a point source located at the origin (see (7.2.1) and (7.2.2)).

*4.5.2.* We are required to prove that

$$n_k(\mathbf{x}_0)\int_{\text{sphere}}^{\mathscr{PV}} T_{ijk}(\mathbf{x},\mathbf{x}_0)\,\mathrm{d}S(\mathbf{x}) = -4\pi\delta_{ij}$$

The validity of this identity emerges from the discussion in problem 4.2.2 bearing in mind that the surface force on a translating sphere is constant.

**Section 4.6**

*4.6.1.* Using the defining expression $(\mathbf{A}-\lambda_1\mathbf{I})\cdot\mathbf{u}=0$ and $\mathbf{v}\cdot(\mathbf{A}-\lambda_2^*\mathbf{I})=0$ where $\mathbf{u}$ and $\mathbf{v}$ are eigenvectors of $\mathbf{A}$ and $\mathbf{A}^{\mathrm{T}}$ corresponding to the eigenvalues $\lambda_1$ and $\lambda_2^*$ respectively, we compute $[\mathbf{v}\cdot(\mathbf{A}-\lambda_2^*\mathbf{I})]\cdot\mathbf{u}=(\lambda_1-\lambda_2^*)\mathbf{v}\cdot\mathbf{u}=0$ which requires that $\mathbf{v}\cdot\mathbf{u}=0$ as long as $\lambda_1\neq\lambda_2^*$.

If $\mathbf{v}$ is the first generalized eigenvector of $\mathbf{A}^{\mathrm{T}}$ then

$$\mathbf{v}\cdot(\mathbf{A}-\lambda_2^*\mathbf{I})^2=0$$

and

$$[\mathbf{v}\cdot(\mathbf{A}-\lambda_2^*\mathbf{I})^2]\cdot\mathbf{u}=(\lambda_1-\lambda_2^*)[\mathbf{v}\cdot(\mathbf{A}-\lambda_1\mathbf{I})]\cdot\mathbf{u}=(\lambda_1-\lambda_2^*)^2\mathbf{v}\cdot\mathbf{u}=0$$

which again requires that $\mathbf{v}\cdot\mathbf{u}=0$ as long as $\lambda_1 \neq \lambda_2^*$. Extending this result, we deduce that all generalized eigenvectors of $\mathbf{A}^T$ are orthogonal to $\mathbf{u}$.

*4.6.2.* First, it is clear that $\mathscr{A}\cdot\mathbf{u}=0$, which indicates that $\mathbf{u}$ is an eigenvector of $\mathscr{A}$ with corresponding eigenvalue equal to zero. Now, recalling that a matrix and its transpose have complex conjugate eigenvalues, we set out to prove that the transpose of $\mathscr{A}$,

$$\mathscr{A}_{ij}^T = A_{ji} - \lambda_1 w_i u_j$$

has the same eigenvalues as $\mathbf{A}^T$ with the exception of $\lambda_1$. Indeed, assuming that $\lambda_2^*$ is an eigenvalue of $\mathbf{A}^T$ (different from $\lambda_1$) with corresponding eigenvector (or generalized eigenvector) $\mathbf{v}$, and using the results of problem 4.6.1, we find that $\mathscr{A}^T\cdot\mathbf{v}=\mathbf{A}^T\mathbf{v}=\lambda_2^*\mathbf{v}$. Thus, $\mathscr{A}$ has the same eigenvalues as $\mathbf{A}$, with the exception of $\lambda_1$ which is replaced by zero.

Bodewig (1959, p. 360) suggests an alternative algebraic proof. Recalling that the determinant of a matrix is equal to the product of its eigenvalues, we deduce that to state that all eigenvalues of $\mathscr{A}$ are equal to those of $\mathbf{A}$ (with the exception of $\lambda_1$) is equivalent to requiring

$$(\lambda_1 - \lambda)\det(\mathscr{A} - \lambda\mathbf{I}) = -\lambda\det(\mathbf{A} - \lambda\mathbf{I})$$

Each side of this equation is an $(n+1)$th degree polynomial in $\lambda$, where $n$ is the order of $\mathbf{A}$. The task is to show that corresponding coefficients of the two polynomials are identical. This may be accomplished by detailed examination of the individual coefficients.

*4.6.3.* The proof is an extension of that of problem 4.6.2. Note that a key observation is that the eigenvectors of the adjoint of a deflated matrix are the same as those of the adjoint of the original matrix.

*4.6.4.* This follows immediately by substituting (4.6.29) into (4.6.28) and carrying out the required computations.

*4.6.5.* This may be shown by direct substitution.

*4.6.6.* This proof proceeds according to the arguments following (4.6.14).

**Section 4.7**

*4.7.1.* Clearly, the boundary integral equation is recovered by setting $\mathbf{q}=(1/8\pi)\mathbf{u}$ and

$$\mathscr{V}_i(\mathbf{x}_0) = -\frac{1}{8\pi\mu}\int_D f_j(\mathbf{x})G_{ji}(\mathbf{x},\mathbf{x}_0)\,dS(\mathbf{x})$$

*4.7.2.* Identifying $\mathscr{V}$ with the flow produced by $N$ point forces and $M$ couplets, we replace (4.7.2) with the more involved expression

$$\mathscr{V}_i(\mathbf{x}_0) = \sum_{n=1}^{N} G_{il}(\mathbf{x}_0,\mathbf{x}_G^n)\mathscr{F}_l^n + \sum_{m=1}^{M} G_{il}^C(\mathbf{x}_0,\mathbf{x}_R^m)\mathscr{L}_l^m \qquad (1)$$

The force and torque with respect to the point $\mathbf{x}_0$ exerted on the distribution domain $D$ are (see problem 2.2.3)

$$\mathbf{F} = -8\pi\mu \sum_{n=1}^{N} \mathscr{F}^n \qquad \mathbf{L} = 8\pi\mu\left[\sum_{n=1}^{N} (\mathbf{x}_0 - \mathbf{x}_G^n) \times \mathscr{F}^n - \sum_{m=1}^{M} \mathscr{L}^m\right] \tag{2}$$

**Section 4.8**

*4.8.1.* As in problem 4.7.2, we stipulate that $\mathscr{V}$ is expressed as in equation (1) of the solution of problem 4.7.2. To complete the boundary integral equation, we must provide a set of $N + M$ vectorial constraints for $\mathscr{F}^n$ and $\mathscr{M}^m$. We may require, for instance, that (4.8.2) applies for the sum of $\mathscr{F}^n$ and $\mathscr{M}^m$ and, in addition, specify the directions of all but one of $\mathscr{F}^n$ and $\mathscr{M}^m$, as well as the relative magnitudes of all $\mathscr{F}^n$ and $\mathscr{M}^m$.

Following the arguments outlined at the beginning of section 4.8 we find that the characteristic solution of the corresponding homogeneous integral equation must be of the form $\mathbf{q}'' = \mathbf{c} + \mathbf{d} \times \mathbf{X}$, where $\mathbf{c}$ and $\mathbf{d}$ are constant vectors. Substituting this expression into (4.8.2) and taking into consideration the imposed constraints, we produce a system of linear homogeneous algebraic equations for $\mathbf{c}$ and $\mathbf{d}$. If it happens that this system has a non-trivial solution, then either (4.8.1) will have an infinity of solutions, or it will have a unique solution.

*4.8.2.* This follows from a mere substitution from (4.8.11).

**Section 4.9**

*4.9.1.* This follows by direct substitution.

*4.9.2.* We wish to investigate the eigenvalues $\mathbf{q}''$ of the homogeneous equation (4.9.16), where $\mathbf{V}''$ and $\mathbf{\Omega}''$ are given by (1) with $\mathbf{q}''$ in place of $\mathbf{q}$. Repeating the analysis that follows (4.9.16) we deduce that $\mathbf{q}''$ must express rigid body motion. Setting $\mathbf{q}'' = \mathbf{U} + \boldsymbol{\omega} \times \mathbf{X}$, substituting into (4.9.16), and using (1) we obtain

$$\mathbf{q}''\frac{4\pi}{\beta^*} = -4\pi\mathbf{q}'' + \varepsilon S_D \mathbf{U} - \zeta \int_D \mathbf{X}' \times (\boldsymbol{\omega} \times \mathbf{X}')\, dS(\mathbf{x}') \times \mathbf{X} \tag{2}$$

which provides us with a homogeneous system of linear equations for $\mathbf{U}$ and $\boldsymbol{\omega}$. Requiring that this system has non-trivial solutions, we obtain

$$\beta = \frac{1}{\dfrac{\varepsilon}{4\pi} S_D - 1} \tag{3}$$

and

$$\det\left\{\mathbf{I}+\beta\left[\left(1+\frac{\zeta}{4\pi}\int_D \mathbf{X}\cdot\mathbf{X}\,\mathrm{d}S(\mathbf{x})\right)\mathbf{I}-\frac{\zeta}{4\pi}\int_D \mathbf{X}\mathbf{X}\,\mathrm{d}S(\mathbf{x})\right]\right\}=0 \tag{4}$$

Equation (4) is a cubic algebraic equation for $\beta$. It will be noted that setting $\varepsilon = 4\pi/S_D$ we recover (4.9.11) and shift the eigenvalue provided by (3) to infinity.

## *Chapter 5*

### Section 5.2

*5.2.1.* This may be shown in a straightforward fashion using (2.3.12).

*5.2.2.* First, we decompose the velocity field $\mathbf{u}$ into the incident field $\mathbf{u}^\infty$ and the disturbance field $\mathbf{u}^D$ owing to the particle. The viscosities of the external and internal fields are $\mu_1$ and $\mu_2$ respectively. Note that the velocity of the incident field is continuous across the interface, but owing to the different viscosities of the external and internal flow, the corresponding stress fields are discontinuous. The discontinuity in the interfacial surface force of the incident field across the interface is given by

$$\Delta\mathbf{f}^\infty = \mathbf{f}^{\infty(1)} - \mathbf{f}^{\infty(2)} = (1-\lambda)\mathbf{f}^{\infty(1)} = \left(\frac{1}{\lambda}-1\right)\mathbf{f}^{\infty(2)} \tag{4}$$

Applying the reciprocal identity (2.3.4) for the internal incident flow $\mathbf{u}^\infty$ at a point $\mathbf{x}_0$ located outside the particle we obtain

$$\int_S f_i^{\infty(2)}(\mathbf{x})G_{ij}(\mathbf{x},\mathbf{x}_0)\,\mathrm{d}S(\mathbf{x}) - \mu_2\int_S u_i^\infty(\mathbf{x})T_{ijk}(\mathbf{x},\mathbf{x}_0)n_k(\mathbf{x})\,\mathrm{d}S(\mathbf{x}) = 0 \tag{5}$$

Using (4) we transform (5) into

$$-\frac{1}{8\pi\mu_1}\int_S \Delta f_i^\infty(\mathbf{x})G_{ij}(\mathbf{x},\mathbf{x}_0)\,\mathrm{d}S(\mathbf{x}) + \frac{1-\lambda}{8\pi}\int_S u_i^\infty(\mathbf{x})T_{ijk}(\mathbf{x},\mathbf{x}_0)n_k(\mathbf{x})\,\mathrm{d}S(\mathbf{x}) = 0 \tag{6}$$

Next, we apply (5.2.3) for the disturbance flow, add to the left-hand side the right-hand side of (6), and finally add the incident velocity field to both sides of the resulting equation to obtain (1).

To derive a boundary integral representation for the internal flow, we apply (2.3.11) for the incident flow at a point $\mathbf{x}_0$ located within the particle obtaining

$$u_j^\infty(\mathbf{x}_0) = \frac{1}{8\pi\mu_2}\int_S f_i^{\infty(2)}(\mathbf{x})G_{ij}(\mathbf{x},\mathbf{x}_0)\,\mathrm{d}S(\mathbf{x}) - \frac{1}{8\pi}\int_S u_i^\infty(\mathbf{x})T_{ijk}(\mathbf{x},\mathbf{x}_0)n_k(\mathbf{x})\,\mathrm{d}S(\mathbf{x}) \tag{7}$$

Using (4) and rearranging we cast (7) into the equivalent form

$$u_j^{\infty}(\mathbf{x}_0) = \frac{1}{\lambda} u_j^{\infty}(\mathbf{x}_0) - \frac{1}{8\pi\mu_1\lambda} \int_S \Delta f_i^{\infty}(\mathbf{x}) G_{ij}(\mathbf{x}, \mathbf{x}_0)\, dS(\mathbf{x})$$
$$+ \frac{1}{8\pi}\left(\frac{1-\lambda}{\lambda}\right) \int_S u_i^{\infty}(\mathbf{x}) T_{ijk}(\mathbf{x}, \mathbf{x}_0) n_k(\mathbf{x})\, dS(\mathbf{x}) \tag{8}$$

Finally, applying (5.2.8) for the internal disturbance flow and adding it to (8) yields the required boundary integral representation. Equation (2) follows from (1) with the help of (2.3.12).

Using (1) and repeating the analysis of section 2.5 we find that the far flow may be represented by the approximate expansion (3).
*5.2.3.* The validity of this expression may readily be shown using the identity

$$\int_{S_P} [f_i^{(2)} x_j - \delta_{ij} \tfrac{1}{3} f_l^{(2)} x_l - \mu_2 (u_i^{(2)} n_j + u_j^{(2)} n_i)]\, dS = 0$$

whose truth is readily established using the divergence theorem.

**Section 5.3**

*5.3.1.* The proof is straightforward. Note that for a force-free and torque-free particle the right-hand sides of these equations vanish, and the far flow is similar to that generated by a symmetric point force dipole.
*5.3.2.* This equation may be derived in a manner completely analogous to that indicated in section 5.3, with the aid of (4.1.7) and (4.1.8).

**Section 5.4**

*5.4.1.* The formulation is considerably simplified by using a two-dimensional Green's function that vanishes over the plane wall (see section 3.5). The result is an equation identical to (5.2.9) or (5.3.9), except that the coefficients $1/4\pi$ in front of the integrals must be replaced by $1/2\pi$. If the interface is periodic, it is appropriate to use a periodic Green's function that vanishes over the wall, as discussed in section 3.5. In this case, the domain of the integral equations is reduced simply to one period of the interface.
*5.4.2.* We label the upper and lower fluid using the indices 1 and 2, respectively. The boundary integral formulation leads to two integral equations

for the interfacial velocity and the surface force on the particle, namely

$$u_j(\mathbf{x}_0) = -\frac{1}{4\pi\mu_1}\left(\frac{1}{\lambda+1}\right)\int_{\text{interface}} \Delta f_i(\mathbf{x}) G_{ij}(\mathbf{x}, \mathbf{x}_0)\, dS(\mathbf{x})$$

$$+\frac{\kappa}{4\pi}\int_{\text{interface}}^{\mathcal{PV}} u_i(\mathbf{x}) T_{ijk}(\mathbf{x}, \mathbf{x}_0) n_k(\mathbf{x})\, dS(\mathbf{x})$$

$$-\frac{1}{4\pi\mu_1}\left(\frac{1}{\lambda+1}\right)\int_{\text{particle}} f_i(\mathbf{x}) G_{ij}(\mathbf{x}, \mathbf{x}_0)\, dS(\mathbf{x})$$

when $\mathbf{x}_0$ is on the interface and

$$U_i + \varepsilon_{ijk}\Omega_j x_{0,k} = -\frac{1}{8\pi\mu_1\lambda}\int_{\text{interface}} \Delta f_i(\mathbf{x}) G_{ij}(\mathbf{x}, \mathbf{x}_0)\, dS(\mathbf{x})$$

$$+\frac{1}{8\pi}\left(\frac{1-\lambda}{\lambda}\right)\int_{\text{interface}} u_i(\mathbf{x}) T_{ijk}(\mathbf{x}, \mathbf{x}_0) n_k(\mathbf{x})\, dS(\mathbf{x})$$

$$-\frac{1}{8\pi\mu_1\lambda}\int_{\text{particle}} f_i(\mathbf{x}) G_{ij}(\mathbf{x}, \mathbf{x}_0)\, dS(\mathbf{x})$$

when $\mathbf{x}_0$ is on the particle surface. Here $\mathbf{U}$ and $\boldsymbol{\Omega}$ are the velocities of translation and rotation of the particle respectively.

*5.4.3.* To formulate this problem we decompose the total flow into the undisturbed parabolic flow that prevails far upstream, and the disturbance component due to the projection. The disturbance component satisfies the equations of unforced Stokes flow. Using the boundary integral formulation we derive an integral representation of the disturbance flow in terms of boundary integrals over the free surface and the projection. The unknowns are the velocity at the free surface and the surface force over the projection. The integrals over the wall vanish because the disturbance velocity as well as the Green's function vanish over the wall.

Considering next flow over a wall with a depression, we find that it is not possible to use the Green's function that vanishes on the wall. The reason is that this Green's function introduces singularities within the domain of flow when its pole is located beneath the wall.

**Section 5.5**

*5.5.1.* Simply substitute $u_k = \varepsilon_{kij} F_j$ into Stokes' theorem.

## *Chapter 6*

**Section 6.2**

*6.2.1.* This is a computer exercise.

*6.2.2.* This becomes evident by noting that the radius of curvature of the trace of the interface in a plane perpendicular to an azimuthal plane as well as to the interface is simply equal to the length of the straight segment subtended between a point on the trace of the interface and the intersection between the normal to the trace of the interface and the $x$-axis (see (2.6.34)).
*6.2.3.* This is a computer exercise.
*6.2.4.* We introduce a parametric representation in which $\eta = \phi$ and $\xi = x$, and write $\mathbf{x} = (x, g(x)\cos\phi, g(x)\sin\phi)$. Using (6.2.12) we compute the unit normal vector

$$\mathbf{n} = \frac{1}{(1 + g'^2)^{1/2}}[-g', \cos\phi, \sin\phi]$$

Furthermore, using (6.2.18) and (6.2.23) we compute $a_{\eta\eta} = g^2$, $a_{\eta\xi} = 0$, $a_{\xi\xi} = 1 + g'^2$, and

$$k(0) = \frac{1}{g}\frac{1}{(1 + g'^2)^{1/2}} \qquad k(\infty) = -\frac{g''}{(1 + g'^2)^{3/2}}$$

The mean curvature of the interface is equal to the average value of $k(0)$ and $k(\infty)$.
*6.2.5.* Considering (6.2.14) we compute

$$\frac{\partial \mathbf{v}}{\partial \eta} = \dot{a}\frac{\partial \mathbf{x}}{\partial \eta} \qquad \frac{\partial \mathbf{v}}{\partial \xi} = \dot{a}\frac{\partial \mathbf{x}}{\partial \xi}$$

where $\dot{a} = \mathrm{d}a/\mathrm{d}t$. Thus,

$$\Delta = \frac{\partial h_n}{\partial t} = 2\dot{a}\,\mathbf{n}\cdot\left(\frac{\partial \mathbf{x}}{\partial \eta} \times \frac{\partial \mathbf{x}}{\partial \xi}\right) = 2\dot{a}h_n$$

from which we conclude that

$$h_n = h_n(0)\mathrm{e}^{2\dot{a}t}$$

*6.2.6.* An ellipsoid is described in parametric form by the equations $x = a\cos\eta$, $y = b\sin\eta\cos\xi$, $z = c\sin\eta\sin\xi$, where $0 < \eta < \pi$, $0 < \xi < 2\pi$, and $a$, $b$, $c$ are the three semi-axes. Note that when the ellipsoid reduces to a sphere, $a = b = c$ and $\eta$ and $\xi$ represent the meridional and azimuthal angle respectively. The system $(\eta, \xi)$ is orthogonal only in the special case where the ellipsoid degenerates into a spheroid, i.e. $b = c$. Using (6.2.12) we compute

$$\mathbf{n} = \frac{\left[\dfrac{\cos\eta}{a}, \dfrac{\sin\eta}{b}\cos\xi, \dfrac{\sin\eta}{c}\sin\xi\right]}{\left[\left(\dfrac{\cos\eta}{a}\right)^2 + \left(\dfrac{\sin\eta}{b}\right)^2\cos^2\xi + \left(\dfrac{\sin\eta}{c}\right)^2\sin^2\xi\right]^{1/2}}$$

The computation of the mean curvature may be effected in a tedious yet straightforward fashion using (6.2.25).

*6.2.7.* If we divide a sphere into eight triangles defined by the cross-section of the sphere with the planes $\phi = 0$, $\phi = \pi/2$, and $\theta = \pi/2$, we shall count $F = 8$, $E = 12$, $V = 6$ and thus, $\chi = F - E + V = 2$. Similarly, if we divide a torus into four cylindrical manifolds, and divide further each manifold into four triangles, we shall count $F = 16$, $E = 24$, $V = 8$ and thus, $\chi = F - E + V = 0$.

*6.2.8.* These are simple applications of the various formulae presented in the text.

*6.2.9.* This is a computer exercise.

*6.2.10.* The sum of the curvature of two perpendicular lines that form angles $\alpha$ and $\alpha + \pi/2$ with respect to the line of maximum curvature is

$$k_{max} \cos^2\alpha + k_{min} \sin^2\alpha + [k_{max} \cos^2(\alpha + \pi/2) + k_{min} \sin^2(\alpha + \pi/2)]$$
$$= k_{max} + k_{min}$$

which is equal to twice the mean curvature.

**Section 6.4**

*6.4.1.* We simply set $f(\mathbf{x}) = 1$ and thus $g(\eta, \xi) = h_n$ (see (6.2.3)), and apply the Gaussian quadrature illustrated in Figure 6.4.1.

*6.4.2.* To desingularize this integral we write

$$I_j^S(\mathbf{x}_0) = \int_{P_{i-1}}^{P_{i+1}} [q_i(\mathbf{x}) G_{ij}(\mathbf{x}, \mathbf{x}_0) + q_j(\mathbf{x}_0) \ln|\mathbf{x} - \mathbf{x}_0|] \, dl(\mathbf{x})$$
$$- q_j(\mathbf{x}_0) \int_{P_{i-1}}^{P_{i+1}} \ln|\mathbf{x} - \mathbf{x}_0| \, dl(\mathbf{x})$$

The last integral is equal to

$$\int_{P_{i-1}}^{P_{i+1}} \ln|\mathbf{x} - \mathbf{x}_0| \, dl(\mathbf{x}) = |\mathbf{x}_{i+1} - \mathbf{x}_0| \ln|\mathbf{x}_{i+1} - \mathbf{x}_0|$$
$$+ |\mathbf{x}_{i-1} - \mathbf{x}_0| \ln|\mathbf{x}_{i-1} - \mathbf{x}_0| - |\mathbf{x}_{i+1} - \mathbf{x}_{i-1}|$$

## Chapter 7

**Section 7.2**

*7.2.1.* The validity of this relation may be shown in a straightforward manner using (7.2.26) and carrying out the required integrations. Note that it will be necessary to use the identity

$$\int_{\text{sphere}} \hat{x}_i \hat{x}_j \hat{x}_k \hat{x}_l \, dS = \frac{4\pi}{15} a^6 (\delta_{ij}\delta_{kl} + \delta_{ik}\delta_{jl} + \delta_{il}\delta_{jk})$$

where $\hat{\mathbf{x}} = \mathbf{x} - \mathbf{x}_0$, $\mathbf{x}_0$ is the center and $a$ is the radius of the sphere.

The symmetry of the first moment tensor implies that the torque exerted on a spherical surface centered at the pole of the stresslet (and thus, the torque exerted on any other surface) is equal to zero.
*7.2.2.* The proof requires a straightforward computation using (7.2.23).
*7.2.3.* See problem 2.5.1.
*7.2.4.* The velocity produced by a singularity of bounded flow is infinite at the pole and must vanish over the solid boundaries of the flow. Placing the pole on a solid boundary makes these two restrictions incompatible and imposes the requirement that the flow due to the singularity vanishes throughout the flow.
*7.2.5.* Considering the asymptotic limit of (7.2.32) as the point $\mathbf{x}$ moves away from the pole, i.e. $|\mathbf{x} - \mathbf{x}_0| \gg h_0$, we find

$$\Sigma_i^{W} \approx 6 \frac{\hat{X}_i \hat{X}_1^2}{R^5}$$

Physically, this velocity field represents the flow produced by the suction or discharge of fluid through a hole of infinitesimal dimensions.

**Section 7.3**

*7.3.1.* The flow due to the radial expansion of an air bubble may be represented simply by a point source. If $a(t)$ is the radius of the bubble, then the velocity field is

$$\mathbf{u} = \frac{\mathrm{d}a}{\mathrm{d}t}\left(\frac{a}{r}\right)^2 \frac{\mathbf{x}}{r}$$

Using (7.2.2) we find that the normal stress on the external surface of the bubble is

$$\mathbf{f} \cdot \mathbf{n} = -\frac{4\mu}{a}\frac{\mathrm{d}a}{\mathrm{d}t}$$

If $P_B$ is the pressure within the bubble, then $\mathbf{f} \cdot \mathbf{n} = -P_B + 2\gamma/a$, where $\gamma$ is the surface tension, and we obtain

$$\frac{\mathrm{d}a}{\mathrm{d}t} = \frac{1}{4\mu}(P_B a - 2\gamma)$$

*7.3.2.* We consider the disturbance flow $\mathbf{u}'$ due to a spherical bubble of radius $a$ embedded in the purely straining flow $\mathbf{u}^\infty = \mathbf{E} \cdot \mathbf{x}$, where $\mathbf{E}$ is a symmetric traceless matrix. The total flow is $\mathbf{u} = \mathbf{u}^\infty + \mathbf{u}'$. For convenience, we set the origin at the center of the bubble.

First, we require that the normal component of the velocity and the shear stress on the surface of the bubble are equal to zero. Thus, we obtain

$$u_i'n_i = -\frac{1}{a}E_{ij}x_ix_j \tag{1}$$

$$\sigma'_{ik}n_k(\delta_{im} - n_in_m) = -2\frac{\mu}{a}E_{ij}x_j\left(\delta_{im} - \frac{x_ix_m}{a^2}\right) \tag{2}$$

at the surface of the bubble.

Next, we postulate that the external disturbance velocity field may be represented by a Stokeslet doublet and a potential quadruple, as in (7.3.11). After some algebra we find that the left-hand sides of (1) and (2) are equal to

$$u_i'n_i = \frac{3}{a^4}\left(d_{ij} + \frac{3}{a^2}\varphi_{ij}\right)x_ix_j - \frac{1}{a^2}\left(d_{ii} + \frac{3}{a^2}\varphi_{ll}\right) \tag{3}$$

$$\sigma'_{ik}n_k(\delta_{im} - n_in_m) = \frac{6}{a^4}\left(d_{ij} + \frac{8}{a^2}\varphi_{ij}\right)x_j\left(\delta_{im} - \frac{x_ix_m}{a^2}\right) \tag{4}$$

Substituting these expressions into (1) and (2) yields $\mathbf{d} = -a^3\mathbf{E}/3$ and $\varphi = 0$ stating that the flow may be represented simply in terms of a Stokeslet doublet. Straightforward application of (2.5.32) yields the Faxen relation for the stresslet

$$\mathcal{S}_{ij} = \frac{4}{3}\pi\mu a^3\left(\frac{\partial u_j^\infty}{\partial x_i} + \frac{\partial u_i^\infty}{\partial x_j}\right)_0 \tag{5}$$

where the subscript 0 indicates evaluation at the center of the bubble.

It will be noted that (5) may be considered as a special case of the Faxen relation for a spherical viscous drop, arising in the limit as the viscosity of the drop tends to zero. The more general expression is

$$\mathcal{S}_{ij} = \frac{2}{3}\pi\mu a^3\left(\frac{2+5\lambda}{1+\lambda}\right)\left[1 + \frac{\lambda}{2(5\lambda+2)}\nabla^2\right]_0\left(\frac{\partial u_j^\infty}{\partial x_i} + \frac{\partial u_i^\infty}{\partial x_j}\right) \tag{6}$$

where $\lambda$ is the ratio of the viscosities of the drop and the ambient fluid (Rallison 1978). The functional form of (6) suggests that the disturbance flow exterior to a drop placed in a purely straining incident flow may be represented in terms of a Stokeslet doublet and a potential quadruple with strengths

$$\mathbf{d} = -\frac{1}{6}\left(\frac{2+5\lambda}{1+\lambda}\right)a^3\mathbf{E} \qquad \varphi = \frac{1}{6}a^3\frac{\lambda(2+5\lambda)}{(1+\lambda)(5\lambda+2)}\mathbf{E}$$

*7.3.3.* This requires a straightforward computation.

*7.3.4.* A straightforward computation reveals that

$$\mathbf{F} = -8\pi\mu\mathbf{c}\frac{1}{\pi^2}\int_D \frac{1}{(a^2-\sigma^2)^{1/2}}\,dA = -16\pi\mu a\,\mathbf{c}$$

*7.3.5.* This flow may be represented by a ring of rotlets oriented in the azimuthal direction. Introducing cylindrical polar coordinates $(x, \sigma, \phi)$, we write

$$\mathbf{u}(\mathbf{x}_0) = am \int_0^{2\pi} \frac{\mathbf{e}_\phi \times \hat{\mathbf{x}}}{|\mathbf{x}|^3} \, d\phi$$

where $a$ is the radius of the ring and $m$ is the strength of the rotlets. The axial and radial components of the velocity are

$$u_x = \frac{m}{a}\left(I_0 - \frac{\sigma}{a} I_1\right) \qquad u_\sigma = \frac{m}{a^2} \hat{x} I_1$$

where

$$I_0 = \frac{4}{[\hat{x}^2 + (\sigma + a)^2]^{3/2}} \left(\frac{1}{1 - k^2}\right) E(k),$$

$$I_1 = \frac{4}{[\hat{x}^2 + (\sigma + a)^2]^{3/2}} \left[\frac{2 - k^2}{k^2(1 - k^2)} E(k) - \frac{2}{k^2} F(k)\right]$$

and $k^2$, $F$, $E$ were defined in (2.4.8) and (2.4.10). Close to the center line, the flow is similar to that produced by the rolling of a slender torus of circular cross-section with radius

$$\left[\frac{2m}{\omega}\left(1 + \frac{5m}{8\omega a^2}\right)\right]^{1/2}$$

where $\omega$ is the angular velocity of the rolling motion.

**Section 7.4**

*7.4.1.* An elementary computation shows that

$$\langle \mathbf{u} \rangle \equiv \frac{1}{4\pi a^2} \int_{\text{sphere}} \mathbf{u}(\mathbf{x}) \, dS(\mathbf{x}) = \frac{4}{3a} \mathbf{g}$$

where $a$ is the radius of the sphere. Setting $\langle \mathbf{u} \rangle = \mathbf{V}$ yields $\mathbf{g} = 3a\mathbf{V}/4$. The force exerted on the sphere is $\mathbf{F} = -8\pi\mu\mathbf{g} = -6\pi\mu a\mathbf{V}$ in perfect agreement with Stokes' law.

For a sphere held in the paraboloidal flow $\mathbf{u}^P = U/a^2[y^2 + z^2, 0, 0,]$ we require $\langle \mathbf{u} \rangle + \langle \mathbf{u}^P \rangle = 0$. Noting that $\langle \mathbf{u}^P \rangle = -2U/3\ [1, 0, 0]$ we obtain $\mathbf{g} = -aU/2\ [1, 0, 0]$, and compute $\mathbf{F} = -8\pi\mu\mathbf{g} = 4\pi\mu a U\ [1, 0, 0]$ in perfect agreement with Faxen's law (see problem 1.4.4).

*7.4.2.* Combining Burgers' approximate solution for flow due to a translating sphere with the generalized Faxen relation (2.5.28) we obtain the approximate expression $\mathbf{F} = 6\pi\mu\mathbf{u}^\infty(\mathbf{x}_0)$. Comparing this relation with the exact relation given in (2.5.36) we notice the absence of the term involving the Laplacian of the velocity.

**Section 7.5**

*7.5.1.* This may be shown by performing a straightforward asymptotic expansion.

*7.5.2.* This flow may be represented by an unsteady Stokeslet and an unsteady symmetric Stokeslet quadruple with strengths

$$\mathbf{g} = \frac{1}{4}(3 + 3\lambda + \lambda^2)\mathbf{V}, \qquad \mathscr{d} = \frac{1}{2\lambda^2}(3 + 3\lambda + \lambda^2 - 3e^{\lambda})\mathbf{V}$$

where $\mathbf{V}$ is the velocity of the oscillation, and the frequency parameter $\lambda$ is defined with respect to the radius $a$ of the sphere. Using (2.5.28) we derive the Faxen relation for the force on a sphere in dimensional form:

$$\mathbf{F} = 2\pi\mu a\left[3 + 3\lambda + \lambda^2 - \frac{1}{2\lambda^2}(3 + 3\lambda + \lambda^2 - 3e^{\lambda})\nabla^2\right]_0 \mathbf{u}^\infty$$

*7.5.3.* This may be proven in a straightforward manner using the defining expression (7.2.22).

The flow due to a sphere oscillating with angular velocity $\Omega$ may be represented simply by a rotlet of strength $\Omega e^{\lambda}/(\lambda + 1)$ placed at the center of the sphere, where $\lambda$ is defined with respect to the radius $a$ of the sphere. Using (2.5.30) we derive the Faxen relation for the torque on a sphere. In dimensional form:

$$\mathbf{F} = 4\pi\mu a^3\left(\frac{e^{\lambda}}{1 + \lambda}\right)\nabla \times \mathbf{u}^\infty$$

*7.5.4.* This is a consequence of Faxen's law (2.5.28). In a frame of reference moving with the particle, an observer sees a uniform incident flow $\mathbf{u} = -\mathbf{V}$. According to the dimensional version of (2.5.28), the force exerted on the particle is equal to $-8\pi\mathbf{g}$. Adding to this the inertial force $\lambda^2 V_p \mathbf{V}$ due to the acceleration of the frame of reference we obtain the desired result.

# *References*


Abramowitz, M. & Stegun, I. A. (1972), *Handbook of Mathematical Functions*, Dover.

Aris, R. (1989), *Vectors, Tensors, and the Basic Equations of Fluid Mechanics*, Dover.

Atkinson, K. E. (1973), Iterative variants of the Nyström method for the numerical solution of integral equations, *Numer. Math.* **22**, 17–31.

Atkinson, K. E. (1976), *A Survey of Numerical Methods for the Solution of Fredholm Integral Equations of the Second Kind*, SIAM.

Atkinson, K. E. (1980), The numerical solution of Laplace's equation in three dimensions II. In *Numerical Treatment of Integral Equations*, Bikhäuser Verlag.

Atkinson, K. E. (1990), A survey of boundary integral methods for the numerical solution of Laplace's equation in three dimensions. In *Numerical Solution of Integral Equations*, ed. M. A. Goldberg, Plenum Press.

Avudainayagam, A. & Jothiram, B. (1987), No-slip images of certain line singularities in a circular cylinder, *Int. J. Eng. Sci.* **25**, 1193–205.

Baker, C. T. H. (1977), *The Numerical Treatment of Integral Equations*, Clarendon Press.

Bakr, A. A. (1986), The boundary integral equation method in axisymmetric stress analysis problems, *Lecture Notes in Engineering*, **14**, Springer-Verlag.

Banerjee, P. K. & Butterfield, R. (1981), *Boundary Element Methods in Engineering Science*, McGraw-Hill.

Barshinger, R. N. & Geer, J. F. (1984), Stokes flow past a thin oblate body of revolution: axially incident uniform flow. *SIAM J. Appl. Math.* **44**, 19–32.

Batchelor, G. K. (1967), *An Introduction to Fluid Dynamics*, Cambridge University Press.

Batchelor, G. K. (1970a), The stress system in a suspension of force-free particles, *J. Fluid Mech.* **41**, 545–70.

Batchelor, G. K. (1970b), Slender body theory for particles of arbitrary cross-section in Stokes flow, *J. Fluid Mech.* **44**, 419–40.

Batchelor, G. K. (1972), Sedimentation in a dilute suspension of spheres, *J. Fluid Mech.* **52**, 245–68.

Batchelor, G. K. & Green, J. T. (1972), The hydrodynamic interaction of two small freely moving spheres in a linear flow field, *J. Fluid Mech.* **56**, 375–400.

Bathe, K-J. (1982), *Finite Element Procedures in Engineering Analysis*, Prentice-Hall.

Bézine, G. & Bonneau, D. (1981), Integral equation method for the study of two-dimensional Stokes flow, *Acta Mech.* **41**, 197–209.

Bhattacharya, P. K. & Symm, G. T. (1984), New formulation and solution of the plane elastostatic traction problem, *Appl. Math. Modelling* **8**, 226–30.

Blake, J. R. (1971) A note on the image system for a Stokeslet in a no-slip boundary, *Proc. Camb. Phil. Soc.* **70**, 303–10.

Blake, J. R. (1972), A model for the micro-structure in ciliated organisms, *J. Fluid Mech.* **55**, 1–23.

Blake, J. R. (1979), On the generation of viscous toroidal eddies in a cylinder, *J. Fluid Mech.* **95**, 109–22.

Blake, J. R. & Chwang, A. T. (1974), Fundamental singularities of viscous flow, *J. Engin. Math.* **8**(1), 23–9.

Bodewig, E. (1959), *Matrix Calculus*, North-Holland.

Brady, J. F. & Bossis, G. (1988), Stokesian dynamics, *Ann. Rev. Fluid Mech.* **20**, 111–57.

Brenner, H. (1963), The Stokes resistance of an arbitrary particle, *Chem. Eng. Science* **18**, 1–25.

Brenner, H. (1964a), The Stokes resistance of an arbitrary particle, II. An extension, *Chem. Eng. Science* **19**, 599–629.

Brenner, H. (1964b), The Stokes resistance of an arbitrary particle, IV. Arbitrary fields of flow, *Chem. Eng. Science* **19**, 703–27.

Burgers, J. M. (1938), On the motion of small particles of elongated form, suspended in a viscous liquid. In *Second Report on Viscosity and Plasticity*, Kon. Ned. Akad. Verhand. (Eerste sectie), DI. XVI, No. 4.

Burgess, G. & Mahajerin, E. (1984), Rotational fluid flow using a least squares collocation technique, *Computers & Fluids* **12**, 311–17.

Carnahan, B., Luther, H. A. & Wilkes, J. O. (1969), *Applied Numerical Methods*, Wiley.

Chan, P. C., Leu, R. J. & Zargar, N. H. (1986), On the solution for the rotational motion of an axisymmetric rigid body at low Reynolds number with application to a finite length cylinder. *Chem. Eng. Commun.* **49**, 145–63.

Chwang, A. T. (1975), Hydromechanics of low-Reynolds-number flow, Part 3. Motion of a spheroidal particle in quadratic flows, *J. Fluid Mech.* **72**, 17–34.

Chwang, A T. & Hwang, W.-S. (1990), Rotation of a torus, *Phys. Fluids* **A 1**, 1309–11.

Chwang, A. T. & Wu, T. Y.-T. (1974), Hydromechanics of low-Reynolds-number flow. Part 1, Rotation of axisymmetric prolate bodies, *J. Fluid Mech.* **63**, 607–22.

Chwang, A. T. & Wu, T. Y.-T. (1975), Hydromechanics of low-Reynolds-number flow. Part 2, Singularity methods for Stokes flows, *J. Fluid Mech.* **67**, 787–815.

Coleman, C. J. (1980), A contour integral formulation of plane creeping Newtonian flow, *Quart. J. Mech. Appl. Math.* **34**, 453–64.

Collins, W. D. (1962), The forced torsional oscillations of an elastic half-space and an elastic stratum, *Proc. London Math. Soc.* **12**, 226–44.

Cox, R. G. (1970), The motion of long slender bodies in a viscous fluid. Part 1, General theory, *J. Fluid Mech.* **44**, 791–810.

Dąbroś, T. A. (1985), Singularity method for calculating hydrodynamic forces and particle velocities in low-Reynolds-number flows, *J. Fluid Mech.* **156**, 1–21.

Davey, K. & Hinduja, S. (1988), Analytical integration of linear three-dimensional triangular elements in BEM, *Appl. Math. Modelling* **13**, 450–61.

Davis, H. T. (1962), *Introduction to Nonlinear Differential and Integral Equations*, Dover.

Dean, W. R. & Montagnon, P. E. (1949), On the steady motion of viscous liquid in a corner, *Proc. Camb. Phil. Soc.* **45**, 389–94.

De Boor, C. (1978), *A Practical Guide to Splines*, Springer-Verlag.

Delves, L. M. & Mohamed, J. L. (1985), *Computational Methods for Integral Equations*, Cambridge University Press.

Dorrepaal, J. M., O'Neill, M. E., & Ranger, K. B. (1984), Two-dimensional Stokes flows with cylinders and line singularities, *Mathematika* **31**, 65–75.

Dritschel, D. G. (1989), Contour dynamics and contour surgery: numerical algorithms for extended, high-resolution modelling of vortex dynamics in two-dimensional, inviscid, incompressible flows, *Comp. Phys. Rep.* **10**, 77–146.

Einstein, A. (1906), Eine neue Bestimmung der Moleküldimensionen, *Ann. Phys.* **69**, 1352–60.

Einstein, A. (1956), *The Theory of Brownian Movement*, Dover.

Evans, E. A. & Skalak, R. (1980), *Mechanics and Thermodynamics of Biomembranes*. CRC Press.

Fairweather, G., Rizzo, F. G. & Shippy, D. J. (1979), Computation of double integrals in the boundary integral equation method, *Adv. Comp. Meth. Partial Diff. Eq. III*, IMACS.

Faxen, H. (1924), Der Widerstand gegen die bewegung einer starren Kugel in einer zähen Flüssigkeit, die zwischen zwei parallelen, ebenen Wänden eingeschlossen ist, *Arkiv Mat. Astron. Fys.* **18** (29), 1–52.

Garabedian, P. R. (1964), *Partial Differential Equations*, Wiley.

Gavze, E. (1990), A boundary integral equation solution of the Stokes flow due to the motion of an arbitrary body near a plane wall with a hole, *Int. J. Multiphase Flow* **16**, 529–43.

Geer, J. (1976), Stokes flow past a slender body of revolution, *J. Fluid Mech.* **78**, 577–600.

Goldstein, H. (1980), *Classical Mechanics*, Addison-Wesley.

Golub, G. E. & Van Loan, C. F. (1989), *Matrix Computations*, The Johns Hopkins University Press.

Gradshteyn, I. S. & Ryshik, I. M. (1980), *Tables of Integrals, Series, and Products*, Academic Press.

Gray, J. & Hancock, G. J. (1955), The propulsion of sea-urchin spermatozoa, *J. Exp. Biol.* **32**, 802–14.

Green, A. E. & Adkins, J. E. (1960), *Large Elastic Deformations*, Oxford University Press.

Hackborn, W. W. (1990), Asymmetric Stokes flow between parallel planes due to a rotlet, *J. Fluid Mech.* **218**, 531–46.

Hackborn, W. W., O'Neill, M. E. & Ranger, K. B. (1986), The structure of an asymmetric Stokes flow, *Quart. J. Mech. Appl. Math.* **39**, 1–14.

Hancock, G. J. (1953), The self-propulsion of microscopic organisms through liquids, *Proc. Roy. Soc. A* **217**, 96–121.

Hansen, E. B. (1987), Stokes flow down a wall into an infinite pool, *J. Fluid Mech.* **178**, 243–56.

Happel, J. & Brenner, H. (1973), *Low Reynolds Number Hydrodynamics*, Martinus Nijhoff.

Hasimoto, H. (1959), On the periodic fundamental solutions of the Stokes equations and their application to viscous flow past a cubic array of spheres, *J. Fluid Mech.* **5**, 317.

Hasimoto, H. (1976), Slow motion of a small sphere in a cylindrical domain. *J. Phys. Soc. Japan* **41**, 2143–4; errata in **42**, 1047.

Hasimoto, H. & Sano, O. (1980), Stokeslets and eddies in creeping flow, *Ann. Rev. Fluid Mech.* **12**, 335–63.

Hasimoto, H., Kim, M.-U. & Miyazaki, T. (1983), The effect of a semi-infinite plane on the motion of a small particle in a viscous fluid, *J. Phys. Soc. Japan* **52**, 1996–2003.

Hebeker, F.-K. (1986), Efficient boundary element methods for three-dimensional exterior viscous flow, *Num. Meth. PDE* **2**, 273–97.

Hetsroni, G. & Haber, S. (1970), Flow in and around a droplet or bubble submerged in an unbounded arbitrary velocity field, *Rheol. Acta* **9**, 488–96.

Higdon, J. J. L. (1979), A hydrodynamic analysis of flagellar propulsion, *J. Fluid Mech.* **90**, 685–711.

Higdon, J. J. L. (1985), Stokes flow in arbitrary two-dimensional domains: shear flow over ridges and cavities, *J.* Fluid Mech. **159**, 195–226.

Higdon, J. J. L. & Pozrikidis, C. (1985), The self-induced motion of vortex sheets, *J. Fluid Mech.* **150**, 201–31.

Hill, R. & Power, G. (1956), Extremum principles for slow viscous flow and the approximate calculation of drag, *Quart. J. Mech. Appl. Math.* **9**, 313–19.

Hocquart, R. & Hinch, E. J. (1983), The long-time tail of the angular-velocity autocorrelation function for a rigid Brownian particle of arbitrary centrally symmetric shape, *J. Fluid Mech.* **137**, 217–20.

Howell, I. D. (1974), Drag due to the motion of a Newtonian fluid through a sparse random array of small fixed rigid objects, *J. Fluid Mech.* **64**, 449–75.

Huang, L. H. & Chwang, A. T. (1986), Hydromechanics of low-Reynolds-number flow. Part 6. Rotation of oblate bodies. *J. Eng. Math.* **20**, 307–22.

Ingber, M. S. & Mitra, A. K. (1986), Grid optimization for the boundary element method, *Int. J. Num. Meth. Eng.* **23**, 2121–36.

Ingham, D. B. & Kelmanson, M. A. (1984), Boundary integral equation analyses of singular, potential, and biharmonic problems, *Lecture Notes in Engineering* **7**, Springer-Verlag.

Jaswon, M. A. & Symm, G. T. (1977), *Integral Equation Methods in Potential Theory and Elastostatics*, Academic Press.

Jeffrey, D. J. & Sherwood, J. D. (1980), Streamline patterns and eddies in low Reynolds number flow, *J. Fluid Mech.* **96**, 315–34.

Kantorovich, L. V. & Krylov, V. I. (1958), *Approximate Methods of Higher Analysis*, Wiley Interscience.

Kanwal, R. P. (1971), *Linear Integral Equations*, Academic Press.

Karageorghis, A. & Fairweather, G. (1987), The method of fundamental solutions for the numerical solution of the biharmonic equation, *J. Comp. Phys.* **69**, 434–59.

Karageorghis, A. & Fairweather, G. (1988), The Almansi method of fundamental solutions for solving biharmonic problems, *Int. J. Num. Meth. Fluids* **26**, 1665–82.

Karageorghis, A. & Fairweather, G. (1989), The simple layer potential method of fundamental solutions for certain biharmonic problems, *Int. J. Num. Meth. Fluids* **9**, 1221–34.

Karrila, S. J., Fuentes, Y. O. & Kim, S. (1989), Parallel computational strategies for hydrodynamic interactions between rigid particles of arbitrary shape in a viscous fluid, *J. Rheology* **33**, 913–47.

Keller, J. B. & Rubinow, A. I. (1976), Slender-body theory for slow viscous flow, *J. Fluid Mech.* **75**, 705–14.

Kellogg, O. D. (1954), *Foundations of Potential Theory*, Dover.

Kelmanson, M. A. (1983a), Modified integral equation solution of viscous flows near sharp corners, *Computers & Fluids* **11**, 307–24.

Kelmanson, M. A. (1983b), An integral equation method for the solution of singular slow flow problems, *J. Comp. Phys.* **51**, 139–58.

Kelmanson, M. A. (1983c), Boundary integral equation solution of viscous flows with free surfaces, *J. Eng. Math.* **17**, 329–43.

Kennedy, M. (1991), *Numerical Investigations of the Dynamics of Three-dimensional Viscous Drops*, Ph. D. Thesis, University of California at San Diego.

Kennedy, M., Pozrikidis, C. & Skalak, R. (1991) On the behaviour of liquid drops and the rheology of dilute suspensions undergoing simple shear flow, typescript, University of California, San Diego.

Kim, M. (1979), Slow viscous flow due to the motion of a sphere on the axis of a circular cone, *J. Phys. Soc. Japan* **46**, 1929–34.

Kim, S. (1985), A note on Faxen laws for nonspherical particles, *Int. J. Multiphase Flow* **11**, 713–19.

Kim, S. (1986), Singularity solutions for ellipsoids in low-Reynolds-number flows: With applications to the calculation of hydrodynamic interactions in suspensions of ellipsoids, *Int. J. Multiphase Flow* **12**, 469–91.

Kim, S. & Arunachalam, P. V. (1987), The general solution for an ellipsoid in low-Reynolds-number flow, *J. Fluid Mech.* **178**, 535–47.

Kim, S. & Karrila, S. J. (1991), *Microhydrodynamics: Principles and Selected Applications*, Butterworth.

Kim, S. & Lu, S.-Y. (1987), The functional similarity between Faxen relations and singularity solutions for fluid–fluid, fluid–solid and solid–solid dispersions, *Int. J.Multiphase Flow* **13**, 837–44.

Kojima, M., Hinch, E. J. & Acrivos, A. (1984), The formation and expansion of a toroidal drop moving in a viscous fluid, *Phys. Fluids* **27**(1), 19–32.

Kress, R. (1989), *Linear Integral Equations*, Springer-Verlag.

Kupradze, V. (1967), On the approximate solution of problems in mathematical physics. *Russ. Math. Surveys* **22**, 59–107.

Ladyzhenskaya, O. A. (1969), *The Mathematical Theory of Viscous Incompressible Flow*, Gordon & Breach.

Lamb, H. (1932), *Hydrodynamics*, Dover.

Landau, L. D. & Lifshitz, E. M. (1987) *Fluid Mechanics*, Pergamon Press.

Langlois, W. E. (1964), *Slow Viscous Flow*, Macmillan.

Larson, R. E. & Higdon, J. J. L. (1987), Microscopic flow near the surface of two-dimensional porous media. Part 2, Transverse flow, *J. Fluid Mech.* **178**, 119–36.

Lawrence, C. J. & Weinbaum, S. (1988), The unsteady force on a body at low Reynolds number; the axisymmetric motion of a spheroid, *J. Fluid Mech.* **189**, 463–89.

Lean, M. H. & Wexler, A. (1985), Accurate numerical integration of singular boundary element kernels over boundaries with curvature, *Int. J. Num. Meth. Eng.* **21**, 211–28.

Li, Z. Z., Barthes-Biesel, D. & Helmy, A. (1988), Large deformations and burst of a capsule freely suspended in an elongational flow, *J. Fluid Mech.* **187**, 179–96.

Lighthill, M. J. (1958), *An Introduction to Fourier Analysis and Generalised Functions*, Cambridge University Press.

Lighthill, M. J. (1975), *Mathematical Biofluiddynamics*, SIAM.

Liron, N. & Blake, J. R. (1981), Existence of viscous eddies near boundaries, *J. Fluid Mech.* **107**, 109–29.

Liron, N. & Mochon, S. (1976), Stokes flow for a Stokeslet between two parallel flat plates, *J. Eng. Math.* **10**, 287–303.

Liron, N. & Shahar, R. (1978), Stokes flow due to a Stokeslet in a pipe, *J. Fluid Mech.* **86**, 727–44.

Lorentz, H. A. (1907), Ein allgemeiner Satz, die bewegung einer reibenden Flüssigkeit betreffend, nebst einigen anwendungen Desselben, *Abhand. Theor. Phys.* (Leipzig) **1**, 23–42.

Love, A. E. H. (1944), *A Treatise on the Mathematical Theory of Elasticity*, Dover.

Lugt, H. J. & Schwiderski, E. W. (1964), Flows around dihedral angles I. Eigenmode analysis, *Proc. Roy. Soc. London A* **285**, 382–99.

Małysa, K., Dąbroś, T. & Van de Ven, T. G. M. (1986), The sedimentation of one sphere past a second attached to a wall, *J. Fluid Mech.* **162**, 157–70.

Mathon, R. & Johnston, R. L. (1977), The approximate solution of elliptic boundary-value problems by fundamental solutions, *SIAM J. Numer. Anal.* **14**, 638–50.

Mayr, M., Drexler, W. & Kuhn, G. (1980), A semianalytical boundary integral approach for axisymmetric elastic bodies with arbitrary boundary conditions, *Int. J. Solids Structures* **16**, 863–71.

Meiburg, E. & Homsy, G. M. (1988), Vortex methods for porous media flows, in *Numerical Simulation in Oil Recovery*, Institute of Mathematics and Its Applications, Vol. 2, Springer-Verlag.

Michael, D. H. (1958), The separation of a viscous liquid at a straight line, *Mathematika* **5**, 82–4.

Mikhlin, S. G. (1957), *Integral Equations and Their Applications to Certain Problems in Mechamics, Mathematical Physics and Technology*, Pergamon Press.

Miyazaki, T. & Hasimoto, H. (1984), The motion of a small sphere in fluid near a circular hole in a plane wall, *J. Fluid Mech.* **145**, 201–21.

Moffat, H. K. (1964), Viscous and resistive eddies near a sharp corner, *J. Fluid Mech.* **18**, 1–18.

Moffat, H. K. & Duffy, B. R. (1980), Local similarity solutions and their limitations, *J. Fluid Mech.* **96**, 299–313.

Muldowney, G. P. (1989), Simulation of time-dependent free-surface Navier-Stokes flow, Ph. D. Thesis, University of Illinois, Urbana-Champaign.

Newhouse, L. & Pozrikidis, C. (1990), The Rayleigh–Taylor instability of liquid layer resting on a plane wall, *J. Fluid Mech.* **217**, 231–54.

Nigam, S. D. & Srinivasan, V. (1975), No-slip images in a sphere, *J. Math. Phys.* Sc. **9**, 389–98.

Nitsche, L. C. & Brenner, H. (1990), Hydrodynamics of particulate motion in sinusoidal pores via a singularity method, *AIChE J.* **36**, 1403–19.

Nyström, E. J. (1930), Über die praktische Auflösung von Integralgleichungen mit Anwendungen auf Randwertaufgaben, *Acta Math.* **54**, 185–204.

Odqvist, F. K. G. (1930), Über die Randwertaufgaben der Hydrodynamik zäher Flüssigkeiten, *Math. Z.* **32**, 329–75.

O'Neill, M. E. (1983), On angles of separation in Stokes flow, *J. Fluid Mech.* **133**, 427–42.

Oseen, C. W. (1927), *Hydrodynamik*, Leipzig.

Pogorzelski, W. (1966), *Integral Equations and Their Applications*, Pergamon Press.

Power, H. & Miranda, G. (1987), Second kind integral equation formulation of Stokes flow past a particle of arbitrary shape, *SIAM J. Appl. Math.* **47** (4) 689–98.

Pozrikidis, C. (1987a), Creeping flow in two-dimensional channels, *J. Fluid Mech.* **180**, 495–14.

Pozrikidis, C. (1987b), A study of peristaltic flow, *J. Fluid Mech.* **180**, 515–27.

Pozrikidis, C. (1988), The flow of a liquid film along a periodic wall, *J. Fluid Mech.* **188**, 275–300.

Pozrikidis, C. (1989a), A singularity method for unsteady linearized flow, *Phys. Fluids A* **1**, 1508–20.

Pozrikidis, C. (1989b), A study of linearized oscillatory flow past particles by the boundary integral method, *J. Fluid Mech.* **202**, 17–41.

Pozrikidis, C. (1990a), The instability of a moving viscous drop, *J. Fluid Mech.* **210**, 1–21.

Pozrikidis, C. (1990b), The deformation of a liquid drop moving normal to a plane solid wall, *J. Fluid Mech.* **215**, 331–63.

Pozrikidis, C. (1990c), The axisymmetric deformation of a red blood cell in uniaxial extensional flow, *J. Fluid Mech.* **216**, 231–54.

Pozrikidis, C. & Thoroddsen, S. T. (1991), The deformation of a liquid film flowing down a plane wall due to a small particle captured on the wall, *Phys. Fluids A*, In press.

Rallison, J. M. (1978), Note on the Faxen relations for a particle in Stokes flow, *J. Fluid Mech.* **88**, 529–33.

Rallison, J. M. (1981), A numerical study of the deformation and burst of a viscous drop in general shear flows, *J. Fluid Mech.* **109**, 465–82.

Rallison, J. M. & Acrivos, A. (1978), A numerical study of the deformation and burst of a viscous drop in an extensional flow, *J. Fluid Mech.* **89**, 191–200.

Ranger, K. B. (1977), The Stokes flow round a smooth body with an attached vortex, *J. Eng. Math.* **11**, 81–8.

Ranger, K. B. (1980), Eddies in two dimensional Stokes flow, *Int. J. Eng. Sci.* **18**, 181–90.

Reinelt, D. A. (1987), The rate at which a long bubble rises in a vertical tube, *J. Fluid Mech.* **175**, 557–65.

Rokhlin, V. (1990), Rapid solution of integral equations of scattering theory in two dimensions, *J. Comp. Phys.* **86**, 414–39.

Sano, O. & Hasimoto, H. (1977), Slow motion of a small sphere in a viscous fluid in a corner I. Motion on and across the bisector of a wedge, *J. Phys. Soc. Japan* **42**, 306–12.

Sano, O. & Hasimoto, H. (1978), The effect of two plane walls on the motion of a small sphere in a viscous fluid, *J. Fluid Mech.* **87**, 673–94.

Sapir, T. & Nir, A. (1985), A hydrodynamic study of the furrowing stage during cleavage, *Physico Chemical Hydr.* **6**, 803–14.

Schwarz, H. R. (1988), *Finite Element Methods*, Academic Press.

Scriven, L. E. (1960), Dynamics of a fluid interface, *Chem. Eng. Science* **12**, 98–108.

Secomb, T. W. & Skalak, R. (1982), Surface flow of viscoelastic membranes in viscous fluids, *Quart. J. Mech. Appl. Math.* **35**, 233–47.

Shail, R. & Townsend, M. S. (1966), On the slow rotation of an oblate spheroid in an unbounded viscous fluid, *Quart. J. Mech. Appl. Math.* **14**, 441–53.

Silliman, W. J. & Scriven, L. E. (1980), Separating flow near a static contact line: slip at a wall and shape of a free surface, *J. Comp. Phys.* **34**, 287–313.

Smith, S. H. (1987a), Low Reynolds number flow resulting from an array of rotlets, *Phys. Fluids A* **30**, 3836–8.

Smith, S. H. (1987b), Unsteady Stokes flow in two dimensions, *J. Eng. Math.* **21**, 271–85.

Stakgold, I. (1967, 1968), *Boundary Value Problems of Mathematical Physics*, 2 volumes, Macmillan.

Stakgold, I. (1979), *Green's Functions and Boundary Value Problems*, Wiley.

Stoker, J. J. (1989), *Differential Geometry*, Wiley.

Stroud, A. H. (1971), *Approximate Calculation of Multiple Integrals*, Prentice Hall.

Stroud, A. H. & Secrest, D. (1966), *Gaussian Quadrature Formulas*, Prentice Hall.

Struik, D. J. (1961), *Lectures on Classical Differential Geometry*, Dover.

Tillett, J. P. K. (1970), Axial and transverse flow past slender axisymmetric bodies, *J. Fluid Mech.* **44**, 401–17.

Tokuda, N. (1972), Viscous flow near a corner in three dimensions, *J. Fluid Mech.* **53**, 129–148.

Tokuda, N. (1975), Stokes solutions for flow near corners in three dimensions, *J. Phys. Soc. Japan* **38**, 1187–94.

Tözeren, H. (1984), Boundary integral equation method for some Stokes problems, *Int. J. Num. Meth. Fluids.* **4**, 159–170.

Tran-Cong, T. & Phan-Thien, N. (1989), Stokes problems of multiparticle systems: A numerical method for arbitrary flows, *Phys. Fluids A* **1**, 453–60.

Tran-Cong, T., Phan-Thien, N. & Graham, A. L. (1990), Stokes problems of multiparticle systems: Periodic arrays, *Phys. Fluids A* **2**, 666–73.

Van Dyke, M. (1975), *Perturbation Methods in Fluid Mechanics*, Parabolic Press.

Varnhorn, W. (1989), Efficient quadrature for a boundary element method to compute three-dimensional Stokes flow, *Int. J. Num. Meth. Fluids* **9**, 185–91.

Wang, H.-C. & Banerjee, P. K. (1989), Multi-domain general axisymmetric stress analysis by BEM, *Int. J. Num. Meth. Eng.* **28**, 2065–83.

Waxman, A. M. (1984), Dynamics of a couple-stress fluid membrane, *Stud. App. Math.* **70**, 63–86.

Weatherburn, C. E. (1961), *Differential Geometry of Three Dimensions*, Cambridge University Press.

Webster, W. C. (1975), The flow about arbitrary, three-dimensional smooth bodies. *J. Ship Eng.* **19**, 206–18.

Weinbaum, S., Ganatos, P. & Yan, Z.-Y. (1990), Numerical multipole and boundary integral equation techniques in Stokes flow, *Ann. Rev. Fluid Mech.* **22**, 275–316.

Wilkinson, J. H. (1965), *The Algebraic Eigenvalue Problem*, Oxford Science Publications.

Williams, W. E. (1962), The reduction of boundary value problems to Fredholm integral equations of the second kind, *ZAMP* **13**, 133–51.

Williams, W. E. (1964), Approximate solution of boundary value problems in potential theory with applications to electrostatics and elastostatics, *Proc. Edinburgh Math. Soc.* **14**, 89–101.

Williams, W. E. (1965), Integral equation formulation of Oseen flow problems, *J. Inst. Maths Applics.* **1**, 339–47.

Williams, W. E (1966a), Boundary effects in Stokes flow, *J. Fluid Mech.* **24**, 285–91.

Williams, W. E. (1966b), A note on slow vibrations in a viscous fluid, *J. Fluid Mech.* **25**, 589–90.

Yiantsios, S. G. & Higgins, B. G. (1989), Rayleigh–Taylor instability in thin viscous films, *Phys. Fluids* **1**, 1484–501.

Youngren, G. K. & Acrivos, A. (1975), Stokes flow past a particle of arbitrary shape: A numerical method of solution, *J. Fluid Mech.* **69**, 377–403.

Youngren, G. K. & Acrivos, A. (1976), On the shape of a gas bubble in a viscous extensional flow, *J. Fluid Mech.* **76**, 433–42.

Zick, A. A. & Homsy, G. M. (1982), Stokes flow through periodic arrays of spheres, *J. Fluid Mech.* **115**, 13–26.

Zienkiewicz, O. C. (1971), *The Finite Element Method in Engineering Science*, McGraw-Hill.

Zinemanas, D. & Nir, A. (1988), On the viscous deformation of biological cells under anisotropic surface tension, *J. Fluid Mech.* **193**, 217–41.

# Index